Reactivity and Structure Concepts in Organic Chemistry

Volume 4

Editors:
Klaus Hafner   Jean-Marie Lehn
Charles W. Rees   P. von Ragué Schleyer
Barry M. Trost   Rudolf Zahradník

W. P. Weber   G. W. Gokel

Phase Transfer Catalysis in Organic Synthesis

Springer-Verlag
Berlin Heidelberg New York 1977

William P. Weber
Professor at the Department of Chemistry
University of Southern California, University Park
Los Angeles, CA 90007/USA

George W. Gokel
Professor at the Department of Chemistry
152 Davey Laboratory, The Pennsylvania State University
University Park, PA 16802/USA

ISBN-3-540-08377-4 Springer-Verlag Berlin Heidelberg New York
ISBN-0-387-08377-4 Springer-Verlag New York Heidelberg Berlin

Library of Congress Cataloging in Publication Data. Weber, William P 1940–. Phase transfer catalysis in organic synthesis. (Reactivity and structure ; v. 4) Includes bibliographies and indexes. 1. Catalysis. 2. Chemistry, Organic—Synthesis. I. Gokel, G. W., 1946– joint author. II. Title. III. Series. QD505.W4. 547'.2. 77-22798

This work is subject to copyright. All rights are reserved, whether the whole or part of the material is concerned, specifically those of translation, reprinting, re-use of illustrations, broadcasting, reproduction by photocopying machine or similar means, and storage in data banks. Under § 54 of the German Copyright Law where copies are made for other than private use, a fee is payable to the publisher, the amount of the fee to be determined by agreement with the publisher.

© by Springer-Verlag Berlin Heidelberg 1977
Printed in Germany

Typesetting: Elsner & Behrens, Oftersheim
Printing and binding: Konrad Triltsch, Würzburg
2152/3140-543210

Dedication
to Heather, Ann and our Mothers

Preface

The field of phase transfer catalysis is a tribute to the chemists involved in process development research. Phase transfer catalysis is a solution to numerous cost and yield problems encountered regularly in industrial laboratories. In fact, much of the early work in this area was conducted by industrial chemists although the work was not labelled phase transfer catalysis at the time. We certainly do not intend to minimize the contributions of academic chemists to this field, but it is an unalterable fact that much of the early understanding and many of the early advances came from industrial laboratories.

A special tribute is due to Dr. Charles Starks of the Continental Oil Company. By the mid sixties, Starks had formulated the principles of phase transfer catalysis and had applied for patents on many reactions that others were later to examine in somewhat greater detail. His mechanistic model of phase transfer catalysis still stands up well today and is a model for much of the thinking in this area. It is fitting that Starks suggested the name "phase transfer catalysis" by which the whole field is now known.

We wish to thank a number of people who have aided us in many ways in the preparation of this volume. We very much appreciate the helpful discussions and insights provided by Drs. Henry Stevens and Andrew Kaman of PPG Industries in Barberton, Ohio. We also thank Dr. L. A. Domeier of Union Carbide for insight on two-phase polymer reactions. We thank Ms. Lorna Blum who assisted in collecting literature in the early stages of this work and Ms. Blanche Garcia who read the entire manuscript and corrected many of the errors. Mr. David Miles, Mr. Mark Clark and Ms. Andrea Saxon also provided valuable assistance.

We are especially indebted to Mrs. Jennifer L. Stevenson Rice who typed the lion's share of the manuscript and Mrs. Debbie Franke who bore a good deal of this burden as well. Finally, we thank Heather for putting up with more drafts than any mortal should have to endure.

March, 1977   W.P.W. and G.W.G.
                                        Los Angeles, California
                                        and University Park, Pennsylvania

Contents

1. Introduction and Principles 1

1.1 Introduction 1
1.2 Early Examples 1
1.3 The Coalescence of Ideas 2
1.4 The Principle of Phase Transfer Catalysis 3
1.5 Evidence for the Mechanism of Phase Transfer Catalysis . 5
1.6 Charged Catalysts: Quaternary Ions 5
1.7 Uncharged Catalysts: The Amines 7
1.8 Uncharged Catalysts: The Crown Ethers 9
1.9 Uncharged Catalysts: The Cryptands 11
1.10 Catalyst Comparisons 12
1.11 Solvents 13
1.12 The Role of Water in Phase Transfer Catalysis 14
1.13 Summary 15
     References 15

2. The Reaction of Dichlorocarbene With Olefins 18

2.1 Introduction 18
2.2 The Mechanism of the Dichlorocyclopropanation Reaction . 19
2.3 Catalytic Cyclopropanation 22
2.4 Dichlorocyclopropanation of Simple Olefins 22
2.5 Cyclopropanation of Enamines 28
2.6 Dichlorocyclopropanation Followed by Rearrangement . . 28
2.7 Carbene Addition to Indoles 31
2.8 Carbene Addition to Furans and Thiophenes 32
2.9 Carbene Addition to Polycyclic Aromatics 33
2.10 Carbene Addition to Conjugated Olefins 33
2.11 Michael Addition of the Trichloromethyl Anion 37
2.12 Dichlorocarbene Addition to Allylic Alcohols: A Cyclopentenone Synthon 41
2.13 Dichlorocarbene to Phenols: Reimer-Tiemann Reactions . . 42
     References 42

3.	*Reactions of Dichlorocarbene With Non-Olefinic Substrates*	44
3.1	Introduction	44
3.2	C–H Insertion Reactions	44
3.3	Reaction With Alcohols: Synthesis of Chlorides	46
3.4	Carbene Addition to Imines	49
3.5	Addition to Primary Amines: Synthesis of Isonitriles	50
3.6	Reaction With Hydrazine, Secondary, and Tertiary Amines	51
3.7	Dehydration With Dichlorocarbene	52
3.8	Miscellaneous Reactions of Dichlorocarbene	55
	References	56
4.	*Dibromocarbene and Other Carbenes*	58
4.1	Introduction	58
4.2	Dibromocarbene Addition to Simple Olefins	58
4.3	Dibromocarbene Addition to Strained Alkenes	61
4.4	Dibromocarbene Addition to Indoles	62
4.5	Dibromocarbene Addition to Michael Acceptors	62
4.6	Other Reactions of Dibromocarbene	65
4.7	Other Halocarbenes	65
4.8	Phenylthio- and Phenylthio(chloro)carbene	66
4.9	Unsaturated Carbenes	67
	References	71
5.	*Synthesis of Ethers*	73
5.1	Introduction	73
5.2	Mixed Ethers: The Mechanism	73
5.3	Rate Enhancement in the Williamson Reaction	75
5.4	Methylation	76
5.5	Phenyl Ethers	76
5.6	Methoxymethyl Ethers of Phenol	78
5.7	Diethers From Dihalomethanes	79
5.8	The Koenigs-Knorr Reaction	80
5.9	Epoxides	81
	References	84
6.	*Synthesis of Esters*	85
6.1	Introduction	85
6.2	Tertiary Amines and Quaternary Ammonium Salts	85
6.3	Noncatalytic Esterification in the Presence of Ammonium Salts	87
6.4	Polycarbonate Formation	88
6.5	Crown Catalyzed Esterification	89
6.6	Crown Catalyzed Phenacyl Ester Synthesis	90
6.7	Crown Catalyzed Esterification of BOC-Amino Acid to Chloromethylated Resins	92
6.8	Cryptate and Resin Catalyzed Esterifications	93

6.9	Synthesis of Sulfonate and Phosphate Esters by PTC	94
	References	95
7.	***Reactions of Cyanide Ion***	**96**
7.1	Introduction	96
7.2	The Mechanism and General Features of the Cyanide Displacement Reaction	96
7.3	The Formation of Alkyl Cyanides	98
7.4	Formation of Acyl Nitriles	101
7.5	Synthesis of Cyanoformates	102
7.6	Cyanohydrin Formation	103
7.7	The Benzoin Condensation	104
7.8	Hydrocyanation, Cyanosilylation, and Other Reactions	106
	References	107
8.	***Reactions of Superoxide Ions***	**109**
8.1	Introduction	109
8.2	Reactions at Saturated Carbon	109
8.3	Additions to Carbonyl Groups	112
8.4	Reactions With Aryl Halides	114
	References	116
9.	***Reactions of Other Nucleophiles***	**117**
9.1	Introduction	117
9.2	Halide Ions	117
9.3	Azide Ions	124
9.4	Nucleophile Induced Elimination Reactions	125
9.5	Nitrite Ion	129
9.6	Hydrolysis Reactions	130
9.7	Anionic Polymerization Initiation	130
9.8	Organometallic Systems	132
9.9	Isotopic Exchange	134
	References	134
10.	***Alkylation Reactions***	**136**
10.1	Introduction	136
10.2	The Substances Alkylated	137
10.3	Phase Transfer Alkylating Agents	139
10.4	Alkylation of Reissert's Compound	144
	References	204
11.	***Oxidation Reactions***	**206**
11.1	Introduction	206
11.2	Permanganate Ion	206
11.3	Chromate Ion	209

Contents

11.4	Hypochlorite Ion	209
11.5	Catalytic Oxidation	210
11.6	Singlet Oxygen	212
11.7	Oxidation of Anions	212
11.8	Phosphorylation	213
	References	214

12.	*Reduction Techniques*	215
12.1	Introduction	215
12.2	Borohydrides	215
12.3	Stoichiometric Reduction Systems	217
12.4	Other Catalytic Reductions	219
12.5	Altered Reactivity	220
	References	220

13.	*Preparation and Reactions of Sulfur Containing Substrates*	221
13.1	Introduction	221
13.2	Preparation of Symmetrical Thioethers	221
13.3	Preparation of Mixed Sulfides	221
13.4	Preparation of Sulfides From Thiocyanates	225
13.5	Preparation of Alkylthiocyanates	226
13.6	Sulfides Resulting From Michael Additions	227
13.7	Synthesis of α, β-Unsaturated Sulfur Compounds	228
13.8	Other Phase Transfer Reactions of Sulfur Containing Substances	229
	References	233

14.	*Ylids*	234
14.1	Introduction	234
14.2	Phase Transfer Wittig Reactions	234
14.3	The Wittig-Horner-Emmons Reaction	237
14.4	Sulfur Stabilized Ylids	240
	References	241

15.	*Altered Reactivity*	242
15.1	Introduction	242
15.2	Cation Effects	243
15.3	Affected Anions	246
15.4	Ambident Nucleophiles	249
	References	250

16.	*Addendum: Recent Developments in Phase Transfer Catalysis*	252

Author Index	267
Subject Index	274

Phase Transfer Catalysis in Organic Synthesis
List of Abbreviations

Abbreviation	Meaning
aliquat 336	see TCMAC
BMEB	N-benzyl-N-methylephedrinium bromide
BPB	butylpyridinium bromide
Brändström's catalyst	tetrabutylammonium bisulfate
BTEAB	benzyltriethylammonium bromide
BTEAC	benzyltriethylammonium chloride
BTMAC	benzyltrimethylammonium chloride
BTMAF	benzyltrimethylammonium fluoride
18-C-6	18-crown-6
crown	any macrocyclic polyether, usually 18-crown-6
cryptate	any of the macrotricyclic diaminopolyethers
[2.1.1]-cryptate	the macrotricyclic diaminopolyether with three bridges, two of the bridges having one oxygen and the remaining bridge having two
[2.2.1]-cryptate	as above except two two-oxygen bridges and one one-oxygen bridge
[2.2.2]-cryptate	the macrotricyclic diaminopolyether with three bridges, each having two oxygen atoms
[2.2.B]-cryptate	the third bridge in the cryptate has a fused benzo group
[$2_o2_o2_s$]-cryptate	the subscript designates whether the bridge contains oxygen or sulfur
[3.2.2]-cryptate	as above except the third bridge is a three-oxygen bridge
CTEAB	hexadecyltriethylammonium bromide
CTEPB	hexadecyltriethylphosphonium bromide
CTMAB	hexadecyltrimethylammonium bromide
CTMAC	hexadecyltrimethylammonium chloride
DBDMA	dibutyldimethylammonium chloride
DB-18-C-6	dibenzo-18-crown-6
DC-18-C-6	see DCH-18-C-6
DCH-18-C-6	dicyclohexyl-18-crown-6
di	the disubstituted products was obtained
DME	1,2-dimethoxyethane
DMEB	N,N-dimethylephedrinium bromide
DMEBr	N-dodecyl-N-methylephedrinium bromide
DMF	dimethylformamide

List of Abbreviations

DMSO	dimethyl sulfoxide
DTEAB	decyltriethylammonium bromide
E	*trans*
HA	n-hexylamine
HDTBP	hexadecyltributylphosphonium bromide
HMPA	see HMPT
HMPT	hexamethylphosphoric triamide
HPB	heptylpyridinium bromide
HTBPC	hexadecyltributylphosphonium chloride
HTEAB	hexyltriethylammonium bromide
LPB	dodecylpyridinium bromide
LTEAB	dodecyltriethylammonium bromide
Makosza's catalyst	see BTEAC
MeCN	acetonitrile
mono	the monosubstituted product was obtained
MTNAC	methyltrinonylammonium chloride
MTPAB	methyltriphenylammonium bromide
NBP	N-butylpiperidine
nr	not reported
[O]	oxidation
OTEAB	octyltriethylammonium bromide
polypode	see explanations in chapter 9
PTC	phase transfer catalyst or phase transfer conditions
quat	a quaternary cation
QX, Q^+X^-	any quaternary ammonium, phosphonium or arsonium cation paired with an anion
Resin	usually a polystyrene based resin with many sites for functional groups, only one of which is designated in the drawing
seed	an oligomer used as initiator in a polymerization reaction
st, *st*	stoichiometric
Stark's catalyst	see TCMAC
TBA	tributylamine
TBA+	tetrabutylammonium ion
TBA cyanide	tetrabutylammonium cyanide
TBAB	tetrabutylammonium bromide or bisulfate, usually the latter
TBAC	tetrabutylammonium chloride
TBACN	tetrabutylammonium cyanide
TBAF	tetrabutylammonium fluoride
TBAI	tetrabutylammonium iodide
$TBA^+N_3^-$	tetrabutylammonium azide
TBAOH	tetrabutylammonium hydroxide
TBA^+OH^-	see TBAOH
TBPC	tetrabutylphosphonium chloride
TCMAC or TOMAC	tricaprylylmethylammonium chloride
TEAC	tetraethylammonium chloride
TEBAC	see BTEAC
THF	tetrahydrofuran
TIAA	triisoamylamine

List of Abbreviations

TMAB	tetramethylammonium bromide
TMEDA	*N,N,N',N'*-tetramethylethylenediamine
TMS	trimethylsilyl
(TMS)$_2$	hexamethyldisilane
TMS-CN	trimethylsilyl cyanide
TOEPB	trioctylethylphosphonium bromide
TOMAC or TCMAC	trioctylmethylammonium chloride
TOPAC	trioctylpropylammonium chloride
TPAB	tetrapropylammonium bromide
TPAsC	tetraphenylarsonium chloride
TPPB	tetraphenylphosphonium bromide
TPPC	tetraphenylphosphonium chloride
tri	the trisubstituted product was obtained
Triton B	benzyltrimethylammonium hydroxide
Ts	toluenesulfonyl (tosyl)
Z	benzyloxycarbonyl
Z	*cis*

1. Introduction and Principles

1.1 Introduction

With the probable exception of Wöhler's isomerization of ammonium isocyanate to urea, most of the major advances in organic chemistry have been preceded and presaged by a number of diverse and perhaps minor advances. In some cases the early work was recognized, extended and built upon. In other cases, early examples of particular phenomena were recognized only after a general statement of principles had been offered. Phase transfer catalysis (ptc) is a major advance which is preceded by earlier examples of related phenomena but most if not all of these examples were recognized as examples of phase transfer catalysis only after the principles had coalesced in several minds across the world.

1.2 Early Examples

There are undoubtedly many early examples of the phenomenon now known as phase transfer catalysis and a presentation of all of them is neither valuable nor feasible. Some of the early work does merit presentation, however, if only to provide a sense of perspective on the field. Wittig and coworkers demonstrated in 1947 the value of utilizing tetramethylammonium cations paired with trityl and fluorenide ions for alkylation in dry alcohol solution [1]. Jarrousse found that cyclohexanol and phenylacetonitrile could both be alkylated in a two phase system and the catalytic effect of quaternary ammonium salts was clearly recognized (see Eqs. 1.1 and 1.2) [2] Sarrett, in his classic cortisone synthesis, found that either Triton B or the Mannich base

$$c\text{-}C_6H_{11}\text{-}OH + C_6H_5CH_2Cl \longrightarrow c\text{-}C_6H_{11}\text{-}O\text{-}CH_2\text{-}C_6H_5 \tag{1.1}$$

$$C_6H_5\text{-}CH_2\text{-}CN + C_6H_5CH_2Cl \xrightarrow{\text{base}} C_6H_5\text{-}CH(CN)\text{-}CH_2\text{-}C_6H_5 \tag{1.2}$$

derived quaternary salt in the presence of hydroxide would yield the annulated product equally well (Eq. 1.3) [3]. Maerker, Carmichael and Port found in 1960 that

Introduction and Principles

(1.3)

the sodium salts of fatty acids could be alkylated by epichlorohydrin more readily in the presence of benzyltrimethylammonium chloride (Eq. 1.4) [4].

$$\text{n-}C_{17}H_{35}COO^- Na^+ + CH_2-\overset{O}{\overset{|\diagdown}{CH}}-CH_2Cl \rightarrow \text{n-}C_{17}H_{35}COO-CH_2-\overset{O}{\overset{|\diagdown}{CH}}-CH_2 + NaCl$$

(1.4)

Resin bound quaternary ammonium hydroxide groups (on anion exchange resins) were reported as catalysts as early as 1952 to effect cyanohydrin formation, the benzoin condensation, and cyanoethylation [5].

The examples presented above are isolated and, one might argue, desultory examples of early literature in phase transfer catalysis. One example which is unquestionably important and must be included in any discussion of early phase transfer catalysis is the method reported by Gibson and Hosking in 1965 [6]. These Australian analytical chemists found that triphenylmethylarsonium permanganate could be prepared, isolated and dissolved in chloroform where it constituted an excellent oxidizing reagent. This work is of particular moment in the authors' opinion because these workers recognized that the reaction could be made catalytic. In fact, the paper contains rather a lucid albeit succinct exposition of the principles of phase transfer catalysis. The paper concluded with the statement, "This application could be important in preparative organic oxidations . . ."

1.3 The Coalescence of Ideas

Thinking along the lines described by Gibson and Hosking had already begun around the world by the time they first delineated the principles of biphasic catalytic oxidation. Mieczyslaw Makosza of the Technical University in Warsaw published the first of a long series of papers in 1965 [7] on two phase reactions which he called "extractive alkylation". Almost simultaneously, patents on the "catalysis of heterogeneous reactions" began to issue to Charles M. Starks of the Continental Oil Company in Ponca City, Oklahoma [8]. At about the same time Arne Brandström of AB Hassle (Sweden) was closely examining the reactions of quaternary ammonium salts in nonpolar media [9]. Brandström called his non-catalytic technique "ion pair extraction". By the time the first non-patent reports of Starks' work appeared [10] he had termed the process "phase transfer catalysis" and this name has been widely adopted and is in general use.

It is very difficult and probably of little value to try to separate the work of these three pioneers. At the time Makosza reported the two-phase carbene reaction [11] the process was described in a patent issued to Starks [8]. While Brandström was pushing back the frontiers of alkylation in nonpolar homogeneous solution [12], Makosza had alkylated a massive number of carbon acids under catalytic conditions [13]. Hennis (Dow, Midland, Michigan) working in a more limited way, was successfully alkylating carboxylates under conditions favorable for *in situ* quaternary salt formation [14]. And so it goes. It is, in our opinion, of value to credit and compliment these early workers in the field now known as phase transfer catalysis and draw attention to the important work each, in his own way, accomplished.

1.4 The Principle of Phase Transfer Catalysis

The fundamental requirement for a bimolecular reaction to occur is collision. No amount of kinetic energy contained by one species can make it react with another species if the two do not come into proximity. Thus Starks reported that octyl bromide could be heated with sodium cyanide for two weeks to no avail [10a]. Such problems have traditionally been overcome by utilizing a solvent or cosolvent which exhibited both lipophilic and hydrophilic properties. For example, methanol, ethanol, acetone and dioxane have all been used as solvents in reactions involving salts and organic substrates. The difficulty has been that the salts were less soluble in these solvents than in water and the organic substrates were generally less soluble in them than in hydrocarbons. More recently, the problem has been remedied, at least in part, by the utilization of dipolar aprotic solvents. Such cation solvating solvents as dimethylsulfoxide, dimethylformamide, acetonitrile and hexamethylphosphoramide have made possible the mutual dissolution of both salts and organic substrates. An additional advantage implicit in the use of a solvent which is primarily a cation solvator is that the anion associated with the solvated cation will be relatively unencumbered by solvation, and therefore quite reactive. The principal difficulty with such solvents has been that they are often costly, frequently hard to purify, dry and maintain in an anhydrous state and they are difficult to recover once the reaction is complete.

An alternative to the use of a dipolar aprotic solvent is to use a nonpolar medium and a cation solvating additive. The use of beta-diamines to solvate and enhance the reactivity of organolithium compounds is well known and documented [15]. Polyethylene glycol derived bases were known to be self-solvating as early as 1963 [16]. A suggestion is recorded in a review in 1965 that the hexamethylether of all *cis*-inositol should be an anion activator by virtue of cation solvation [17].

The principle of specific solvation was well established in the literature by the time phase transfer catalysis became a recognized technique. The great beauty of the phase transfer method, however, resides in the fact that the method is general, mild and catalytic. Phase transfer catalysis utilizing either a quaternary ammonium or phosphonium salt as a catalyst works in the following way. In general, there are two immiscible phases. One of these phases (usually aqueous) contains a reservoir of the

Introduction and Principles

salt expected to function either as base or nucleophile. The second phase is organic and contains the substrate which is expected to react with the salt. Because the salt-containing phase is insoluble in the substrate-containing phase, there will be no reaction observed in the absence of interfacial phenomena [18]. A phase transfer catalyst is added to the mixture. This is ordinarily a quaternary ammonium or phosphonium halide or bisulfate which contains a lipophilic cation. The lipophilic cation enjoys solubility in both aqueous and organic phases and when in contact with the aqueous reservoir of salt, exchanges anions with the excess of anion in the salt solution. The quaternary ion is often given the cognomen "quat" and is frequently represented by the symbol "Q". The anion exchange is represented by the equilibrium shown in equation 1.5.

$$Q^+X^-(aq) + M^+Nu^-(aq) \rightleftharpoons Q^+Nu^-(aq) + M^+X^-(aq) \tag{1.5}$$

The exchange of anion is of little import, however, if nothing further than that which is formulated occurs. Not only must the anion which will function as nucleophile be paired with Q^+, but it must find its way into the organic solution. A second equilibrium is therefore a requirement for phase transfer catalysis to be successful: namely the phase transfer equilibrium. This is formulated in equation 1.6.

$$Q^+Nu^-(aq) \rightleftharpoons Q^+Nu^-(org). \tag{1.6.}$$

Once the nucleophile or base (represented by Nu) is in solution in nonpolar (organic) media, the displacement or deprotonation can take place with product formation. In the case of a nucleophilic displacement reaction, Q^+ would ultimately be ion-paired with the nucleofuge. If the leaving group were X^-, the ion pair QX would be generated and would be subject to the equlibria formulated above. Starks has offered a now classic diagram of the phase transfer catalytic cycle [10a].

$$QNu + R-X \longrightarrow R-Nu + QX \quad \text{Organic phase}$$

$$QNu + MX \rightleftharpoons MNu + QX \quad \text{Aqueous phase}$$

Starks phase transfer catalytic cycle diagram

Note that in this cycle it is not necessary that the ion pair, QX, generated in the organic phase be identical to the ion pair originally added as phase transfer catalyst. It is only necessary that there be present in solution a lipophilic cation, Q^+, or some equivalent cation solvator (see below) and that whatever the identity of X, it must be exchangable with Nu. Probably the most common choice of catalyst has been a quaternary ammonium or phosphonium chloride. The chloride ion readily exchanges with such diverse nucleophiles as hydroxide and cyanide and therefore allows the cycle to be complete.

1.5 Evidence for the Mechanism of Phase Transfer Catalysis

We have discussed above in general terms the requisites and expectations associated with the basic principles of phase transfer catalysis. In fact, these ideas accord well with what is now known about the mechanism of many phase transfer processes. Detailed work has been carried out by several groups, and their conclusions are in substantial agreement.

Starks [10] examined the reaction of cyanide ion with n-octyl bromide (Eq. 1.7) and found that 1) the reaction occurred in the organic phase; 2) the displacement was first order in alkyl halide and first order in catalyst (QX); 3) the rate of reaction

$$Q^+CN^- + C_8H_{17}Br \rightarrow C_8H_{17}CN + Q^+Br^- \qquad (1.7)$$

was shown to be directly proportional to the catalyst concentration; and 4) reaction rate was independent of stirring rate. Regarding the latter point, it was found that at low stirring speed, mass transfer was retarded. Beyond a minimal stirring rate which ensured effective intermixing of the phases, there was no further variation in the reaction rate.

Herriott and Picker [19] addressed the same question in a somewhat different fashion. They examined the two phase reaction of secondary-octyl bromide with hydroxide ion. Based on Ingold's prediction, if the reaction occurred in the organic phase, elimination products would predominate, whereas reaction in the aqueous phase would favor substitution. From the predominance of elimination products, it was inferred that reaction occurred in the organic phase. Beyond a minimum value, stirring rate was found not to affect the reaction rate, a fact which allows one to exclude interfacial phenomena as important factors. The independence of catalytic effectiveness on the general shape of the catalyst helped exclude micellar effects from consideration.

1.6 Charged Catalysts: Quaternary Ions

Herriott and Picker have also reported a careful study of catalyst efficiencies [19b]. The system examined was the reaction of thiophenoxide ion with *n*-bromooctane (Eq. 1.8). The second order rate constants for the reaction as a function of catalyst are shown in Table 1.1. Note that the comparison is valid only for the benzene-water

$$C_6H_5S^-Q^+ + Br\text{-}C_8H_{17} \rightarrow C_6H_5\text{-}S\text{-}C_8H_{17} + QBr \qquad (1.8)$$

system which was used in the study. It is expected, however, that the general trends will apply to other heterogeneous systems such as chloroform-, dichloromethane-, or *ortho*-dichlorobenzene-water. Herriott and Picker offer several generalizations based largely on the data reported in their table and reproduced here (Table 1.1).

Table 1.1. Catalytic efficiencies in benzene-water [19b]

Catalyst	Abbreviation	Relative rate
1. $(CH_3)_4NBr$	TMAB	$< 2.2 \times 10^{-4}$
2. $(C_3H_7)_4NBr$	TPAB	7.6×10^{-4}
3. $(C_4H_9)_4NBr$	TBAB	0.70
4. $(C_4H_9)_4NI$	TBAI	1.000[a]
5. $(C_8H_{17})_3NCH_3Cl$	TOMAC	4.2
6. $C_6H_5CH_2N(C_2H_5)_3Br$	BTEAB	$< 2.2 \times 10^{-4}$
7. $C_5H_5NC_4H_9Br$	BPB	$< 2.2 \times 10^{-4}$
8. $C_5H_5NC_7H_{15}Br$	HPB	3.1×10^{-3}
9. $C_5H_5NC_{12}H_{25}Br$	LPB	0.012
10. $C_6H_{13}N(C_2H_5)_3Br$	HTEAB	2.0×10^{-3}
11. $C_8H_{17}N(C_2H_5)_3Br$	OTEAB	0.022
12. $C_{10}H_{21}N(C_2H_5)_3Br$	DTEAB	0.032
13. $C_{12}H_{25}N(C_2H_5)_3Br$	LTEAB	0.039
14. $C_{16}H_{33}N(C_2H_5)_3Br$	CTEAB	0.065
15. $C_{16}H_{33}N(CH_3)_3Br$	CTMAB	0.020
16. $(C_6H_5)_4PBr$	TPPB	0.34
17. $(C_6H_5)_4PCl$	TPPC	0.36
18. $(C_6H_5)_3PCH_3Br$	MTPAB	0.23
19. $(C_4H_9)_4PCl$	TBPC	5.0
20. $(C_8H_{17})_3PC_2H_5Br$	TOEPB	5.0
21. $C_{16}H_{33}P(C_2H_5)_3Br$	CTEPB	0.25
22. $(C_6H_5)_4AsCl$	TPAsC	0.19
23. Dicyclohexyl-18-crown-6	DCH-18-C-6	5.5

[a] TBAI arbitrarily set equal to 1.00.

They are as follows: 1) The larger quaternary ions are more effective than smaller ones. 2) The catalytic efficiency increases as the length of the longest chain increases. 3) The more symmetrical ions are more effective than those with only one long chain. A certain minimum carbon number is required for the catalyst to be soluble and effective in the organic medium. The better catalysts seem to be those which have the minimal lipophilicity and also have bulky groups surrounding the quaternary positively charged heteroatom. Tetrabutylammonium is a much more efficient catalyst than is hexadecyltrimethylammonium although the latter has three more carbon atoms than the former. The associated anion probably forms a tighter ion pair with the quaternary cation when the quat is less hindered than when the charge is buried. It is interesting to note that the very widely used "Makosza catalyst", benzyltriethylammonium chloride, does not show high efficiency in this study. 4) Phosphonium ions are somewhat more effective and thermally stable than the corresponding ammonium catalysts and both are better than arsonium systems. 5) Substitution of the quaternary ion by alkyl rather than aryl groups yields more effective catalysts. 6) Reaction rates are generally greater in *ortho*-dichlorobenzene (and presumably in other chlorocarbon media) than in benzene, and both are better than heptane. In connection with this latter point, Ugelstad and coworkers have studied the reactions of quaternary ammonium phenoxide ions with alkyl halides in a variety of media and concluded that the reactivity of phenoxide under these conditions can be attributed to the lack of ion pairing with the cation (see Sect. 5.3) [20].

The catalysts which have been used most commonly to date are benzyltriethylammonium chloride ("Makosza's catalyst, BTEAC, TEBAC), trioctylmethylammonium chloride ("Starks' catalyst", aliquat 336, tricaprylylmethylammonium chloride, TOMAC) and tetrabutylammonium hydrogen sulfate (Brandström's catalyst, TBAB). Makosza's catalyst is quite easy to prepare; it can be recrystallized to purity and kept in a reasonably anhydrous state with little difficulty. Aliquat 336 is readily available as a yellow oil from General Mills and is quite inexpensive. It is more efficient than either Brandström's or Makosza's catalyst according to the Herriott-Picker study. Tetrabutylammonium hydrogen sulfate is more expensive than either of the other catalysts singled out, but has two important advantages. The hydrogen sulfate anion is highly hydrophilic and readily partitions into the aqueous phase where it plays no further role in the reaction. Moreover, the crystalline hydrogen sulfate can be treated with a variety of salts and anion exchange can be effected. Brandström and coworkers have prepared numerous salts by this anion exchange reaction and details are given in his book on ion pair extraction [12].

Resin bound quaternary ammonium ions have also been used as catalysts in a variety of reactions conducted in nonpolar media. Such reactions as cyanohydrin formation, cyanoethylation, and the benzoin condensation, were all achieved in the early 1950's [5, 21] and the method was well established by the 1960's [22]. In most of the anionic reactions, a resin bound quaternary ammonium cation chloride anion pair was converted to the hydroxide form and then used directly in the reaction. The method has recently attracted new attention [23, 24].

One fact about the use of quaternary ions as phase transfer catalysts should be noted. In general, the large, lipophilic quaternary ions are soft in the HSAB sense [25]. As a consequence, the quat tends to pair with the softest anion available in solution. If both iodide ions and hydroxide ions were present, for example, the quat would pair with iodide. If reaction with hydroxide was desired, the catalyst would be "poisoned" by the presence of iodide. The source of an ion such as iodide could be from the catalyst originally added or it could be the leaving group in the substitution reaction. The choice of reaction conditions should therefore include a consideration of cation, anion, nucleophile and nucleofuge.

1.7 Uncharged Catalysts: The Amines

There are several reports in the literature of tertiary amine catalysis of reactions which appear to be of the phase transfer type. The first such example is the reaction of potassium benzoate with benzyl chloride to give benzyl benzoate, reported in a German patent issued in 1913 to be catalyzed by triethylamine [26]. Merker and Scott in 1961 utilized *in situ* quaternary ammonium carboxylate formation to facilitate the same esterification reaction [27]. Hennis and coworkers rediscovered and clarified amine catalysis in the reaction of benzyl chloride with potassium acetate

[14]. They found that the reactions were actually catalyzed by quaternary ammonium salts generated *in situ*. In these reactions, sodium iodide was added as a cocatalyst and reacted with the alkyl chloride to yield an alkyl iodide which alkylated the tertiary amine. Once generated, the quaternary ammonium ion served as catalyst in the normal sense of a phase transfer catalyst. (see Eqs. 1.9–1.11).

$$R-Cl + NaI \rightarrow R-I + NaCl \tag{1.9}$$

$$R'_3N + R-I \rightarrow RNR'^{+}_3 I^- \ (\equiv Q^+I^-) \tag{1.10}$$

$$CH_3-COO^-K^+ + Cl-CH_2-C_6H_5 \xrightarrow{Q^+I^-} CH_3-COO-CH_2C_6H_5 \tag{1.11}$$

Reeves and coworkers have recently published examples of tertiary amine catalyzed nitrile formation [29], thiocyanate formation [30], alkylation [31], and carbene formation [31]. These reactions almost certainly involve catalysis by quaternary ammonium ions generated *in situ*.

More recently, Normant and coworkers have reported catalysis of the reaction between potassium acetate and benzyl chloride in acetonitrile by polyamines in a two phase system [28]. It seems likely that the catalytic activity reported by Normant *et al.* is related to the earlier alkylations discussed above. The authors state in their communication, however, their finding that "... quaternary ammonium salts corresponding to the diamines do not activate anions under the experimental conditions used...". In the reaction referred to here, it seems likely that the diamines are playing a dual role. The diamines are probably assisting in the solubilization of the solid (and relatively insoluble) potassium acetate by chelation of the potassium cation and the homogeneous reaction is then probably catalyzed by the quaternary ion formed *in situ*. That the catalytic activity of the amine depends on the hardness of the cation (the harder the cation, the less catalytic activity) [28] seems to accord with this interpretation although lattice energy differences cannot be discounted.

Most of Normant's catalysts are 1,2-diamines [28]. The early work of Jarrousse [2] involved not only a quaternary ion but a 1-amino-2-ether system which could coordinate cations just as the diamines do. Both the coordination effect and quaternary ion effect may be operative in that early example and also in Normant's work. Moreover, the phosphonate catalysts recently developed by Mikolajczky and coworkers [32], as a result of their earlier finding that Wittig-Horner-Emmons reactions require no additional catalyst [33], all contain two potential coordination sites. The combination of coordination properties and nucleophilic nitrogen is quite apparent in the so-called polypode ligands developed by Montanari [34].

An interesting case of catalysis by tertiary amines was reported by Isagawa and coworkers [35] (see Sect. 2.2). These investigators found that the two-phase carbene reaction originally reported almost simultaneously and independently by both Starks [36] and Makosza [11] was catalyzed not only by quaternary ammonium cations, but by amines as well. Makosza has offered an explanation of this phenomenon in terms of *in situ* quaternary ammonium ion formation [37]. He suggests that deprotonation of chloroform occurs at the aqueous-organic interface [38], and that the dichlorocarbene generated by decomposition of the trichloromethide ion coordinates with trialkylamine to give an ylid which functions as a base in the bulk organic phase. The ylid reacts with chloroform to give a quaternary ammonium trichloromethide ion pair. He further suggests a mechanism by which the reaction with olefin to give dichlorocyclopropanes might occur (see Eqs. 1.12–1.14).

$$CHCl_3 + NaOH\ (aq) \rightleftharpoons CCl_3^-\ Na^+ + H_2O \rightleftharpoons NaCl + :CCl_2 + H_2O \quad (1.12)$$

$$:CCl_2 + R_3N\ (org) \rightleftharpoons R_3\overset{+}{N}-\overset{-}{C}Cl_2\ (org) \quad (1.13)$$

$$R_3\overset{+}{N}-\overset{-}{C}Cl_2 + CHCl_3 \rightleftharpoons R_3\overset{+}{N}-CHCl_2CCl_3^- \rightarrow$$

$$R_3\overset{+}{N}-CHCl_2Cl^- + :CCl_2 \xrightarrow{\rightleftharpoons} R_3N + CHCl_3 + \underset{Cl}{\overset{Cl}{\triangleright\!\!\!<}} \quad (1.14)$$

In general, then, most of the catalytic activity attributed to tertiary amines has been shown to be or can probably be explained by *in situ* 'onium salt formation. The catalytic activity of the phosphorylated sulfoxides bespeak the need for further exploration in this area [34].

1.8 Uncharged Catalysts: The Crown Ethers

The macrocyclic polyethers were prepared by Pedersen a decade ago and shown to complex a variety of cationic substrates [39]. Among these substrates are alkali metal cations, alkaline earth cations and ammonium ions [40]. It has been shown that hindered esters could be saponified by KOH in toluene solution in the presence of dicyclohexyl-18-crown-6 [41] or cryptate [42]. Potassium hydroxide is both solubilized in toluene and activated by the crown ether [43]. Sam and Simmons, however, first clearly demonstrated the potential of crowns as phase transfer reagents by solubilizing potassium permanganate in benzene solution (see Eq. 1.15) and using the solubilized reagent in a variety of oxidation processes [44].

$$KMnO_4 + \text{(dicyclohexyl-18-crown-6)} \rightleftharpoons [\text{K}^+\text{(crown)}]\ MnO_4^- \quad (1.15)$$

Phase transfer catalysis utilizing crown ethers differs slightly from the approach which utilizes quaternary ions [45]. Probably the most notable point is that crown ether catalyzed reactions are frequently of the so-called solid-liquid type. In these reactions, an organic substrate is dissolved in an organic solvent. This solution is then placed in contact with a solid reagent (or less commonly an immiscible solution of the reagent, i.e., an aqueous reservoir). Crown ether is added and this forms a complex with the salt. The cation-complexed ion pair is then soluble in the organic phase where the reaction takes place.

The success of crown ether catalyzed phase transfer processes depends on several factors [45]. The crown must be capable of forming a complex with the solid reagent. In this situation, the crown is competing with the lattice energy of the solid. The presence of small amounts of water is probably very important in this process and will be mentioned later. Second, the crown-complexed salt must be soluble in the reaction medium. As a result, the more lipophilic crown ethers will tend, other factors being equal, to be more efficient catalysts for phase transfer processes. Finally, the equilibrium between the metal-cation-crown complex and the nucleophile and nucleofuge must be such that anions can readily be exchanged. If the nucleofuge is selectively paired with the crown-cation complex, the only reagent available in solution will simply lead to ligand exchange and not to product. The requisites are set forth in equations 1.16–1.18.

$$M^+Nu^-(\text{solid}) + \text{Crown (sol'n)} \rightleftharpoons \text{Crown} \cdot M^+Nu^- (\text{sol'n}) \quad (1.16)$$

$$\text{Crown} \cdot M^+Nu^- + R-Y \longrightarrow \text{Crown} \cdot M^+Y^- + R-Nu \quad (1.17)$$

$$\text{Crown} \cdot M^+Y^- (\text{sol'n}) \rightleftharpoons \text{Crown (sol'n)} + M^+Y^- (\text{solid}) \quad (1.18)$$

The three crown ethers which have been most widely used are dibenzo-18-crown-6 [39], dicyclohexyl-18-crown-6 [39] and 18-crown-6 [46]. The dibenzo crown (*1*) owes its popularity to the fact that a straightforward and efficient preparation of this stable and readily handled substance was published early and a detailed procedure is available in Organic Syntheses [43]. The difficulties with this compound are two-fold. First, it is not one of the better cation complexing crowns [40, 47],

1 *2* *3*

and secondly, its solubility in hydrocarbon media is marginal. More popular than *1* is dicyclohexyl-18-crown-6 (*2*) which is prepared from *1* by hydrogenation [39]. Compound *2* is quite soluble in a variety of organic media, is a strong cation complexer and is a stable white solid. Unfortunately, it is a skin irritant [39] and it is relatively expensive to make. The most widely used crown catalyst has been 18-crown-6 [39]. This compound is somewhat less lipophilic than *2*, but much more so than *1*. Moreover, 18-crown-6 can be prepared quite readily by several methods and can be obtained in a pure form [45]. Finally, 18-crown-6 is an excellent cation complexer and is effective in a wide variety of phase transfer processes [45].

Herriott and Picker, in their catalyst evaluation report [19b], indicated that dicyclohexyl-18-crown-6 was as efficient a catalyst in a two-phase system (see above) as any quaternary compound. Unfortunately, 18-crown-6 was not studied. It is the authors' expectation that 18-crown-6 (*3*) will be nearly as good a catalyst in most applications, both solid-liquid and liquid-liquid as the dicyclohexyl compound will be, but its ready availability makes it the catalyst of choice in most applications.

The crown ether called 18-crown-6 is prepared by reaction of triethylene glycol with triethylene glycol dichloride in the presence of base [46] or by the cyclo-oligomerization of ethylene oxide [51]. Purification is effected by distillation and crystallization. The sequence is shown in equation 1.19.

$$HO(CH_2CH_2O)_3H + Cl(CH_2CH_2O)_2CH_2CH_2Cl \xrightarrow{base}$$

$$H_2C\!-\!\!\overset{O}{\underset{}{\diagdown}}\!\!CH_2 \xrightarrow{M^+BF_4^-}$$

(1.19)

One final note regarding the use of crown ethers as phase transfer catalysts: there is little literature which directly compares quaternary ammonium catalysts with crown ethers in liquid-liquid processes (see Sect. 1.10) [48]. There are examples where both have been tried and are effective. In general, however, it appears that for solid-liquid phase transfer processes, the crowns are far better catalysts than are the quaternary ammonium ions. In order for a solid-liquid phase transfer process to succeed, the catalyst must remove an ion pair from a solid matrix. The quaternary catalysts have no chelating heteroatoms with available lone pairs which would favor such a process. The combination of a quaternary catalyst and some simple coordinating amine or ether would probably succeed [28, 32, 34]. It seems likely, as mentioned above, that it is the combination of diamine and quaternary catalyst generated *in situ* which accounts for the success of Normant's catalysts [28]. It is interesting to speculate on the possibility of using a quaternary ammonium compound and a drop of water as a catalytic system.

1.9 Uncharged Catalysts: The Cryptands

One of the most interesting developments in recent years is the preparation of a variety of macrobicylic aminoethers by Lehn and his coworkers [49]. Their similarity to the crown ethers and to Simmons in-out bicyclic amines [50] is apparent. In a sense these cryptates combine the properties of both in that they can complex and encapsulate species within their cavities. Their binding constants with metal ions,

Introduction and Principles

protons, etc. are enormous and they are of value as phase transfer catalysts just as the crowns are. In fact, most of the discussion regarding crowns as phase transfer catalysts applies equally well to cryptates. That the cryptates are less extensively used reflects the fact that there is not yet a one-or two-step synthesis which produces them, and as a consequence, they are expensive.

The commonly used 2.2.2-cryptate is prepared by condensing triethylene glycol dichloride with tosylamide [49]. This affords the doubly N-protected macrocyclic aminoether *4*. Detosylation yields the compound containing secondary nitrogen atoms (*5*). Double amide formation at high dilution affords the bicyclic structure *6* which can be reduced to the desired cryptate, *7*. The sequence is formulated in equations 1.20–1.22.

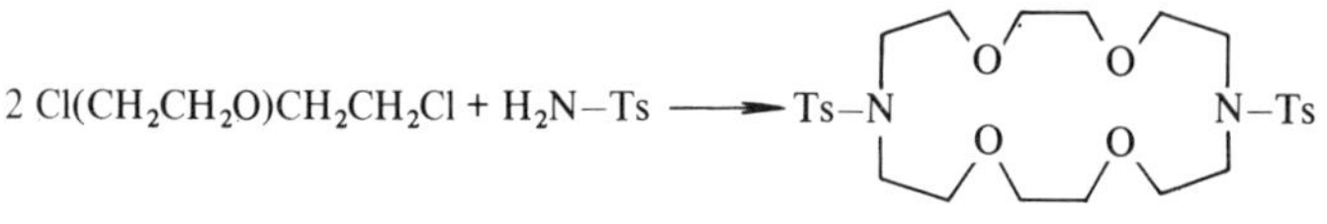

$$2\ Cl(CH_2CH_2O)CH_2CH_2Cl + H_2N-Ts \longrightarrow Ts-N\underset{4}{\diagdown\diagup}N-Ts \qquad (1.20)$$

$$\xrightarrow{-Ts} HN\underset{5}{\diagdown\diagup}NH \xrightarrow{(ClCOCH_2OCH_2)_2} \underset{6}{\text{bicyclic diamide}} \qquad (1.21)$$

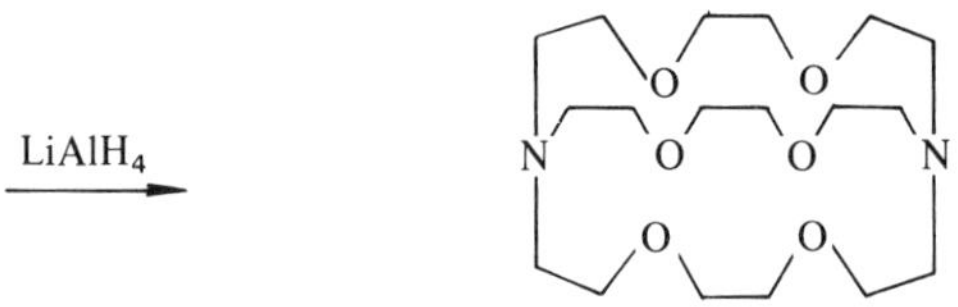

$$\xrightarrow{LiAlH_4} \underset{7}{\text{cryptate}} \qquad (1.22)$$

It should be apparent from an examination of the reaction sequence that the preparation of these compounds, although not complicated, is cumbersome. There are differences in the methods by which these and related compounds are prepared, but they all require two ring formation steps and the second of these (5 → 6, 1.21) has generally been accomplished at high dilution. As a consequence, these compounds remain expensive (even for catalytic requirements) and their use is still infrequent despite their obvious importance and potential.

1.10 Catalyst Comparisons

It should be clear from the preceding discussion that substances which can ion-pair with anions or complex the cationic half of a salt molecule may function as phase transfer catalysts. We have dealt above with quaternary ammonium ions, phosphonium ions, amines (which are quat sources), crown ethers and cryptands. Other substances have been prepared which show catalytic activity as phase transfer reagents. These include the aminopolyethers [34], phosphorylsulfoxides [32], and certain naturally occurring ionophores [52].

Although numerous substances have been utilized as phase transfer reagents in specific cases, very little comparative work is available. Such a study [48] is reported for what might almost be called the standard reaction for catalyst evaluation: the displacement of chloride from benzyl chloride by acetate ion. Included in this study are yield and rate data (half lives are compared) for (*inter alia*) several crown ethers, aminopolyethers, cryptates, an "octopus" molecule and nonactin [48]. Several generalizations are offered which will not be reiterated here. We note that such comparisons can be a valuable guide to catalyst selection.

1.11 Solvents

Many phase transfer processes are conducted in the absence of solvent. The early ester formation reactions, for example, were carried out with solid carboxylate salt in the presence of a mixture of alkyl halide and a small amount of tertiary amine [14a]. (It was not until the efficacy of sodium iodide as a cocatalyst was demonstrated that 2-butanone was added as cosolvent [14b].) In Starks' synthesis of alkyl cyanides by direct displacement of halide, no organic cosolvent was present [10]. Numerous other examples are recorded. Nevertheless, it is common to conduct a phase transfer reaction in the presence of an organic solvent or cosolvent, particularly if the substrate is a solid.

In solid-liquid phase transfer processes, i.e., those reactions in which a solid reagent is phase transferred by a crown [46] or occasionally by a tertiary amine, a cosolvent is ordinarily used, regardless of whether or not the substrate is a solid. In principle, any solvent which does not itself undergo reaction (unless this is the desired end) is acceptable. The most commonly used solvents for solid-liquid phase transfer processes have been benzene (and other hydrocarbons), dichloromethane and chloroform (and other chlorocarbons) and acetonitrile. The latter solvent can be successfully utilized in solid-liquid systems whereas it should be unacceptable in liquid-liquid systems because of its miscibility with water. Chloroform and dichloromethane are commonly and successfully used, although both undergo reactions; the former being readily deprotonated to yield either trichloromethide anion or carbene [38], and the latter suffering nucleophilic displacement [19b, 53, 54].

In liquid-liquid phase transfer systems, the quality of water immiscibility is of particular consequence. Again, hydrocarbons and chlorocarbons have been used most commonly, with the aforementioned *caveat* applying to the latter. A general notion of which solvent will be best for a particular application can be gained by considering the type of reaction to be conducted. If dichlorocarbene is to be generated, for example, the choice of solvent is obvious. It appears that for most other applications, the chlorocarbons are somewhat better solvents than are the hydrocarbons. Brandström has reported data which can also serve as a helpful guide [55]. He determined the amount of 1-(2-allylphenoxy)-3-isopropylaminopropan-2-ol (alprenolol) hydrochloride which was extracted from aqueous solution by several different solvents. The data are reproduced in Table 1.2, below. Note that chloroform seems to be the best solvent, but recall also that it can readily be deprotonated. Dichloromethane and 1,2-dichloroethane seem to be about equally effective in solvating the salts, but methylene chloride is less prone to undergo substitution than dichloroethane, although the latter's higher boiling point can be an advantage. We also note that in at least one case, special solvent properties have been attributed to ethyl acetate [56].

Table 1.2. Extraction of alprenolol hydrochloride

Solvent	Percent extracted
Chloroform	100
1,2-dichloroethane	92
Dichloromethane	88
Carbon tetrachloride	0.6
Trichloroethylene	1.5
Ethyl acetate	16
Ether	0.6
Benzene (formed 3 phases)	–

1.12 The Role of Water in Phase Transfer Catalysis

The role of water in phase transfer catalysis is, like the weather, much discussed and little understood. Several groups have examined the organic phase in two-phase aqueous-organic systems and found that water is present to the extent of a few molecules per ion [10, 19, 57, 58]. The amount of water present depends on the catalyst, the cation, the anion, and the quaternary 'onium salt and concentration.

In liquid-liquid phase transfer systems, water is obviously accompanying the quaternary ammonium ion pair and is likely hydrogen bonding the associated anion or otherwise solvating the system. The need for water in cyanoformate formation [59] and the indifference of nitrile formation to its presence [60] suggests a variable role which complicates the situation even further. In light of the presence of water, it seems inappropriate to designate phase transfer nucleophiles "naked" or "bare" anions. In fact, conclusive evidence has been presented for an $S_N 2$ case [19b]. The reactivity of tetrabutylammonium thiophenoxide when prepared in advance and dissolved in anhydrous benzene was found to be approximately an order of magnitude more reactive than the same ion pair formed *in situ* in a mixture of benzene and water [19b]. It seems likely that the thiophenoxide anion is encumbered by water molecules and its reactivity correspondingly reduced.

Water probably plays an important role in solid-liquid phase transfer processes as well as in liquid-liquid reactions. Freedman has noted that anhydrous potassium permanganate in the presence of an organic solvent and dicyclohexyl-18-crown-6 imparts no color to the organic phase. Upon addition of a drop of water, the organic phase quickly turns purple [61]. The solvation of the ion pair by water is surely competing with the lattice energy of the solid and creating a local liquid-liquid interface. Once the lattice energy is overcome, normal phase transfer processes then intercede. It seems likely that the combination of a quaternary ammonium catalyst and a small amount of water or one of Normant's diamines might make a very economical and useful catalyst system. Careful work is needed in this area to clarify the role of water in both liquid-liquid and solid-liquid phase transfer processes.

1.13 Summary

Phase transfer processes rely on the catalytic effect of quaternary 'onium or crown type compounds to solubilize in organic solutions otherwise insoluble anionic nucleophiles and bases. The solubility of the ion pairs depends on lipophilic solvation of the ammonium or phosphonium cations or crown ether complexes and the associated anions (except for small amounts of water) are relatively less solvated. Because the anions are remote from the cationic charge and are relatively solvation free they are quite reactive. Their increased reactivity and solubility in nonpolar media allows numerous reactions to be conducted in organic solvents at or near room temperature. Both liquid-liquid and solid-liquid phase transfer processes are known; the former ordinarily utilize quaternary ion catalysts whereas the latter have ordinarily utilized crowns or cryptates. Crowns and cryptates can be used in liquid-liquid processes, but fewer successful examples of quaternary ion catalysis of solid-liquid processes are available. In most of the cases where amines are reported to catalyze phase transfer reactions, *in situ* quat formation has either been demonstrated or can be presumed.

References

1. Wittig, G., Heintzeler, M., Wetterling, M.-H.: Annalen *557*, 201 (1947).
2. Jarrousse, M. J.: C. R. Acad. Sci., Ser. C. *232*, 1424 (1951).
3. Poos, G. I., Arth, G. E., Beyler, R. E., Sarrett, L. H.: J. Amer. Chem. Soc. *75*, 422 (1953).
4. Maerker, G., Carmichael, J. F., Port, W. S.: J. Org. Chem. *26*, 2681 (1961).
5. Schmidle, C. J., Mansfield, R. C.: Ind. Eng. Chem. *44*, 1388 (1952).
6. Gibson, N. A., Hosking, J. W.: Aust. J. Chem. *18*, 123 (1965).
7. Makosza, M., Serafin, B.: Rocz. Chem. *39*, 1223 (1965).
8. Starks, C. M., Napier, D. R.: Fr. Demande 1,573,164. See Chem. Abstr. *72*, 115271t (1970).
9. Brandström, A., Gustavii, K.: Acta Chem. Scand. *23*, 1215 (1969).

10. a) Starks, C. M.: J. Amer. Chem. Soc. *93*, 195 (1971).
    b) Starks, C. M., Owens, R. M.: J. Amer. Chem. Soc. *95*, 3613 (1973).
11. Makosza, M., Wawrzyniewicz, M.: Tetrahedron Let. *1969*, 4659.
12. Brandström, A.: Preparative Ion Pair Extraction. Lakemedel: Apotekarsocieteten, AB Hassle *1974*.
13. See Chapter 10.
14. a) Hennis, H. E., Easterly, J. P., Collins, L. R., Thompson, L. R.: Ind. Eng. Chem. Prod. Res. Dev. *6*, 193 (1967).
    b) Hennis, H. E., Thompson, L. R., Long, J. P.: Ind. Eng. Chem. Prod. Res. Dev. *7*, 96 (1968).
15. Agami, C.: Bull. Soc. Chim. Fr. *1970*, 1619.
16. Ugelstadt, J., Mork, P. C., Jensen, B.: Acta Chem. Scand. *17*, 1455 (1963).
17. Jackman, L. M. in Parker, A. J.: Adv. Org. Chem. Meth. Res. *5*, 1 (1965).
18. Menger, F. M.: Chem. Soc. Rev. *1*, 229 (1972).
19. a) Herriott, A. W., Picker, D.: Tetrahedron Let. *1972*, 4517.
    b) Herriott, A. W., Picker, D.: J. Amer. Chem. Soc. *97*, 2345 (1975).
20. Ugelstadt, J., Ellingsen, T., Berge, A.: Acta Chem. Scand. *20*, 1593 (1966).
21. Galat, A.: J. Amer. Chem. Soc. *70*, 3945 (1948).
22. Shimo, K., Wakamatsu, S.: J. Org. Chem. *28*, 504 (1963).
23. Cainelli, G., Maneschalchi, F.: Synthesis *1975*, 723.
24. Cinquini, M., Colonna, S., Molinari, H., Montanari, F., Tundo, P.: J. C. S. Chem. Commun. *1976*, 394.
25. Pearson, R. G.: Hard and Soft Acids and Bases. Stroudsburg, Pa.: Dowden, Hutchinson and Ross, Inc. 1973.
26. German Patent, 268,261 to BASF, Dec. 22, 1913.
27. Merker, R. L., Scott, M. J.: J. Org. Chem. *26*, 5180 (1961).
28. Normant, H., Cuvigny, N. T., Savignac, P.: Synthesis *1975*, 805.
29. Reeves, W. P., White, M. R.: Synth. Comm. *6*, 193 (1976).
30. Reeves, W. P., White, M. R., Hilbrich, R. G., Biegert, L. L.: Synth. Comm. *6*, 509 (1976).
31. Reeves, W. P., Hilbrich, R. G.: Tetrahedron *32*, 2235 (1976).
32. Mikolajczyk, M., Grzejszczak, S., Zatorski, A., Montanari, F., Cinquini, M.: Tetrahedron Let. *1975*, 3757.
33. Mikolajczyk, M., Grzejszczak, S., Midura, W., Zatorski, A.: Synthesis *1975*, 278.
34. Fornasier, R., Montanari, F., Podda, G., Tundo, P.: Tetrahedron Let. *1976*, 1381.
35. Isagawa, K., Kimura, Y., Kwon, S.: J. Org. Chem. *39*, 3171 (1974).
36. Starks, C. M., Napier, D. R.: British Pat. 1,227,144, filed April 5, 1967.
37. Makosza, M., Kacprowicz, A., Fedorynski, M.: Tetrahedron Let. *1975*, 2119.
38. Makosza, M.: Pure and Applied Chemistry *43*, 439 (1975).
39. Pedersen, C. J.: J. Amer. Chem. Soc. *89*, 2495, 7017 (1967).
40. Christensen, J. J., Eatough, D. J., Izatt, R. M.: Chem. Rev. *74*, 351 (1974).
41. Pedersen, C. J., Bromels, M. J.: U.S. Patent 3,847,949 to E. I. duPont de Nemours Co., Inc., Chem. Abstr. *82*, 73049 (1975).
42. Dietrich, B., Lehn, J. M.: Tetrahedron Let. *1973*, 1225.
43. Pedersen, C. J.: Org. Synth. *52*, 66 (1972).
44. Sam, D. J., Simmons, H. E.: J. Amer. Chem. Soc. *94*, 4024 (1972).
45. Gokel, G. W., Durst, H. D.: Synthesis *1976*, 168.
46. Gokel, G. W., Cram, D. J., Liotta, C. L., Harris, H. P., Cook, F. L.: J. Org. Chem. *39*, 2445 (1974).
47. Pedersen, C. J., Frensdorff, H. K.: Angew. Chem. Int. Ed. *11*, 16 (1972).
48. Knochel, A., Oehler, J., Rudolph, G.: Tetrahedron Let. *1975*, 3167.
49. Lehn, J. M.: Structure and Bonding *16*, 1 (1973).
50. Simmons, H. E., Park, C. H.: J. Amer. Chem. Soc. *90*, 2429 (1968).
51. Dale, J., Daasvatn, K.: J. C. S. Chem. Commun. *1976*, 295.
52. Hurd, C. D.: J. Org. Chem. *39*, 3144 (1974).
53. Herriott, A. W., Picker, D.: Synthesis *1975*, 447.

References

54. Holmberg, K., Hansen, B.: Tetrahedron Let. *1975*, 2303.
55. Brandström, A., Gustavii, K.: Acta Chem. Scand. *23*, 1215 (1969).
56. Lee, G. A., Freedman, H. H.: Tetrahedron Let. *1976*, 1641.
57. Landini, D., Montanari, F., Pirisi, F. J.: J. C. S. Chem. Commun. *1974*, 879.
58. Kheifets, V. C., Yakorlear, N. A., Krasil'shchik, B. Ya.: Zh. Prickl. Chim. *46*, 549 (1973).
59. Childs, M. E., Weber, W. P.: J. Org. Chem. *41*, 3486 (1976).
60. Zubrick, J. W., Dunbar, B. I., Durst, H. D.: Tetrahedron Let. *1975*, 71.
61. Freedman, H. H.: private communication.

2. The Reaction of Dichlorocarbene With Olefins

2.1 Introduction

Although phase transfer catalysis is a many faceted technique, it was the observation that dichlorocarbene could be generated in a two-phase aqueous-organic system in which sodium hydroxide was used as base that first captured the attention of the organic chemical community. Both Starks [1, 2] and Makosza [3] reported the dichlorocyclopropanation of cyclohexene in the late 1960's. The reaction was conducted as shown in equation 2.1.

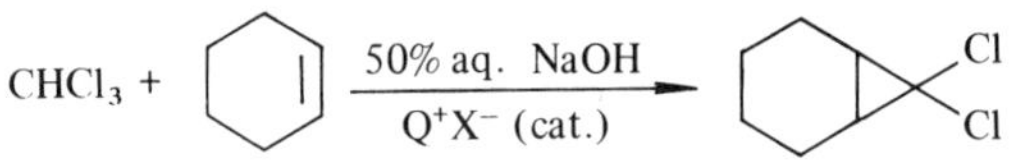

$$CHCl_3 + \text{cyclohexene} \xrightarrow[Q^+X^- \text{ (cat.)}]{50\% \text{ aq. NaOH}} \text{dichloronorcarane} \quad (2.1)$$

The convenience and potential of this method was immediately apparent and its disclosure was followed by numerous examples and extensions. The method was of particular interest because previous methods used to generate dichlorocarbene all required rigorous exclusion of moisture. Among these methods are treatment of chloroform with potassium *t*-butoxide in pentane, pyrolysis of anhydrous sodium trichloroacetate, and the thermal decomposition of phenyl(bromodichloromethyl)-mercury.

The availability of 1,1-dichlorocyclopropanes by phase transfer processes combined with reduction methods such as sodium in ethanol, lithium in *t*-butanol [4], or sodium in liquid ammonia [5] provides a high yield two-step alternative to the Simmons-Smith reaction (Eq. 2.2).

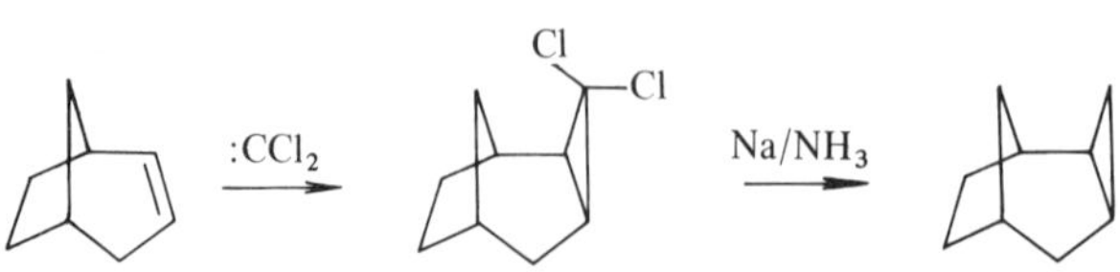

$$\text{norbornene} \xrightarrow{:CCl_2} \text{dichloro adduct} \xrightarrow{Na/NH_3} \text{tricyclic product} \quad (2.2)$$

2.2 The Mechanism of the Dichlorocyclopropanation Reaction

The principles and certain mechanistic features of typical phase transfer catalyzed reactions are described in Sect. 1.4 in the preceding chapter. We reiterate here some of the discussion found in that section so that a discussion of the mechanism of cyclopropanation will be more easily followed and compared with other phase transfer processes.

Careful kinetic work has made it clear that the nucleophilic addition of cyanide to primary alkyl chlorides occurs under liquid-liquid phase transfer conditions according to the following mechanism. The quaternary ammonium cation paired with some anion is soluble in both the aqueous phase (because of its charge) and in the organic phase (due to its lipophilicity). The organic substrate, in this case an alkyl chloride, is present in the organic phase. In the aqueous phase, an equilibrium exists between the quaternary alkyl ammonium cation/chloride ion pair and the sodium cyanide ion pair, on the one hand, and the ammonium cation/cyanide ion pair and the sodium chloride ion pair on the other. Both of these quaternary ammonium cation-anion pairs are in equilibrium with their counterparts in the organic phase, and their distribution depends on their partition coefficients between the aqueous and organic phases. The quaternary ammonium cyanide ion pair in the organic phase, however, is in the presence of a primary alkyl chloride (which is insoluble in water) and a nucleophilic displacement reaction can readily occur, leading to a quaternary ammonium cation/chloride ion pair and primary alkyl cyanide. The sequence is shown schematically below.

$$Q^+CN^- + R-Cl \rightarrow R-CN + Q^+Cl^- \quad \text{(organic phase)}$$
$$Q^+CN^- + NaCl \rightleftarrows NaCN + Q^+Cl^- \quad \text{(aqueous phase)}$$

The quaternary ammonium cation/cyanide anion pair is quite reactive when in the organic phase. This is so because the bulky alkyl groups surrounding the charged nitrogen, phosphorus (etc.) keep the cyanide anion remote from the positive charge. The anionic nucleophile is therefore not well solvated by the cation, nor is it solvated by the organic solvent which is lipophilic. The anion does enjoy some solvation, however, by a few molecules of water per ion which are apparently pulled into the organic phase from the aqueous reservoir [6]. Nevertheless, it is clearly less solvated and more reactive than it would be in aqueous or alcoholic media. A discussion of other factors relating to this basic mechanism can be found in Chapter 1.

The basic mechanistic principles set forth above apply to the cyclopropanation reaction as well as to nucleophilic displacements with some modification [7]. The dichlorocyclopropanation reaction differs principally from the nucleophilic displacement scheme (Eq. 2.2) in that the reactive species is not phase transferred, but rather is generated *in situ* in the organic phase. Dichlorocarbene forms in the organic phase from trichloromethyl anion, which in turn is formed from chloroform. It is the strong quaternary ammonium hydroxide base which initiates the sequence by deprotonating chloroform in the organic phase. It is also this base which is involved in the phase transfer. The scheme is shown in equations 2.3–8, below.

$$\text{NaOH(aq)} + \text{QCl(aq)} \rightleftharpoons \text{QOH(aq)} + \text{NaCl(aq)} \tag{2.3}$$

$$\text{QOH(aq)} \rightleftharpoons \text{QOH(org)} \tag{2.4}$$

$$\text{QOH(org)} + \text{CHCl}_3\text{(org)} \rightleftharpoons \text{H}_2\text{O} + \text{Q}^+\text{Cl}_3\text{C}^-\text{(org)} \tag{2.5}$$

$$\text{Q}^+\text{Cl}_3\text{C}^- \rightleftharpoons \text{QCl} + :\text{CCl}_2 \tag{2.6}$$

$$\text{R}_2\text{C}=\text{CR}_2 + :\text{CCl}_2 \rightarrow \underset{\text{R}_2\text{C}}{\overset{\text{R}_2\text{C}}{\diagdown}}\!\!\!\diagup\!\!\text{CCl}_2 \tag{2.7}$$

$$\text{QCl(org)} \rightleftharpoons \text{QCl(aq)} \tag{2.8}$$

Note that a molecule of water is generated in the organic phase (Eq. 2.5) for each molecule of chloroform which is deprotonated. In addition, some water must accompany the quaternary ammonium hydroxide ion as it undergoes phase transfer from the aqueous to the organic phase (Eq. 2.4). The dichlorocarbene which is generated in the organic phase reacts with the water present in this phase only to a minor extent, an advantage of the phase transfer technique [3]. In contrast, dichlorocarbene generated in the presence of bulk water is hydrolyzed so rapidly that cyclopropanation products are observed to a minor extent, if at all.

Despite the many advantages of the phase transfer method for generating dichlorocarbene, it should be noted that the reactive species arises from the trichloromethide anion as it does in most other methods. Because of this, attempted reactions with electron poor olefins will yield products arising from Michael addition of Cl_3C^- to the olefin instead of, or in addition to, cyclopropanation products. The thermal decomposition of trihalomethyl metal compounds remains the unique method for generating dihalocarbenes without prior formation of a trihalomethyl anion [8].

The question of where the acid-base equilibration (QOH + $CHCl_3$, Eq. 2.5) actually occurs has been the subject of much serious discussion. Equation 2.5 represents the proton transfer as a reaction which takes place in the bulk organic phase. The alternative suggestion is that the reaction occurs at the aqueous-organic interface. While the proton transfer reaction may well occur at the interface, it seems unlikely that the quaternary ammonium trichloromethide undergoes decomposition to chloride and dichlorocarbene at the interface rather than in the bulk organic phase since the latter would be expected to undergo extensive hydrolysis at the interface.

Makosza has argued that the proton transfer occurs at the interface, and that the remainder of the reaction sequence occurs in the bulk organic phase [9]. He has drawn attention to the following facts. First, hydroxide is a harder ion than either trichloromethide or chloride and the latter two would tend to pair with the soft quaternary ion rather than the former. As a consequence, the base concentration in the organic phase should be low. In addition, numerous examples of isotopic (C-D for C-H) exchange are known for weak carbon acids. These exchange reactions are frequently accomplished under biphasic conditions in the absence of a phase transfer catalyst. Finally, the observation that tertiary amines are effective catalysts for the dichlorocarbene generation reaction seems to be in accord with the proton transfer reaction being interfacial [10, 11].

The mechanism of tertiary amine catalysis can be explained as follows [11]. Hydroxide anion reacts with chloroform at the phase boundary to generate the trichloromethyl anion which then decomposes to yield dichlorocarbene and chloride anion. Dichlorocarbene reacts at the phase boundary with the tertiary amine present rather than with water resulting in ylid formation. This ylid is presumably stable enough to be transported into the bulk organic phase where it can react as a base with chloroform to yield the trichloromethyl anion and the trialkyldichloromethylammonium cation. The trichloromethyl anion decomposes to yield dichlorocarbene and chloride ion. The trialkyldichloromethylammonium cation-chloride ion pair then returns to the interface where it reacts with more hydroxide ion to regenerate the ylid which again serves as lipophilic base. Nucleophilic attack by chloride anion could also occur on the trialkyldichloromethylammonium cation to yield trialkylamine and chloroform. The alternative mechanism in which the ylid serves as a dichlorocarbene transfer agent is ruled out by the fact that if this occurred, dichlorocarbene could be transferred directly to electron-poor olefins to form 1,1-dichlorocyclopropanes instead of these olefins undergoing Michael addition of trichloromethide ion. That the Michael additions occur indicates that the ylid serves principally as a lipophilic base and not as a dichlorocarbene transfer reagent. Makosza's scheme is shown in equations 2.9–11 [11].

$$CHCl_3 + NaOH(aq) \rightleftharpoons CCl_3^- Na^+ + H_2O \rightleftharpoons NaCl + :CCl_2 + H_2O \quad (2.9)$$

$$:CCl_2 + R_3N(org) \rightarrow R_3\overset{+}{N}-\overset{-}{C}Cl_2(org) \quad (2.10)$$

$$R_3\overset{+}{N}-\overset{-}{C}Cl_2 + CHCl_3 \rightleftharpoons R_3\overset{+}{N}-CHCl_2 \, Cl_3C^- \rightarrow \quad (2.11)$$

$$R_3\overset{+}{N}-CHCl_2 \, Cl^- + :CCl_2 \xrightarrow{R_2C=CR_2} R_3N + CHCl_3 + \begin{array}{c} R_2C \\ | \\ R_2C \end{array}\!\!\!\!\!\bigtriangleup\!CCl_2$$

Further evidence for this proposal is the finding that if a stoichiometric rather than catalytic amount of trialkylamine is added, the reaction is completely suppressed. This is reasonable if the trialkylamine captures all of the dichlorocarbene to irreversibly form ylid. The reaction of carbenes with trialkylamine to yield ylids capable of deprotonating chloroform was shown independently. Thus, decomposition of phenylchlorodiazine in a mixture of vinyl acetate and chloroform led to the formation of 2-chloro-2-phenylcyclopropyl acetate (see Eq. 2.12). On the other hand, the same reaction in the presence of triethylamine gave 2,2,2-trichloroisopropyl acetate (see Eq. 2.13). Formation of this product can be explained as follows: reaction of chloro-

$$\underset{ClN}{\overset{C_6H_5N}{\diagdown\!\!\diagup\!\!|\!|}} \xrightarrow[\Delta]{OAc} \underset{OAc}{\overset{C_6H_5Cl}{\diagdown\!\!\diagup}} \quad (2.12)$$

phenylcarbene with triethylamine yields an ylid which can deprotonate chloroform to give trichloromethide anion which adds in the Michael sense to vinyl acetate [11].

$$\underset{Cl}{\overset{C_6H_5}{\diagdown}}\!\!\!\underset{N}{\overset{N}{\diagup}} \xrightarrow[CHCl_3]{(n-Bu)_3N, \, \diagup\!\!\diagdown OAc} \underset{Cl}{\overset{C_6H_5}{\diagdown}}\!\!C\text{-}\overset{+}{N}(n\text{-}Bu)_3 + CHCl_3 \longrightarrow$$

$$C_6H_5\text{-}CHCl\text{-}\overset{+}{N}(n\text{-}Bu)_3 + Cl_3C^- \longrightarrow \tag{2.13}$$

$$Q^+ \, CCl_3^- + \diagup\!\!\diagdown OAc \longrightarrow \underset{OAc}{\overset{CCl_3}{\diagdown\!\!\diagup}}$$

2.3 Catalytic Cyclopropanation

Quaternary alkylammonium salts, tertiary amines, and crown ethers have all been utilized as catalysts in the reaction of hydroxide with chloroform to yield dichlorocarbene. The most commonly utilized catalyst has been benzyltriethylammonium chloride (see Sect. 1.7) but other quaternary ammonium chloride catalysts have proved effective. Cetyltrimethylammonium chloride and tricaprylmethylammonium chloride (Aliquat 336) have both been used effectively in the cyclopropanation of simple alkenes. The use of *beta*-hydroxyethyltrialkylammonium hydroxides as phase transfer catalysts results in increased regioselectivity in the addition of dichlorocarbene to olefins [12]. Crown ethers such as dibenzo and dicyclohexyl-18-crown-6 have both been utilized in place of quaternary ammonium compounds. 18-Crown-6 has also been used as a catalyst in the phase transfer thermal decomposition of sodium trichloroacetate to yield dichlorocarbene [13].

The facility of the reaction and the economy of the reagents led a number of workers to examine the cyclopropanation of a large number of olefins. We present here a comprehensive listing of the reaction products grouped according to increasing complexity of the substrate.

2.4 Dichlorocyclopropanation of Simple Olefins

The dichlorocyclopropanation of simple olefins is characterized by good yields, convenient reaction conditions and inexpensive reagents. When there is more than one isolated double bond in a substrate, products of both mono and multiple cyclopropanation are isolated unless a selective catalyst is used (see above). Examples of the simple dichlorocyclopropanation reaction are presented in Table 2.1. Examples of multiple dichlorocyclopropanation are recorded in Table 2.2. Imines also add dichlorocarbene and are discussed in Sect. 3.4.

Table 2.1. Dichlorocyclopanation of simple olefins

Number of carbons	Olefin	% Yield	Catalyst	References
2	$Cl_2C=CHCl$	9	$NaOCOCCl_3/Q^+$	[13]
3	$CH_2=CH-CH_2Br$	36	BTEAC	[14]
4	$Cl-CH_2CH=CH-CH_2Cl$	45	CTMAB	[15]
	(cyclic sulfone, SO$_2$)	–	–	[16]
5	$(CH_3)_2C=CHCH_3$	60	BTEAC	[3]
	Cyclopentene	16	$NaOCOCCl_3/Q^+$	[13]
	(CH$_3$-substituted cyclic sulfone, SO$_2$)	–	–	[16]
6	1-hexene	60	TCMAC	[1]
	2-methyl-1-pentene	83	NBP	[10]
	n-C_4H_9–O–CH=CH$_2$	71	BTEAC	[3]
	n-C_3H_7–O–CH$_2$–CH=CH$_2$	64	BTEAC	[3]
	Cyclohexene	21	TCMAC	[1]
		79	TIAA	[10]
		77	TBA	[10]
		72	BTEAC	[3]
		98	CTMAC	[17]
		61	DB-18-C-6	[18]
		27	DC-18-C-6	[19]
		52	DB-18-C-6	[19]
		57	$NaOCOCCl_3/Q^+$	[13]
7	Cycloheptene	84	CTMAC	[17]
8	1-octene	60	TCMAC	[2]
	Cyclooctene	73	$NaOCOCCl_3/Q^+$	[13]
	C_6H_5–CH=CH$_2$	85	BTEAC	[20]
		87	DB-18-C-6	[21]
		80	BTEAC	[3]
		70	$HOCH_2CH_2\overset{+}{N}R_3X^-$ [NB: $[\alpha]_D^{18}$ + 3.18 (C = 25.9, CHCl$_3$)]	[12]
		44	DC-18-C-6	[19]
		72	DB-18-C-6	[19]
	(norbornene)	98	BTEAC	[5]
	(chloronorbornene)	61	BTEAC	[5]

Table 2.1 (continued)

Number of carbons	Olefin	% Yield	Catalyst	References
	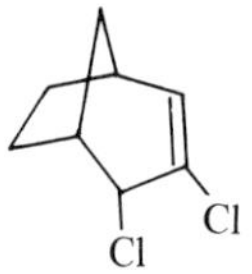 [norbornene with CHCl₂ group structure]	68	BTEAC	[5]
9	Z-n-C₄H₉–CH=CH–Si(CH₃)₃	80	CTMAC	[22]
	α-methylstyrene	71	BTEAC	[20]
	E-propenylbenzene	80	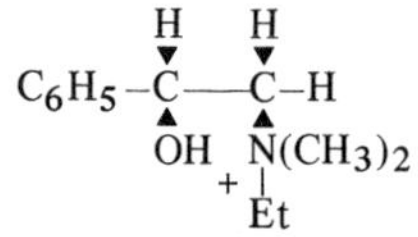 $C_6H_5-\overset{H}{\underset{OH}{C}}-\overset{H}{\underset{N(CH_3)_2}{C}}-H$ $+$ Et [NB: $[\alpha]_D^{28}$ + 1.2 (C = 17.1, CHCl₃)]	[12]
	Indene	75	BTEAC	[23]
10	1,2-dihydronaphthalene	55	BTEAC	[24]
	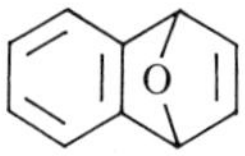 [bicyclic structure with O]	57	BTEAC	[25]
	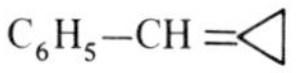 C₆H₅–CH=◁	75	BTEAC/EtOH	[26]
	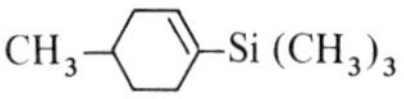 CH₃–⌬–Si(CH₃)₃	72	CTMAC	[22]
11	3-methyl-1,2-dihydronaphthalene	93	BTEAC	[24]
	1,3-dimethylindene	85	BTEAC	[23]
	3,3-dimethylindene	85	BTEAC	[23]
	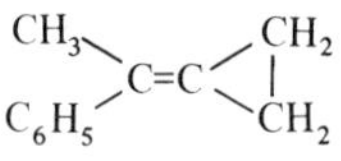 $\underset{C_6H_5}{\overset{CH_3}{\diagdown}}C=C\underset{CH_2}{\overset{CH_2}{\diagup}}$	53	BTEAC/EtOH	[26]
	1,3,3-trimethylindene	85	BTEAC	[23]
	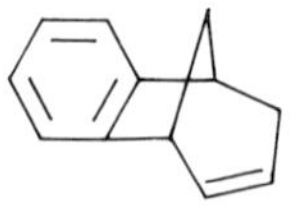 [benzofused bicyclic structure]	92	CTMAC	[27]
	Vinylferrocene	75	BTEAC	[20]
13	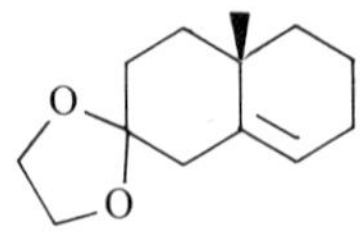 [decalin-type structure with dioxolane]	≃100	BTEAC	[28]
	1-butylindene	75	BTEAC	[23]

Table 2.1 (continued)

Number of carbons	Olefin	% Yield	Catalyst	References
	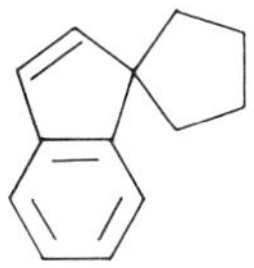 (spiro indene-cyclopentane structure)	70	BTEAC	[23]
	α-methylvinylferrocene	75	BTEAC	[20]
14	Phenanthrene	79	CTMAC	[17]
		79	CTMAC	[29]
		65	BTEAC	[30]
	$(C_6H_5)_2C=CH_2$	70	BTEAC	[20]
		79	BTEAC	[31]
	$E-C_6H_5-CH=CH-C_6H_5$	60	BTEAC	[20]
		75	DB-18-C-6	[18]
		96	BTEAC	[31]
15	$(C_6H_5)_2C=CH-CH_3$	85	BTEAC	[31]
16	$(C_6H_5)_2C=\triangleleft$	95	BTEAC/EtOH	[26]
18	α-phenylvinylferrocene	70	BTEAC	[20]
	β-phenylvinylferrocene	97	BTEAC	[20]
20	$(C_6H_5)_2C=CHC_6H_5$	80	BTEAC	[31]
22	$(C_6H_5)\underset{OCH_3}{C}=CHCH(C_6H_5)_2$	no yield reported	BTEAC	[32]
23	$(C_6H_5)\underset{OCH_3}{C}=CH-CH(C_6H_5)C_6H_4Me-4$	no yield reported	BTEAC	[32]
	$(C_6H_5)\underset{OCH_3}{C}=CH-CH(C_6H_5)C_6H_4OMe-4$	no yield reported	BTEAC	[32]
27	$(C_6H_5)_2C=CH-CH(C_6H_5)_2$	no yield reported	BTEAC	[32]
	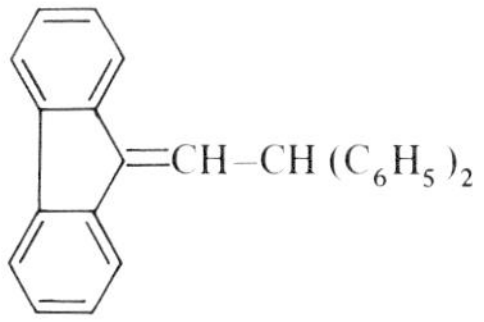 (fluorenylidene)=CH–CH$(C_6H_5)_2$	46	BTEAC	[32]
	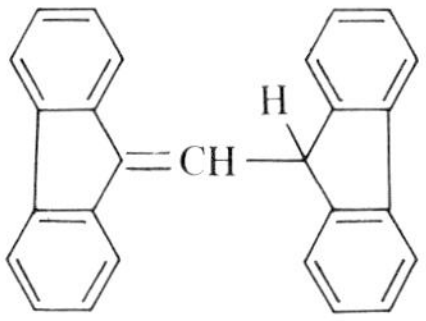 (fluorenylidene)=CH–CH(fluorenyl)	no yield reported	BTEAC	[32]

The Reaction of Dichlorocarbene With Olefins

Table 2.2. Polycyclopropanation of nonconjugated polyolefins

Number of carbons	Olefin	Adduct	% Yield	Catalyst	References
8	$(CH_3)_2Si(CH_2CH=CH_2)_2$	di	77	CTMAC	[17]
8	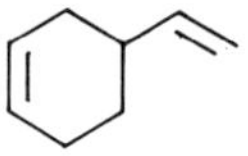	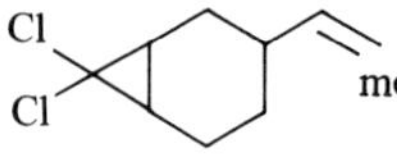 mono	93	$HOCH_2CH_2\overset{+}{N}R_3$	[12]
10	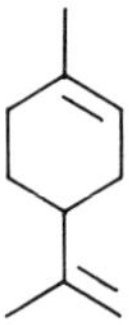	di	78	CTMAC	[17]
		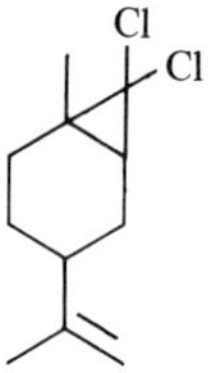 mono	62	$HOCH_2CH_2\overset{+}{N}R_3$	[12]
	1,2-divinylbenzene	di	76	BTEAC	[33]
	1,3-divinylbenzene	di	77	BTEAC	[33]
	1,4-divinylbenzene	di	85	BTEAC	[33]
12	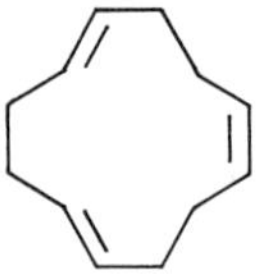	tri	≈100	CTMAC	[12]
		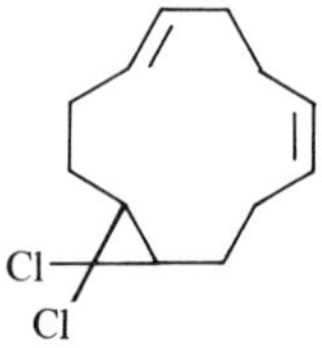 mono	72	$HOCH_2CH_2\overset{+}{N}R_3$	[12]
	1,3-di-(2-propenyl)benzene	di	91	BTEAC	[33]
	1,3,5-trivinylbenzene	tri	75	BTEAC	[33]
15	1,3,5-tri-(2-propenyl)benzene	tri	86	BTEAC	[33]
26	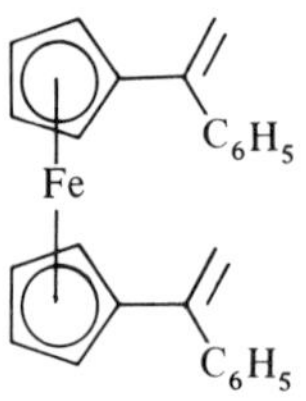	di	71	BTEAC	[20]
polymer	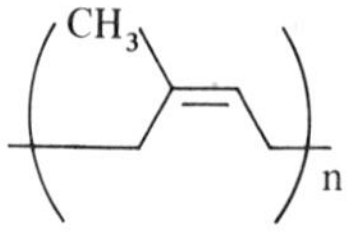	poly	≈100	CTMAC	[17]

There are several phase transfer cyclopropanation reactions which deserve special comment. For example, addition of dichlorocarbene to the conjugated carbon-carbon double bond of styrene systems goes in excellent yield. Addition of dichlorocarbene to 1,2-dihydronaphthalenes [24] and to indenes [23] led to the expected 1,1-dichlorocyclopropane adducts. Heating the 1,1-dichlorocarbene adducts of indenes with ethanolic KOH provides an efficient route to 2-chloronaphthalene derivatives [23] (see Eq. 2.14).

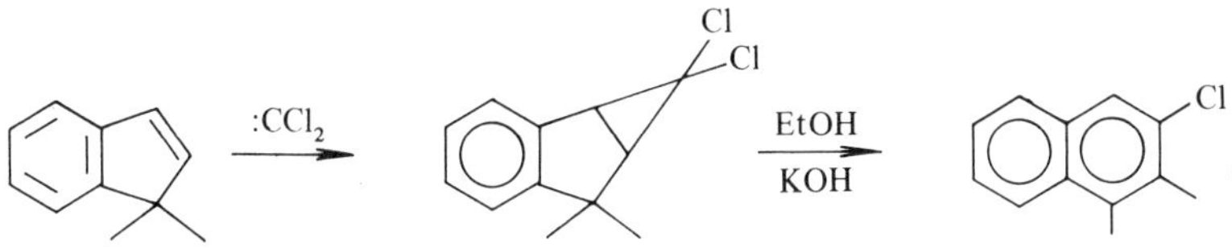

(2.14)

Addition of dichlorocarbene to phenyl substituted methylene cyclopropanes provides an efficient route to 1,1-dichloro-2-phenyl-spiropentanes [26] (see Eq. 2.15).

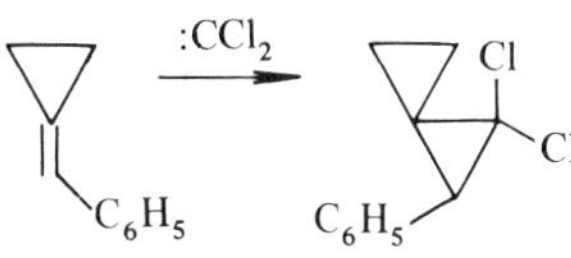

(2.15)

The success of these reactions is somewhat surprising. A phenyl group is known to be inductively electron withdrawing and, as a consequence, the carbon-carbon double bond of a styryl-system is less nucleophilic and therefore might have been expected to react less easily with the electrophilic carbene. Substitution of two or even three phenyl groups on the carbon-carbon double bond does not seem to inhibit the reaction [31]. In fact, dichlorocarbene adds to the C_9-C_{10} double bond of phenanthrene to yield a stable 1,1-dichlorocyclopropane adduct [17]. The reaction is also useful for preparing the difficultly accessible ferrocenyl cyclopropanes from the corresponding ferrocenyl olefins [20]. The reaction of dichlorocarbene with diphenylacetylene under phase transfer conditions gives the anticipated carbene adduct which then undergoes hydrolysis. By this sequence, a low yield of diphenylcyclopropenone was obtained [31].

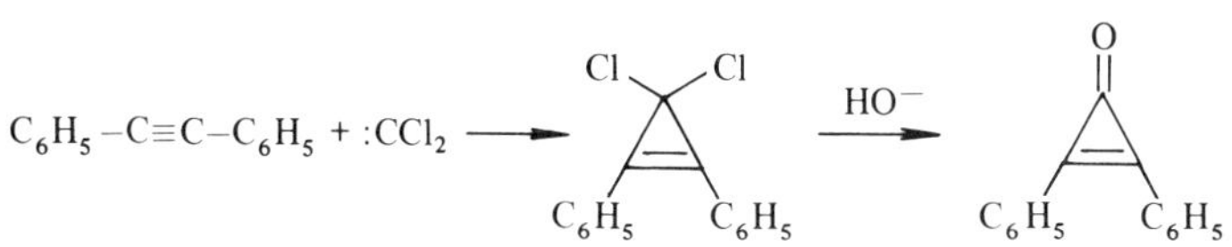

$C_6H_5-C{\equiv}C-C_6H_5 + :CCl_2 \longrightarrow$ (2.16)

Three examples of selective mono adducts of dichlorocarbene to carbon-carbon double bonds in the side chain of steroidal systems have been reported [4]. In each case, the arrow indicates the site of addition.

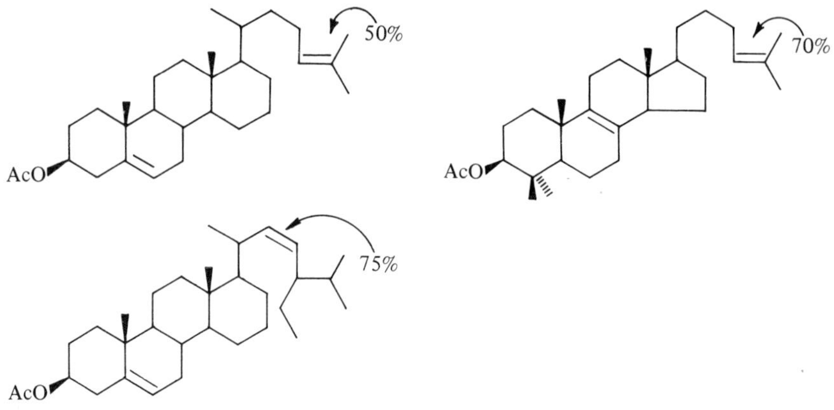

2.5 Cyclopropanation of Enamines

Enamines are readily formed from ketones and therefore constitute an accessible class of substituted olefins. The dichlorocyclopropanation reaction of enamines has not been a preparatively useful one in the past. However, when the reaction is conducted under phase transfer conditions, moderate to excellent yields of the dichlorocarbene adducts can be obtained [34, 35]. A variety of enamines have been subjected to phase transfer cyclopropanation and the results are recorded in Table 2.3.

2.6 Dichlorocyclopropanation Followed by Rearrangement

In cases where the dichlorocarbene adduct of an olefin is either strained or otherwise unstable, the adduct can ionize. The cation thus generated can rearrange and then add back chloride or it can eliminate. In the latter case, if a source of dichlorocarbene is also present, further addition to the system can occur. The addition of dichlorocarbene to norbornadiene, for example, affords only monoaddition products, but each of these results from rearrangement [36–38]. The reaction is formulated in equation 2.17.

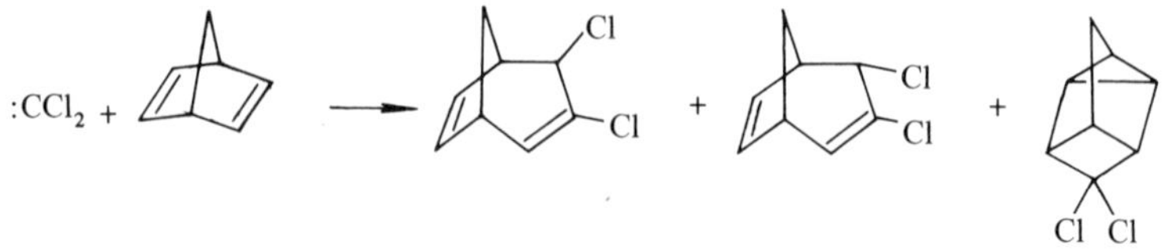

Yield	Catalyst					
80%	CTMAC [38]	80	:	7	:	13
50%	BTEAC [36]	77	:	12	:	11

Table 2.3. Phase transfer dichlorocyclopropanation of enamines

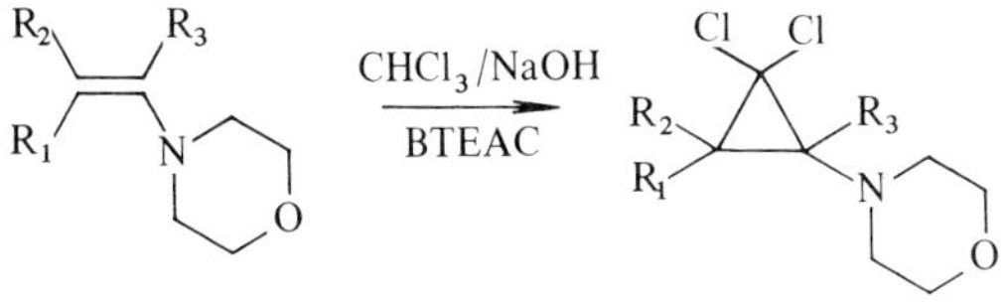

R_1	R_2	R_3	% Yield	References
CH_3	CH_3	H	69	[35]
			64	[34]
CH_3	CH_3	i-C_3H_7	60	[35]
CH_3	C_6H_5	H	86	[35]
H	H	C_6H_5	90	[35]
H	$-(CH_2)_4-$		64	[35]
			62	[34]
H	$-(CH_2)_5-$		75	[35]
H	i-C_3H_7	H	49	[34]
α-tetralone			95	[35]
β-tetralone			92	[35]
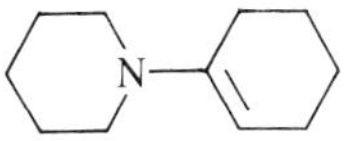			75	[34]
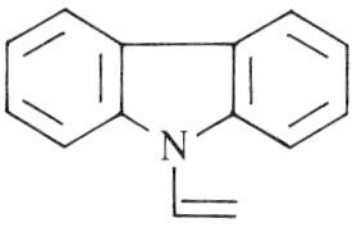			78	[34]

Norbornene [5] and methylnorbornene [39] both undergo phase transfer dichlorocyclopropanation with rearrangement (see Eqs. 2.18 and 2.19). In the former case, a single product was isolated and in the latter case, products of both ring expansion and ion pair collapse and ring expansion followed by dehydrochlorination were isolated.

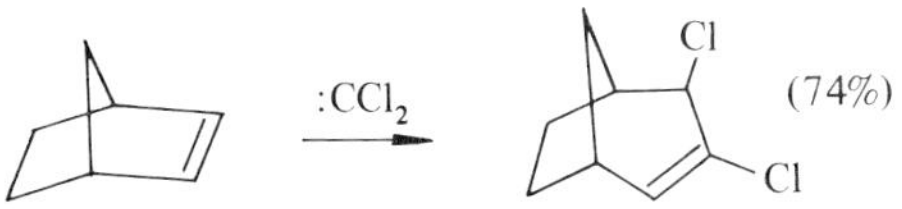

(2.18) [5]

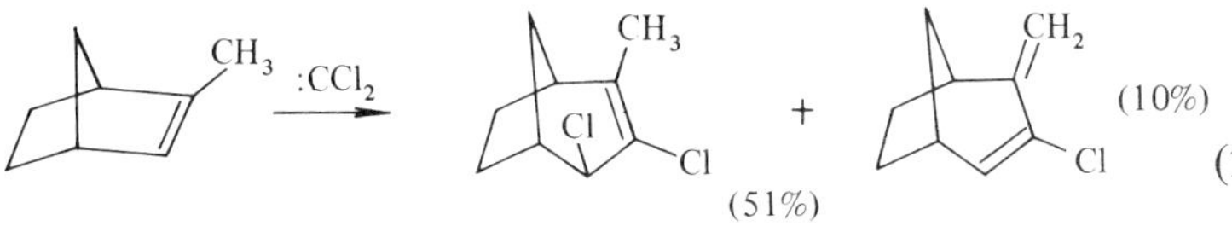

(2.19) [39]

A closely related example is found in the dichlorocyclopropanation of 7-*t*-butoxynorbornadiene [36]. One aspect of this reaction seems surprising. It appears that addition of dichlorocarbene occurs almost exclusively *syn* to the *t*-butoxy group. This

The Reaction of Dichlorocarbene With Olefins

result is unexpected on steric grounds, and probably can be attributed to coordination of the electrophilic species by the oxygen lone pairs, prior to, and perhaps during, addition to the double bond. The reaction is shown in equation 2.20.

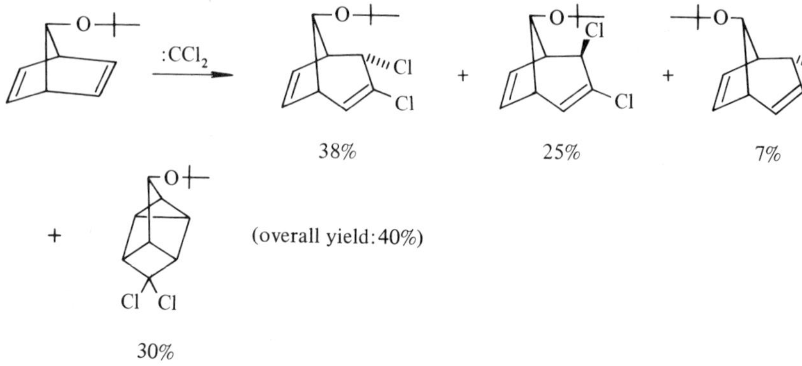

$$\text{(structures: starting material} \xrightarrow{:CCl_2} \text{38\% + 25\% + 7\% + 30\% (overall yield: 40\%)}$$

(2.20)

Reaction of *endo*-5-hydroxymethyl-2-norbornene with dichlorocarbene generated under phase transfer conditions leads to 3-chloro-5-oxatricyclo-[5.2.1.0^{4,8}]-dec-2-ene as the major product (see Eq. 2.21). Formation of this product probably involves initial addition of dichlorocarbene to the carbon-carbon double bond to yield a 1,1-dichlorocyclopropane which ionizes and ring-opens to form a chloro-substituted allylic carbonium ion. This cation is then trapped by the intramolecular nucleophilic alcohol [40].

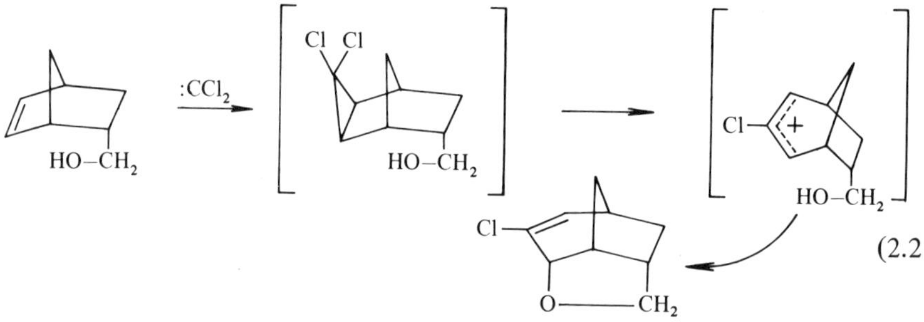

(2.21)

The addition of dichlorocarbene to hexamethyl Dewar benzene has been studied (see Eq. 2.22) [37, 41]. The initial mono-dichlorocarbene adduct undergoes ionization and opening of the cyclopropane ring to yield an allylic carbonium ion which loses a proton from a methyl group to yield adduct I.

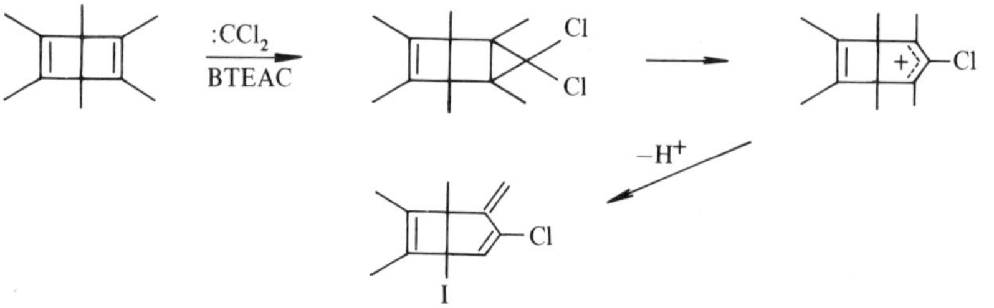

(2.22)

The ring opening ionization is probably favored by relief of strain. Further addition of dichlorocarbene to the chloro-substituted carbon-carbon double bond yields adduct II. Finally, a third addition of dichlorocarbene to the exocyclic olefin yields adduct III [41].

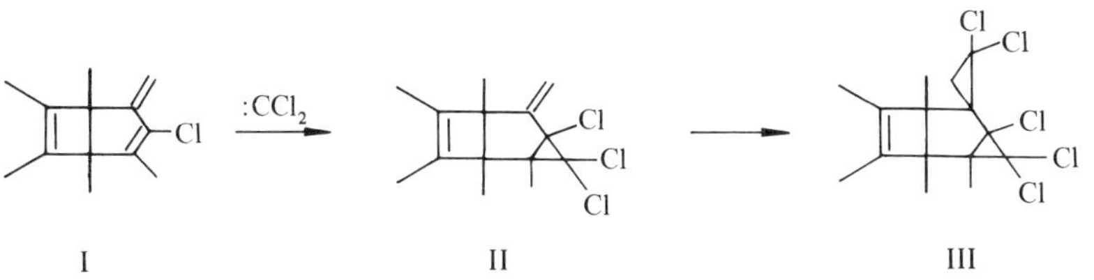

(2.23)

The ratio of the three products depends on the molar ratio of chloroform used as well as on reaction time. Control experiments showed that generation of dichlorocarbene in the presence of I led to II and III, while generation of dichlorocarbene in the presence of II led to high yields of III. The lack of reactivity of the carbon-carbon double bond in the second cyclobutene ring may result from factors similar to those operating in the norbornadiene case.

One further addition-rearrangement sequence which should be mentioned is the reaction of phase transfer generated dichlorocarbene with spirocyclopropylindene. The initial carbene adduct is unstable under the reaction conditions and undergoes ring expansion and dehydrochlorination to yield a mixture of 1-(2-chloroethyl)-3-chloronaphthalene and 1-vinyl-3-chloronaphthalene [23]. The reaction is formulated in equation 2.24.

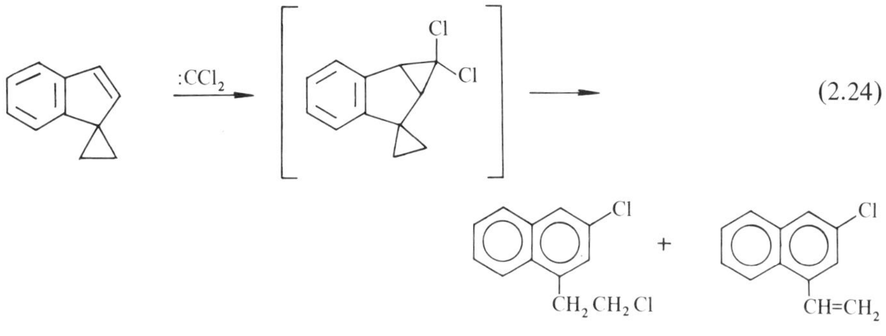

(2.24)

2.7 Carbene Addition to Indoles

Structurally related to, if somewhat more reactive than, indenes are the nitrogen containing heterocycles known as indoles. Just as the indenes will add dichlorocarbene to give an initial adduct which then undergoes rearrangement and dehydrohalogenation, the indoles rearrange and aromatize to form substituted 3-chloroquinolines [42]. 2,3-Dimethylindole, for example, reacts with dichlorocarbene to give, ultimately, 3-chloro-2,4-dimethylquinoline. The sequence is presented in equation 2.25 for the general case and a series of examples is presented in Table 2.4.

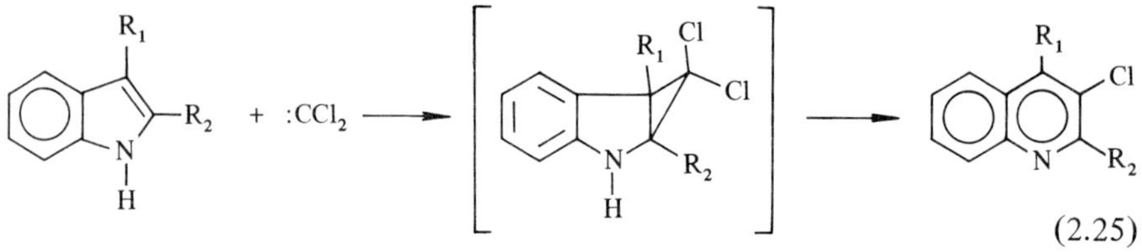

The Reaction of Dichlorocarbene With Olefins

$$(2.25)$$

Table 2.4. Quinoline formation from indoles by carbene addition [42]

R_1	R_2	% Yield	Catalyst
CH_3	H	53	BTEAC
CH_3	H	37	DB-18-C-6
CH_3	H	36	N,N-dimethyl-*n*-dodecylamine
H	C_6H_5	47	BTEAC
H	C_6H_5	68	DB-18-C-6
H	C_6H_5	49	N,N-dimethyl-*n*-dodecylamine
CH_3	CH_3	55	BTEAC
H	CH_3	45	BTEAC

2.8 Carbene Addition to Furans and Thiophenes [43]

The addition of dichlorocarbene to the C_2–C_3 double bonds of 2-methylfuran, 2-methylthiophene, and 2-methylbenzofuran is similar to reaction of dichlorocarbene with indole [42]. Simple mono-adducts of dichlorocarbene are not isolated but rather products arising from these initial adducts by the following reaction sequence [43]. Ionization of this initial adduct followed by ring opening yields a pentadienylic carbonium ion/chloride ion pair. Loss of a proton from the methyl group yields a ring expanded triene with an exocyclic double bond. A similar process has been observed in the addition of dichlorocarbene to 2-methylnorbornene. Finally, addition of a second equivalent of dichlorocarbene to the exocyclic double bond yields product (Eq. 2.26).

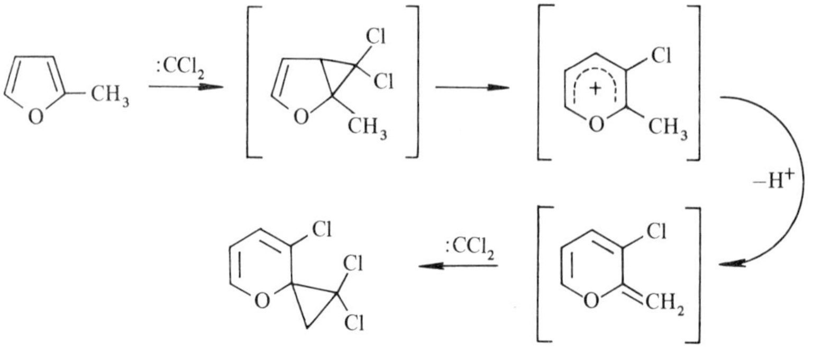

$$(2.26)$$

The phase transfer catalyzed (BTEAC) addition of dichlorocarbene to 2-methylfuran (2.5%), 2-methylthiophene (2.3%) and 2-methylbenzofuran (38%) yields products in accord with equation 2.26 [43].

2.9 Carbene Addition to Polycyclic Aromatics

Analogous additions of dichlorocarbene to the carbon-carbon double bonds of methyl substituted naphthalenes and benzenes also have been reported. The products appear to arise from the following reaction sequence. Initial addition of dichlorocarbene to a methyl substituted carbon-carbon double bond yields a 1,1-dichlorocyclopropane. Ionization of this intermediate with ring-opening yields a substituted tropylium ion. Proton loss from the methyl group yields 1-methylene-2,4,6-cycloheptatriene. Addition of dichlorocarbene to the exocyclic methylene group yields product (see Eq. 2.27). While this reaction is mechanistically interesting, it is of little preparative use since yields are poor [24]. Likewise, the addition of dichlorocarbene to naphthalene [30, 43], anthracene [30], and 1,4,5,8-tetramethoxynaphthalene [44] have all been studied.

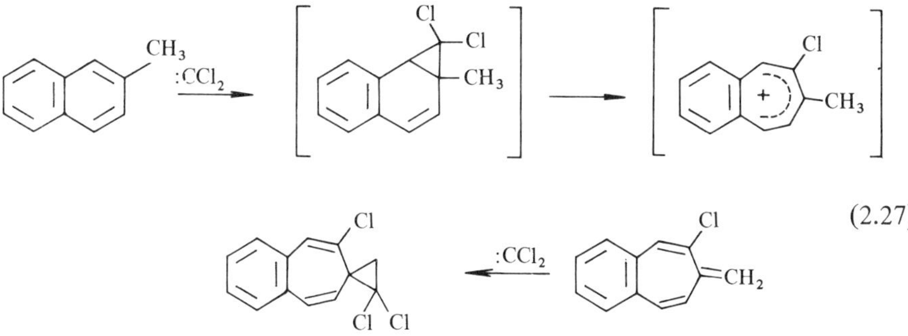

(2.27)

2.10 Carbene Addition to Conjugated Olefins

Dichlorocarbene generated under phase transfer conditions is both an inexpensive reagent and a reactive species. This is true to such an extent that a large excess of the reagent (which is of marginal cost) can convert cyclooctatetraene to the tetra-cyclopropyl compound. The reaction appears to be sequential, the tetrasubstituted compound being favored (up to 55% yield) by a large excess of reagent [46], and the monoadduct being favored by an excess of olefin. The results of dichlorocarbene addition to a number of conjugated dienes are recorded in Table 2.5.

We have previously mentioned that the reaction of dichlorocarbene with diphenylacetylene yields diphenylcyclopropenone. Dichlorocarbene apparently reacts more readily with a carbon-carbon double bond than with a carbon-carbon triple bond.

Table 2.5. Addition of dichlorocarbene to 1,3-dienes

Reactant	Product		Catalyst	References
1,3-butadiene CHCl₃ 2 : 1	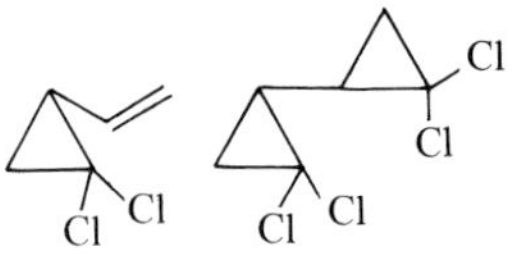 42%	6%	BTEAC	[45]
1,3-butadiene CHCl₃ 1 : 2	13%	38%	BTEAC	[45]
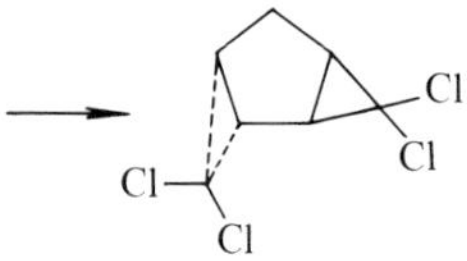 Cyclopentadiene	10%		BTEAC	[45]
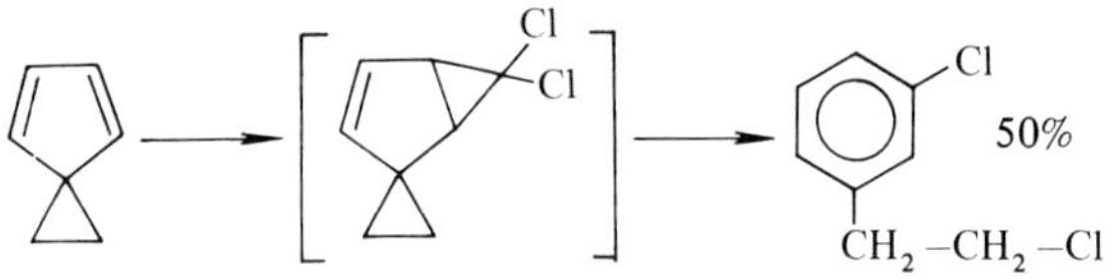 (spiro cyclopentadiene)	50%		BTEAC	[23]
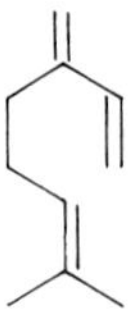 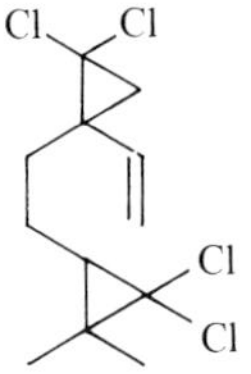 (myrcene-type diene)	42%		BTEAC	[57]
Cyclooctatetraene	mono		BTEAC	[31]
	di		BTEAC	[31]
	tri		BTEAC	[46]
	tetra		BTEAC	[46]
1,4-diphenyl-1,3-butadiene	41% meso diadduct		BTEAC	[31]
	7% racemic diadduct		BTEAC	[31]

Several examples of reactions of dichlorocarbene with conjugated unsaturated systems possessing both C–C double and C–C triple bonds have been reported [47, 48].

Adducts of dichlorocarbene and allenes have also been prepared under phase transfer conditions. The product of dichlorocarbene addition to tetraphenylallene is particularly interesting because addition is reported to occur at the C=C double bonds of a benzene ring rather than to the C=C double bond of the allene system. Likewise, the reaction of dichlorocarbene with phenylallene yields an unusual product. In contrast, tetraphenylbutatriene reacts with dichlorocarbene to give a mixture of normal products [31].

Table 2.6. Addition of dichlorocarbene to conjugated enynes

Reactant	Product	Catalyst	References
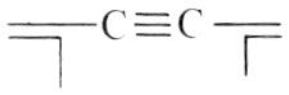 (CH₃)₂C=C≡C–C(CH₃)₂ type alkyne	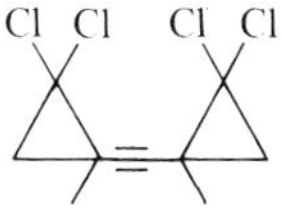 bis-dichlorocyclopropane 62%	BTEAC	[47]
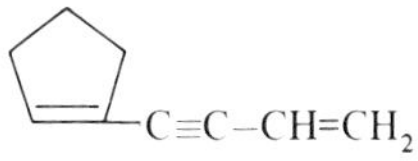 cyclopentenyl–C≡C–CH=CH₂	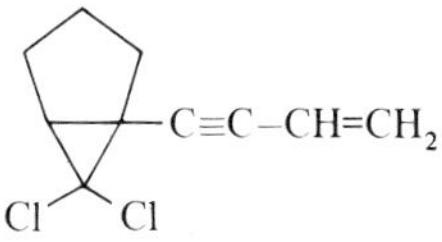 dichlorocyclopropanated cyclopentane–C≡C–CH=CH₂	BTEAC	[48]
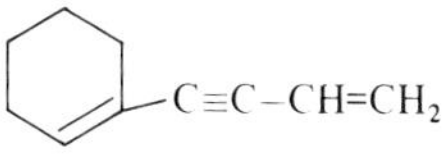 cyclohexenyl–C≡C–CH=CH₂	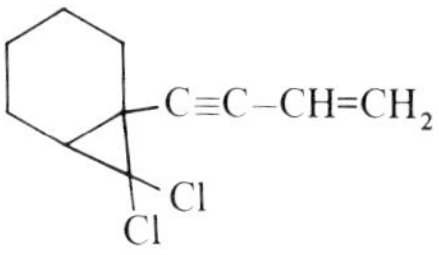 dichlorocyclopropanated cyclohexane–C≡C–CH=CH₂	BTEAC	[48]
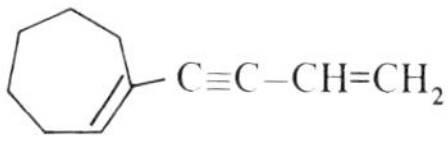 cycloheptenyl–C≡C–CH=CH₂	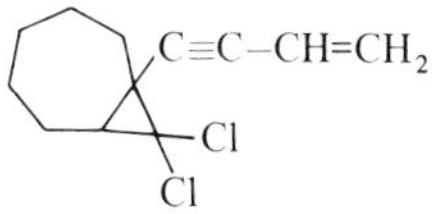 dichlorocyclopropanated cycloheptane–C≡C–CH=CH₂	BTEAC	[48]

Table 2.7. Addition of dichlorocarbene to allenes [49, 50]

Substrate	Product	Catalyst	References
$H_2C=C=CH_2$	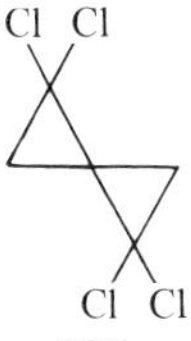  bis-adduct 28%, mono-adduct 6%	BTEAC	[49]
$(CH_3)_2C=C=CH_2$	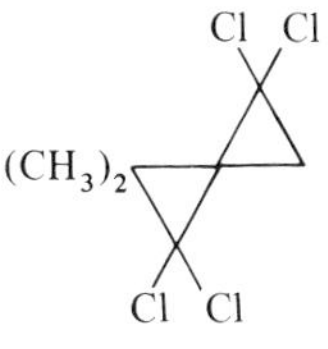 $(CH_3)_2$C–bis-dichlorocyclopropane	BTEAC	[49]
	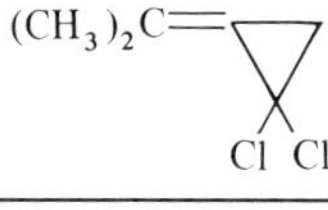 $(CH_3)_2C=$cyclopropane(Cl,Cl) 88%, 5 hrs. at 0°C	BTEAC	[49]

The Reaction of Dichlorocarbene With Olefins

Table 2.7 (continued)

Substrate	Product	Catalyst	References
$(CH_3)_2C=C=C(CH_3)_2$	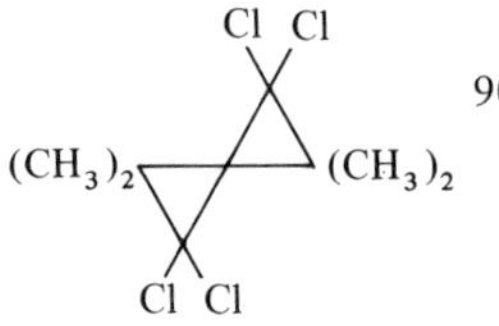 $(CH_3)_2$-[bis-dichlorocyclopropane]-$(CH_3)_2$ 90%	BTEAC	[49]
$(C_6H_5)_2C=C=C\begin{smallmatrix}C_6H_5\\H\end{smallmatrix}$	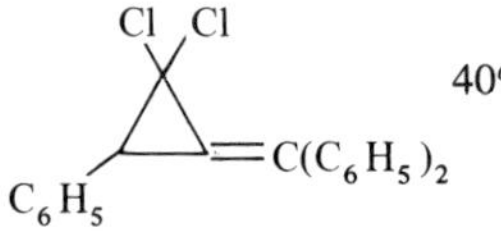 dichlorocyclopropane with C_6H_5, =C(C_6H_5)_2 40%	BTEAC	[49]
$(C_6H_5)_2C=C=C(C_6H_5)_2$	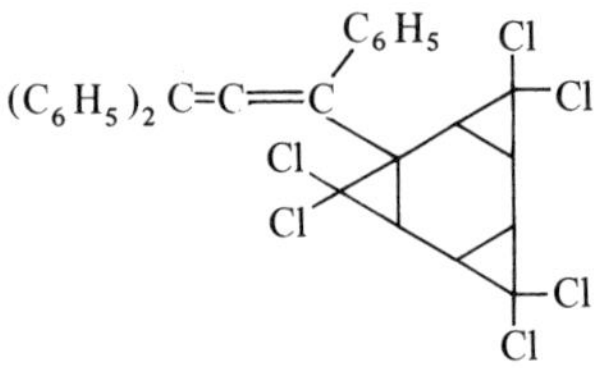 $(C_6H_5)_2C=C=C$ (tris-dichlorocyclopropane adduct with C_6H_5, Cl substituents)	BTEAC	[49]
$\begin{smallmatrix}C_6H_5\\CH_3\end{smallmatrix}>C=C=CH_2$	bis-adduct with CH_3, C_6H_5, Cl, Cl	BTEAC	[50]
	mono-adduct CH_3, C_6H_5, Cl, Cl, =CH_2		
$\begin{smallmatrix}H\\C_6H_5\end{smallmatrix}>C=C<\begin{smallmatrix}CH_3\\H\end{smallmatrix}$	bis-dichlorocyclopropane with H, C_6H_5, CH_3, H	BTEAC	[50]
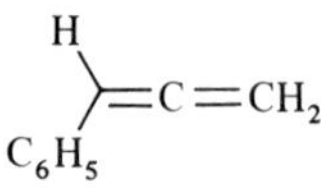 $\begin{smallmatrix}H\\C_6H_5\end{smallmatrix}>C=C=CH_2$	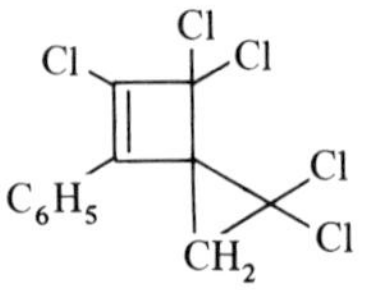 cyclobutene with Cl, Cl, C_6H_5, CH_2, Cl, Cl 21%	BTEAC	[50]

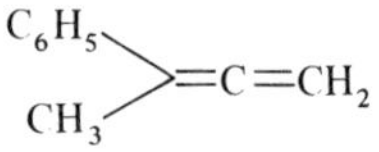

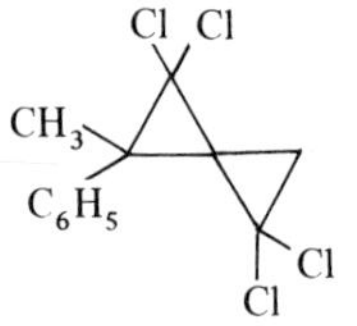

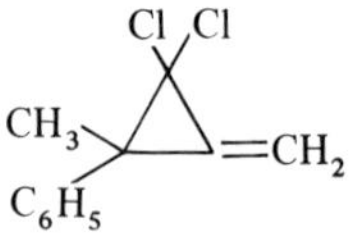

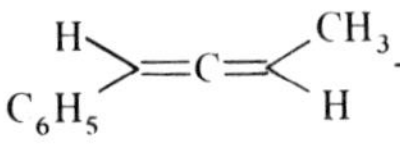

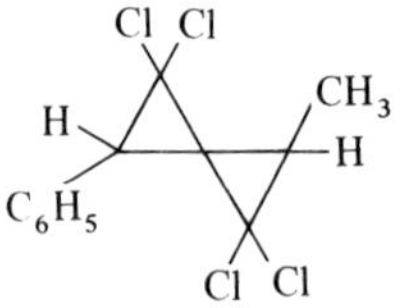

Table 2.7 (continued)

Substrate	Product	Catalyst	References
$(C_6H_5)_2C=C=C=C(C_6H_5)_2$	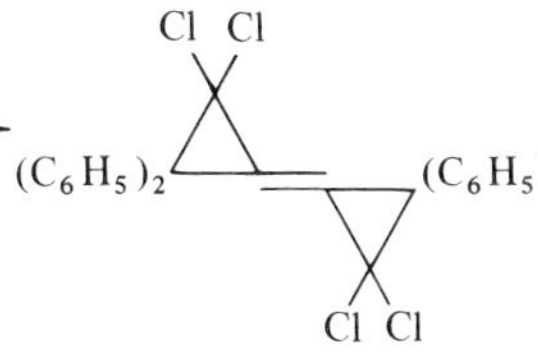 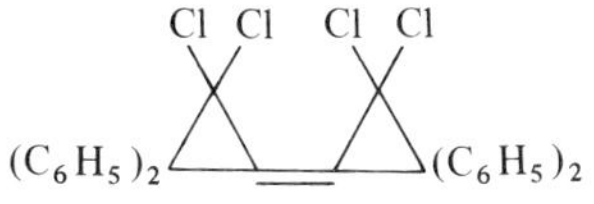 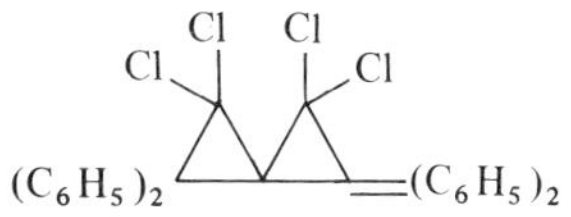 [cyclopropane with $(C_6H_5)_2$ and $=C(C_6H_5)_2$ substituents, tetrachloro], 5%		
	or	BTEAC	[31]
	[bis-cyclopropane product with $(C_6H_5)_2$ groups, tetrachloro]		
	[product with $(C_6H_5)_2$ and $=C(C_6H_5)_2$, tetrachloro], 24%		

2.11 Michael Addition of the Trichloromethyl Anion

At the beginning of this chapter (Sect. 2.2), we mentioned that the generation of dichlorocarbene under phase transfer conditions involved initial formation of the trichloromethyl anion as an intermediate which undergoes alpha-elimination to dichlorocarbene. Attempts to add dichlorocarbene to carbon-carbon double bonds with strongly electron withdrawing groups often result in the Michael addition of trichloromethyl anion rather than formation of the desired dichlorocyclopropane. Thus, acrylonitrile reacted with chloroform and 50% aqueous sodium hydroxide in the presence of either benzyltriethylammonium chloride or dibenzo-18-crown-6 [51], (both of which are normally effective phase transfer catalysts for carbene generation) to yield 4,4,4-trichlorobutyronitrile. Similar results were obtained with α,β-unsaturated esters, nitriles, and sulfones. If the Michael acceptor possesses an α-substituent, the major product is the 1,1-dichlorocyclopropane. These dichlorocyclopropanes may be formed by initial Michael addition of the trichloromethyl anion to the α,β-unsaturated system to yield an anion which then undergoes an intramolecular displacement of chloride to close the three-membered ring. It seems more likely, though, that these systems are less susceptible to Michael addition (because a more substituted carbanion is formed) and more nucleophilic so addition of dichlorocarbene to the double bond becomes more competitive. The two pathways are shown in equation 2.28. A series of results is presented in Table 2.8.

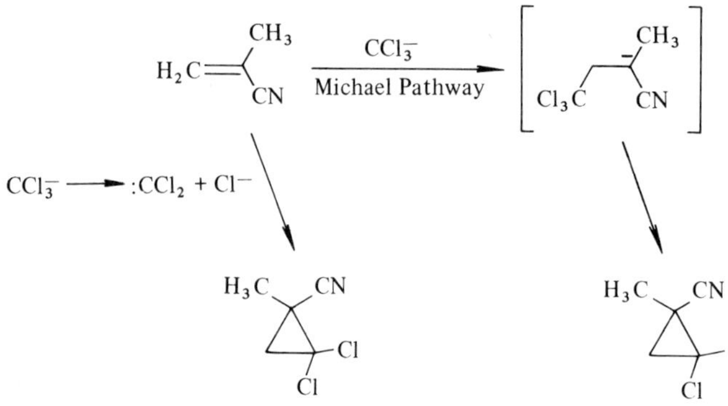

The Reaction of Dichlorocarbene With Olefins

$$CCl_3^- \longrightarrow :CCl_2 + Cl^-$$

(2.28)

Table 2.8. Addition of CCl_3^- and cyclopropanation of potential Michael acceptors

Substrate	Product	% Yield	Catalyst	Ref.
$CH_2=CH-CN$	$CCl_3-CH_2-CH_2-CN$	40	DB-18-C-6	[51]
		72	BTEAC	[53]
$CH_3CH=CH-CN$	$CCl_3-\overset{CH_3}{\underset{}{CH}}-CH_2-CN$	17	BTEAC	[52]
$\underset{H}{\overset{C_6H_5}{\diagdown}}C=\underset{CN}{\overset{H}{\diagup}}$	cyclopropane with C_6H_5, H, CCl_3, CN	14	BTEAC	[52]
$CH_2=CH-CO_2CH_3$	cyclopropane (CCl_3, CO_2CH_3)	43	BTEAC	[53]
$CH_2=CH-CO_2-i-C_3H_7$	cyclopropane (CCl_3, $CO_2-i-C_3H_7$)	43	BTEAC	[52]
$CH_3CH=CH-CO_2-i-C_3H_7$	$CCl_3-\overset{CH_3}{\underset{}{CH}}-CH_2-CO_2-i-C_3H_7$	20	BTEAC	[52]
$CH_2=CH-SO_2-C_6H_5$	$CCl_3-CH_2-CH_2-SO_2-C_6H_5$	70	BTEAC	[52]
$CH_2=C(CH_3)-CN$	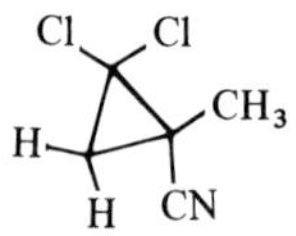 cyclopropane with Cl, Cl, CH_3, H, H, CN	42	BTEAC	[53]
		14	BTEAC	[52]
	cyclopropane with CH_3, H, CCl_3, CN	13	BTEAC	[53]
		6	BTEAC	[52]
$CH_2=C(CH_3)-CO_2C_2H_5$	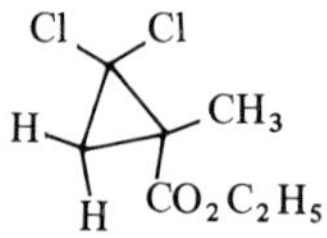 cyclopropane with Cl, Cl, CH_3, H, H, $CO_2C_2H_5$	85	BTEAC	[53]

Table 2.8 (continued)

Alkene	Product	% Yield	Catalyst	Ref.
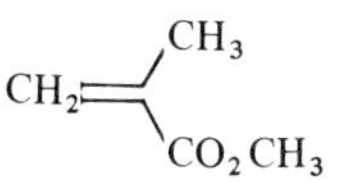 $CH_2=C(CH_3)(CO_2CH_3)$	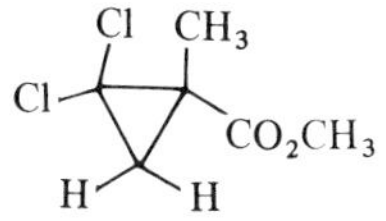 Cl,Cl,CH$_3$,CO$_2$CH$_3$ cyclopropane	20	BTEAC	[52]
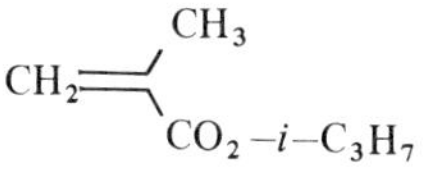 $CH_2=C(CH_3)(CO_2\text{-}i\text{-}C_3H_7)$	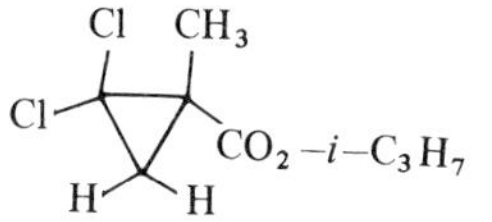 Cl,Cl,CH$_3$,CO$_2$-i-C$_3$H$_7$ cyclopropane	55	BTEAC	[52]
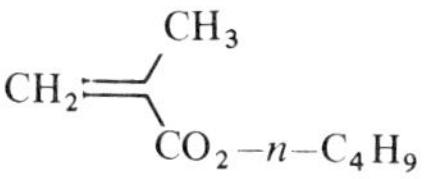 $CH_2=C(CH_3)(CO_2\text{-}n\text{-}C_4H_9)$	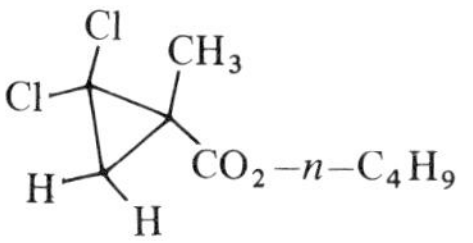 Cl,Cl,CH$_3$,CO$_2$-n-C$_4$H$_9$ cyclopropane	52	BTEAC	[52]
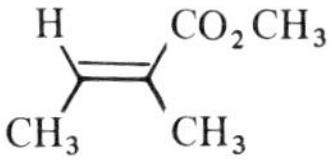 (H)(CH$_3$)C=C(CO$_2$CH$_3$)(CH$_3$)	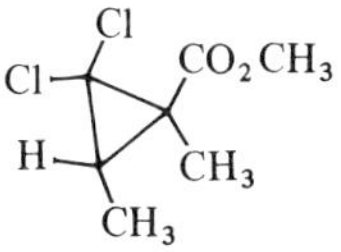 Cl,Cl,CO$_2$CH$_3$,CH$_3$,CH$_3$ cyclopropane	69	BTEAC	[53]

Electron rich enol acetates react in a fashion similar to that described above for other poor Michael acceptors. Vinyl acetate reacts in the Michael sense with chloroform under phase transfer conditions to give 2,2,2-trichloroisopropyl acetate [14, 53], but more highly substituted enol acetates undergo dichlorocyclopropanation. The reaction scheme is similar to that shown in equation 2.28 (above) and the Michael addition of trichloromethide ion to vinyl acetate and the cyclopropanation of bicyclo-[3.2.1] heptanone enol acetate [54] are shown in equations 2.29 and 2.30, respectively.

$$CH_2=CH-O-CO-CH_3 \xrightarrow[\text{(ptc)}]{Cl_3C^-} CH_3-CH(CCl_3)-OCOCH_3 \qquad (2.29)$$

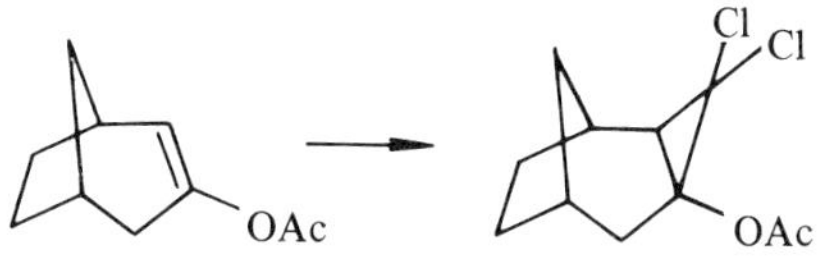

$$\text{(bicyclic-OAc)} \longrightarrow \text{(bicyclic with CCl}_2\text{, OAc)} \qquad (2.30)$$

Reaction of 1,3-butadiene [55] and 1,2-butadiene [53] systems bearing electron withdrawing groups with chloroform under alkaline phase transfer conditions results in low yields of 1,1-dichlorocyclopropanes. In these systems, dichlorocarbene generally adds to the double bond most remote from the electron withdrawing substituent. Examples are shown in equations 2.31–2.33.

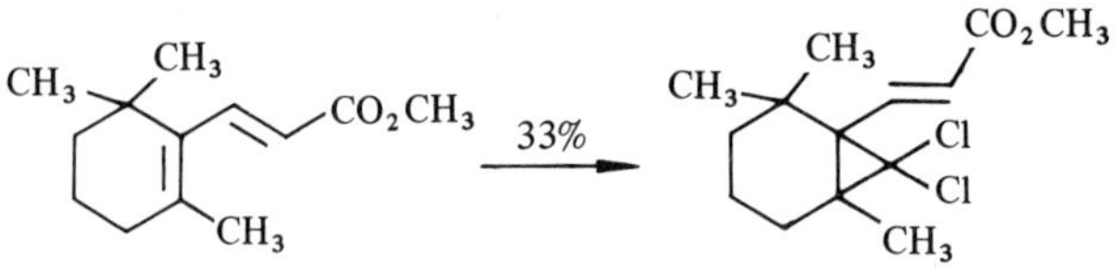

The Reaction of Dichlorocarbene With Olefins

(2.31) [55]

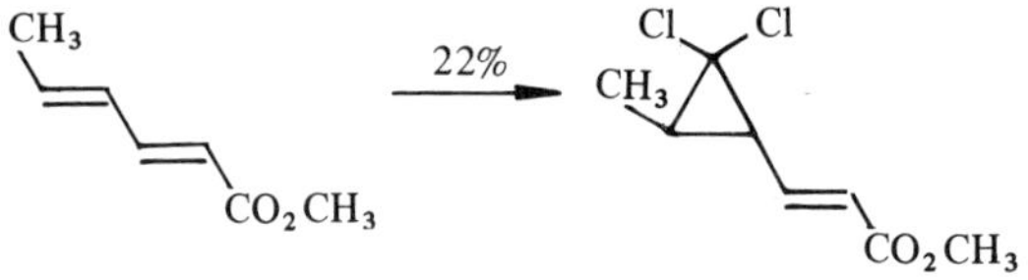

(2.32) [55]

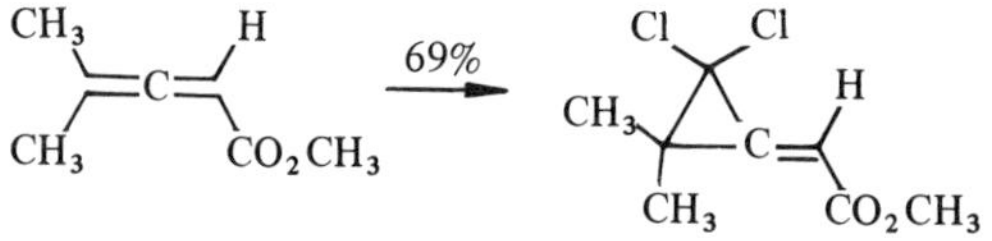

(2.33) [53]

E-Methyl cinnamate, E-methyl crotonate, and methyl 3-methylcrotonate are unusual in that they undergo multiple addition of dichlorocarbene when subjected to phase transfer conditions [53]. A detailed study of this reaction has resulted in evidence for an initial cyclopropanation followed by a series of dehydrochlorinations and trichloromethyl anion additions. The resulting tetrachlorospiropentanes which result from this complex sequence are shown in equations 2.34–2.36. Note that in these cases, there is no electron releasing group on the same carbon as the Michael activating substituent.

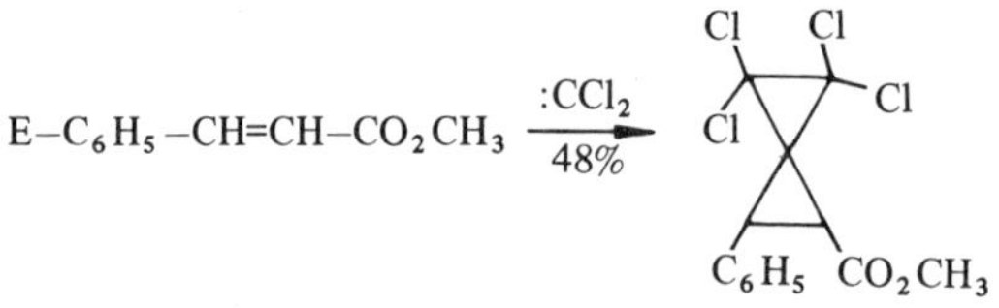

$$E-C_6H_5-CH=CH-CO_2CH_3 \xrightarrow[48\%]{:CCl_2}$$ (2.34)

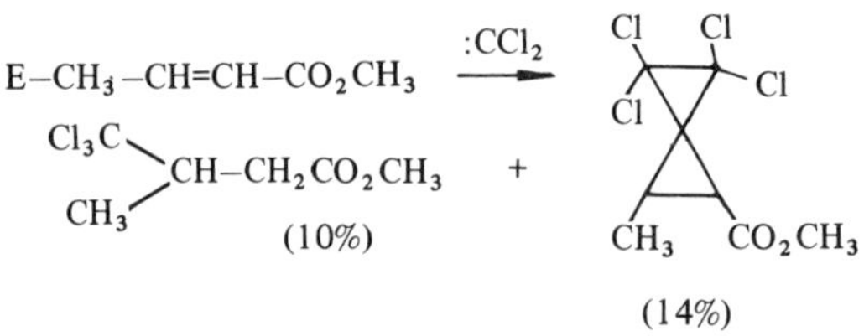

$$E-CH_3-CH=CH-CO_2CH_3 \xrightarrow{:CCl_2}$$ (2.35)

$$(CH_3)_2C=CH-CO_2CH_3 \xrightarrow[35\%]{:CCl_2}$$ (2.36)

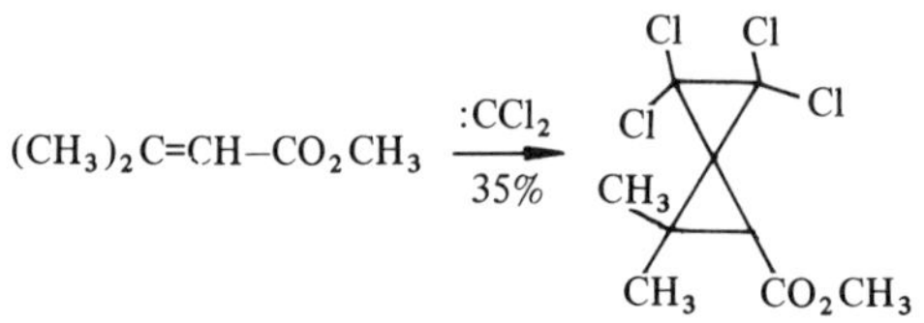

2.12 Dichlorocarbene Addition to Allylic Alcohols: A Cyclopentenone Synthon

The formation of 1,1-dichlorocyclopropanes from allylic alcohols is of particular synthetic value because the initial adducts can undergo rearrangement under acidic conditions to give good yields of cyclopentenones (see Eq. 2.37) [55]. This result is of mechanistic interest as well since alcohols react with dichlorocarbene to yield the corresponding chloride and carbon monoxide. This reaction is discussed in Chapter 3. In the case of allylic alcohols, dichlorocarbene prefers to react at the carbon-carbon double bond rather than at the hydroxyl function.

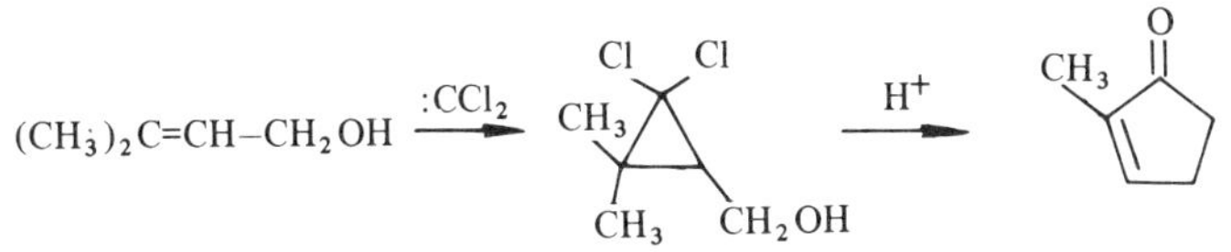

$$(CH_3)_2C=CH-CH_2OH \xrightarrow{:CCl_2} \text{(dichlorocyclopropane)} \xrightarrow{H^+} \text{(methylcyclopentenone)} \quad (2.37)$$

Table 2.9. Cyclopropanation of allylic alcohols [55] according to eq. 2.38

Substrate	% Yield	References	Catalyst
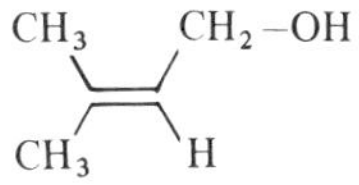 (CH₃)(CH₃)C=CH-CH₂OH (H)	74	[56]	CTMAC
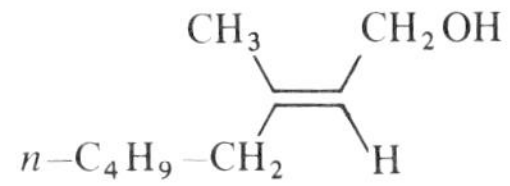 (CH₃)(n-C₄H₉-CH₂)C=CH-CH₂OH (H)	79	[56]	CTMAC
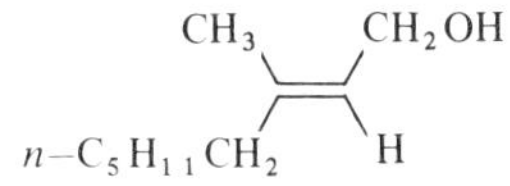 (CH₃)(n-C₅H₁₁CH₂)C=CH-CH₂OH (H)	60	[56]	CTMAC
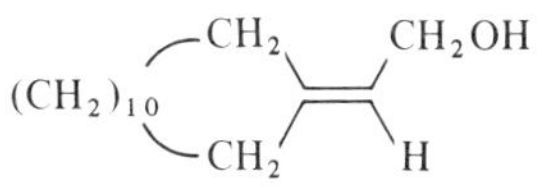 cyclic (CH₂)₁₀ with =CH-CH₂OH	none reported	[56]	CTMAC
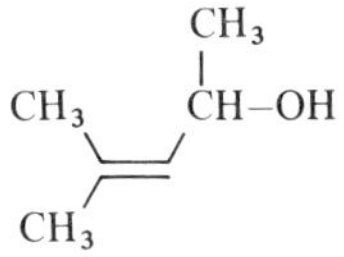 (CH₃)₂CH-OH with CH₃ geminal (dimethyl allylic)	92	[56]	CTMAC
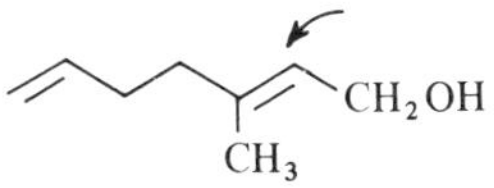 alkenyl-CH(CH₃)-CH₂OH (trisubstituted)	57	[56], [12]	$C_6H_5-CH_2-\overset{+}{N}(CH_3)_2-CH_2-CH_2-OH$ OH^-

The Reaction of Dichlorocarbene With Olefins

Excellent regioselectivity is observed in the reaction of 3-methyl-2,6-heptadien-1-ol with dichlorocarbene generated in the presence of benzyl-2-hydroxyethyldimethylammonium hydroxide as phase transfer catalyst. Dichlorocarbene adds, in this case, exclusively to the trisubstituted double bond bearing the hydroxymethyl function rather than to the terminal double bond (see Eq. 2.38). The results of dichlorocyclopropanation of several allylic olefin systems are recorded in Table 2.9.

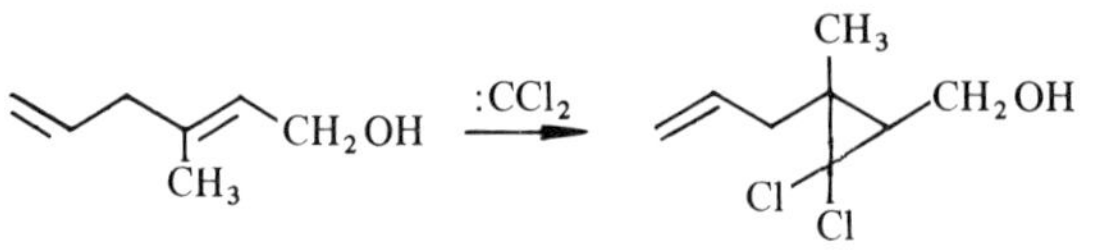

$$\text{(structure)} \xrightarrow{:CCl_2} \text{(structure)} \qquad (2.38)$$

2.13 Dichlorocarbene to Phenols: Reimer-Tiemann Reactions

Despite the long history of the Reimer-Tiemann reaction, little has been reported in the phase transfer literature on this subject. There appears to be a single report of work in which phase transfer generated dichlorocarbene has been added to a phenolic substrate [58]. 4-Methyl-2,6-nonamethylenephenol reacts with dichlorocarbene in chloroform solution to yield the product of ortho insertion (76%) as shown in equation 2.39. The reactions of dichlorocarbene with several other lipophilic phenols are likewise reported [58].

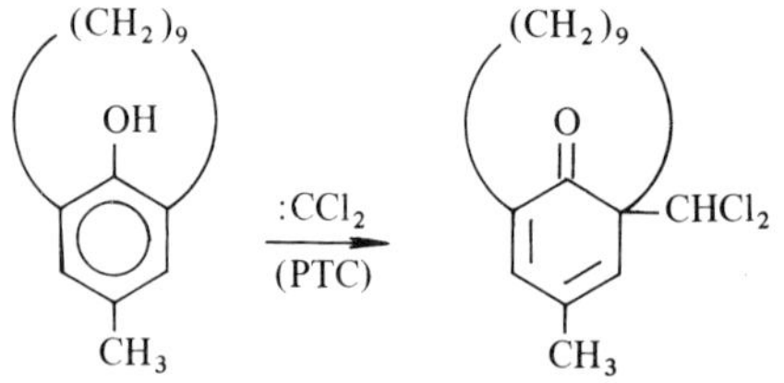

$$\text{(structure)} \xrightarrow[\text{(PTC)}]{:CCl_2} \text{(structure)} \qquad (2.39)$$

Note also that use of chiral *beta*-hydroxyethyltrialkylammonium salts as catalysts led to optically active 1,1-dichlorocyclopropanes with the prochiral substrates styrene and *trans*-propenyl benzene. While the degree of asymmetric induction was low, the feasibility of catalytic asymmetric induction in PTC reactions has certainly been demonstrated [12].

References

1. Starks, C. M., Napier, D. R.: British Patent 1,227,144, filed 4/5/67.
2. Starks, C. M.: J. Am. Chem. Soc. *93*, 195 (1971).
3. Makosza, M., Wawrzyniewicz, M.: Tetrahedron Let. *1969*, 4659.
4. Ikan, R., Markus, A., Goldschmidt, Z.: J. Chem. Soc. Perkin I *1972*, 2423.
5. Kraus, W., Klein, G., Sadlo, H., Rothenwöhrer, W.: Synthesis *1972*, 485.

References

6. Starks, C. M., Owens, R. M.: J. Am. Chem. Soc. *95*, 3613 (1973).
7. Dehmlow, E. V., Lissel, M.: Tetrahedron Let. *1976*, 1783.
8. Seyferth, D.: Acc. Chem. Res. *5*, 65 (1972).
9. Makosza, M.: Pure and Applied Chemistry *43*, 439 (1975).
10. Isagawa, K., Kimura, Y., Kwon, S.: J. Org. Chem. *39*, 3171 (1974).
11. Makosza, M., Kacprowicz, A., Fedorynski, M.: Tetrahedron Let., *1975*, 2119.
12. Hiyama, T., Sawada, H., Tsukanaka, M., Nozaki, H.: ibid. *1975*, 3013.
13. Dehmlow, E. V.: ibid. *1976*, 91.
14. Makosza, M., Fedorynski, M.: Synthesis *1974*, 274.
15. Boswell, R. F., Bass, R. G.: J. Org. Chem. *40*, 2419 (1975).
16. Gaoni, Y.: Tetrahedron Let. *1976*, 2167.
17. Joshi, G. C., Singh, N., Pande, L.: ibid. *1972*, 1461.
18. Kostikov, R. R., Molchanov, A. P.: Zhur. Organ. Khim. *11*, 1767 (1975).
19. Landini, D., Maia, A. M., Montanari, F., Pirisi, F. M.: Gazz. Chim. Ital. *105*, 863 (1975).
20. Gokel, G. W., Shepherd, J. P., Weber, W. P., Boettger, H. G., Holwick, J. L., McAdoo, D. J.: J. Org. Chem. *38*, 1913 (1973).
21. Makosza, M., Ludwikow, M.: Angew. Chem. Int. Ed. *13*, 665 (1974).
22. Miller, R. B.: Synth. Commun. *1974*, 341.
23. Makosza, M., Gajos, I.: Rocz. Chem. *48*, 1883 (1974).
24. Müller, B., Weyerstahl, P.: Tetrahedron *32*, 865 (1976).
25. Ranken, P. F., Harty, B. J., Kapiack, L., Battiste, M.: Syn. Comm. *3*, 311 (1973).
26. Dunkelblum, E., Singer, B.: Synthesis *1975*, 323.
27. Hahn, R. C., Johnson, R. P.: Tetrahedron Let. *1973*, 2149.
28. Moss, R. A., Smudin, D. J.: J. Org. Chem. *41*, 611 (1976).
29. Joshi, G. C., Singh, N., Pande, L. M.: Synthesis *1974*, 317.
30. Blume, G., Neumann, T., Weyerstahl, P.: Ann. *1975*, 201.
31. Dehmlow, E. V., Schönefeld, J.: Ann. *744*, 42 (1971).
32. Kajigaeshi, S., Kuroda, N., Matsumoto, G., Weda, E., Nagashima, A.: Tetrahedron Let. *1971*, 4887.
33. Effenberger, F., Kurtz, W.: Chem. Ber. *106*, 511 (1973).
34. Makosza, M., Kacprowicz, A.: Bull. Acad. Polon. Sci. *22*, 467 (1974).
35. Graefe, J., Adler, M., Mühlstädt, M.: Z. Chem. *15*, 14 (1975).
36. Kwantes, P. M., Klumpp, G. W.: Tetrahedron Let. *1976*, 707.
37. Dehmlow, E. V.: ibid. *1972*, 175.
38. Jefford, C. W., de los Heros, V., Burger, U.: Tetrahedron Let. *1976*, 703.
39. Jefford, C. W., Sweeney, A., Delay, F.: Helv. Chim. Acta. *55*, 2214 (1972).
40. Sasaki, T., Eguchi, S., Kiriyama, T.: J. Org. Chem. *38*, 2230 (1973).
41. Hart, H., Nitta, M.: Tetrahedron Let. *1974*, 2109.
42. Kwon, S., Nishimura, Y., Ikeda, M., Tamura, Y.: Synthesis *1976*, 249.
43. Weyerstahl, P., Blüme, G.: Tetrahedron *28*, 5281 (1972).
44. Tsunetsugu, J., Sato, M., Ebine, S.: J. Chem. Soc., Chem. Commun. *1973*, 363.
45. Dehmlow, E. V.: Tetrahedron Let. *1972*, 175.
46. Dehmlow, E. V., Klabuhn, H., Hass, E. C.: Ann. *1973*, 1063.
47. Dehmlow, S. S., Dehmlow, E. V.: Ann. *1973*, 1753.
48. Tishchenko, I. G., Glazkov, Y. V., Kulinkovich, O. G.: Zhur. Organ. Khim. *12*, 2510 (1973).
49. Greibrokk, T.: Acta Chem. Scand. *27*, 3207 (1973).
50. Dehmlow, E. V.: Tetrahedron Let. *1975*, 203.
51. Makosza, M., Ludwikow, M.: Angew. Chem. Int. Ed. *13*, 665 (1974).
52. Makosza, M., Gajos, I.: Bull. Acad. Polon. Sci. *20*, 33 (1972).
53. Dehmlow, E. V.: Ann. *758*, 148 (1972).
54. Kraus, W., Röthenwöhner, W., Sadlo, H., Klein, G.: Angew. Chem. Int. Ed. *11*, 641 (1972).
55. Dehmlow, E. V., Höfle, G.: Chem. Ber. *107*, 2760 (1976).
56. Hiyama, T., Tsukanaka, M., Nozaki, H.: J. Am. Chem. Soc. *96*, 3713 (1974).
57. DeSmet, A., Anteunis, M., Tavernier, D.: Bull. Soc. Chim. Belg. *84*, 67 (1975).
58. Hiyama, T., Ozaki, Y., Nozaki, H.: Tetrahedron *30*, 2661 (1974).

3. Reactions of Dichlorocarbene With Non-Olefinic Substrates

3.1 Introduction

The reactions of dichlorocarbene with a variety of olefinic and acetylenic substrates have been discussed in Chap. 2. We wish now to turn our attention to the reactions of this species with a number of other substrates which either are non-olefinic or contain double bonds which do not constitute the major reactive function. The substrates considered here are alcohols, imines, amines, amides, thioethers, and hydrocarbons. With the exception of the latter, all of these species appear to react by initial coordination of the electrophilic carbene with a Lewis basic site. Subsequent reactions attributable to differences in the basic function or involvement with other reactive sites lead to differences in the chemistry of each substrate, and each is therefore considered separately.

3.2 C–H Insertion Reactions

One of the most unusual reactions of dichlorocarbene is its insertion into C–H single bonds to yield a dichloromethyl substituted alkane. The yields thus far reported for a variety of substrates are generally poor, even though a large excess of dichlorocarbene seems required. Nevertheless, the ease of generating dichlorocarbene under phase transfer conditions and the inexpensiveness of this method, makes it a useful approach to carbon-hydrogen insertion/substitution.

Adamantane has been reported to react with a large excess of phase transfer generated dichlorocarbene to give a 91% yield (based on recovered starting material) of 1-dichloromethyladamantane (see Eq. 3.1). The fact that no 2-dichloromethyladamantane was obtained (despite a good material balance), suggests that a cationic or cationoid intermediate may be involved since there are three secondary hydrogens per tertiary hydrogen. In addition, electron withdrawing groups such as bromo or dichloromethyl seem to retard further reaction [1]. On the other hand, electron donating groups such as methoxy seem to aid in the C–H insertion. Methoxycyclohexane, for example, reacted with dichlorocarbene to give 1-methoxy-1-dichloromethylcyclohexane in 13% yield whereas methylcyclohexane gave only a 4% yield of the corresponding insertion product [3].

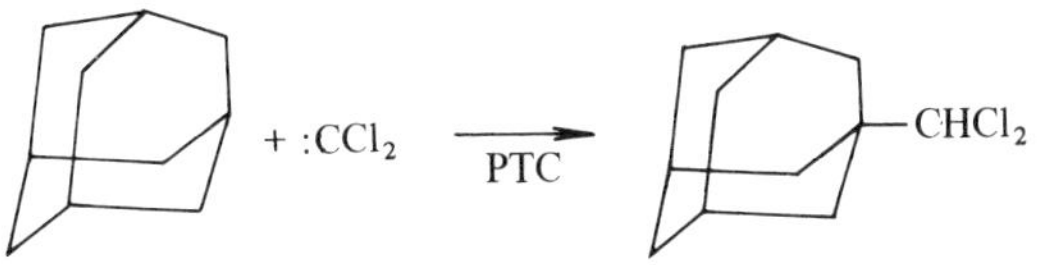

$$\text{adamantane} + :CCl_2 \xrightarrow{PTC} \text{adamantyl-CHCl}_2 \tag{3.1}$$

Table 3.1. Insertion of dichlorocarbene into C–H bonds

Number of carbons	Substrate	Product	% Yield	References
4	Tetrahydrofuran	2-dichloromethyltetrahydrofuran	18	[5]
6	Cyclohexane	(no reaction)	–	[3]
	Diisopropyl ether	Isopropyl-(2-dichloromethylisopropyl)ether	5	[3]
7	Methylcyclohexane	1-dichloromethyl-1-methylcyclohexane	4	[3]
	Methoxycyclohexane	1-dichloromethyl-1-methoxycyclohexane	13	[3]
8	Ethylbenzene	1,1-dichloro-2-phenylpropane	2	[5]
9	Cumene	1,1-dichloro-2-methyl-2-phenylpropane	18, 22	[5, 3]
10	Tetralin	1-dichloromethyltetralin	21	[5]
	cis-decalin	cis-9-dichloromethyldecalin	29	[3]
	trans-decalin	trans-9-dichloromethyldecalin	4	[3]
	Adamantane	1-dichloromethyladamantane	91	[1]
	1-bromoadamantane	(no reaction)	–	[1]
	1-methyl-1,2-dihydroquinoline	2-dichloromethyl-1-methyl-1,2-dihydroquinoline	30	[4]
		4-dichloromethyl-1-methyl-1,4-dihydroquinoline (as reaction intermediate)	20	[4]
11	1-methyladamantane	1-dichloromethyl-3-methyladamantane	≃100	[1]
	1-methoxyadamantane	1-dichloromethyl-3-methoxyadamantane +	poor	[1]
		2′,2′-dichloroethoxyadamantane	poor	[1]
	1-carboxyadamantane	(no reaction)	poor	[1]
12	1,3-dimethyladamantane	1-dichloromethyl-3,5-dimethyladamantane	≃100	[1]
14	Diamantane	1-dichloromethyldiamantane	63	[2]
		4-dichloromethyldiamantane	37	[2]

Insertion of phase transfer generated dichlorocarbene occurs with retention of configuration. Thus cis-decalin is transformed (29% yield) into cis-9-dichloromethyl-decalin [3]. The reaction is formulated in equation 3.2. Note that the ion pair intermediate illustrated cannot be a free carbonium ion/anion pair because the reaction occurs with retention. It is possible, however, that the reaction involves a tight ion pair corresponding to that illustrated.

$$\text{cis-decalin} + :CCl_2 \rightarrow [\text{ion pair with }^-CHCl_2] \rightarrow \text{cis-9-dichloromethyl-decalin} \quad (3.2)$$

The dichlorocarbene insertion reaction has been studied primarily with hydrocarbons and a number of examples are recorded in Table 3.1. Also included in this table are several examples of insertion reactions which occur alpha to ether oxygen or a tertiary nitrogen.

3.3 Reaction With Alcohols: Synthesis of Chlorides

The reactions of phase transfer generated dichlorocarbene with organic molecules possessing such heteroatoms as oxygen, nitrogen, and sulfur, and no other more reactive functionality, have led to a number of useful transformations. Alcohols react under phase transfer conditions to give chlorides [6]. Allylic alcohols which contain not only the hydroxyl group but a reactive double bond react preferentially at the olefinic site [7]. This preference is discussed in Sect. 2.11. In the absence of such complications, alcohols larger than about C_7 react to yield chlorides whereas small, water soluble alcohols generally yield the corresponding orthoformates [8].

The transformation of lipophilic alcohols into chlorides by the action of dichlorocarbene can be explained according to the following sequence [6]. Reaction of electrophilic dichlorocarbene with one of the oxygen lone pairs leads to a zwitterion (structure I in equation 3.3) in which the positive charge is on oxygen and the negative charge is on carbon. Proton loss from I affords alkoxydichlorocarbanion II.

$$R-OH + :CCl_2 \rightarrow R-\overset{+}{\underset{H}{O}}-\bar{C}Cl_2 \rightarrow R-O-\bar{C}Cl_2 \leftarrow RO^- + :CCl_2$$

$$\text{I} \qquad \text{II}$$

$$R-O-\overset{O}{\overset{\|}{C}}H \overset{\text{hydrolysis}}{\longleftarrow}$$

$$R-Cl \leftarrow R^+ C\equiv O\ Cl^- \leftarrow R-O-\overset{..}{C}-Cl \xrightarrow{S_Ni} RCl + CO$$

$$\text{III}$$

(3.3)

An alternative route to II is the direct reaction of alkoxide ion with dichlorocarbene. In any event, once formed, II can undergo protonation and hydrolysis to yield a formate ester or it can lose chloride (α-elimination) to form an alkoxychlorocarbene (III). The alkoxychlorocarbene (III) can then either expel carbon monoxide via an S_Ni type reaction or lose CO sequentially. The alkyl chloride results from ion pair collapse. This sequence is illustrated in equation 3.3.

l-Menthyl alcohol reacts with dichlorocarbene under phase transfer conditions to yield predominantly *l*-menthyl chloride, a fact which suggests that an S_Ni pathway is important although not unique. A significant amount of leakage into the sequential or carbonium ion pathway probably accounts for the production of almost equal amounts of *exo* and *endo*-2-norbornyl chlorides from *endo*-2-norbornanol. Furthermore, small amounts of formate esters are isolated in these reactions, suggesting that hydrolysis of II occurs but is not a major process. Several examples are recorded in Table 3.2.

Table 3.2. Reaction of alcohols with dichlorocarbene to yield chlorides [6]

Substrate	Product	% Yield
1-adamantyl alcohol	1-adamantyl chloride	94
l-methyl alcohol	*l*-menthyl chloride	no yield reported
	d-menthyl chloride	no yield reported
	$[\alpha]_d$ −23.6°	
Benzyl alcohol	Benzyl chloride	90
2-*exo*-norbornyl alcohol	2-*exo*-norbornyl chloride	90
2-*endo*-norbornyl alcohol	2-*exo*-norbornyl chloride	47
	2-*endo*-norbornyl chloride	44
1-adamantyl methyl alcohol	1-adamantyl methyl chloride	40
	Homo-adamantyl chloride	13
	1-adamantyl methyl formate	35

The phase transfer carbene reaction with methanol, 1- and 2-propanol, *t*-butanol and cyclohexanol is not successful, leading to complex product mixtures. Ethanol and 2,2,2-trifluoroethanol, on the other hand, yield the corresponding orthoformates in 36% and 33% yields respectively [8]. The transformation is formulated in equation 3.4.

$$CH_3CH_2OH \xrightarrow[PTC]{CHCl_3} (CH_3CH_2O)_3CH \qquad (3.4)$$

The reaction of an allylic alcohol with dichlorocarbene is noted above and in Sect. 2.11 to give products of cyclopropanation rather than the corresponding chloride [7]. In certain other cases, cyclopropanation of a molecule containing both a hydroxyl group and an olefin may lead to products of rearrangement and intramolecular reaction (see Sect. 2.6). The reaction of *endo*-5-hydroxymethyl-2-norbornene, a homoallylic alcohol, is of interest in this regard. The major product of reaction of this alcohol with dichlorocarbene is 3-chloro-5-oxatricyclo[5.2.1.0^{4,8}]dec-2-ene and probably arises from cyclopropanation [9]. The unstable dichlorocyclopropane ionizes with ring opening to yield a carbonium ion which is intercepted by the intramolecular hydroxyl group. Formation of the major product according to this sequence is shown in equation 3.5. The formation of the minor products in this reaction can be accounted for by an application of similar principles.

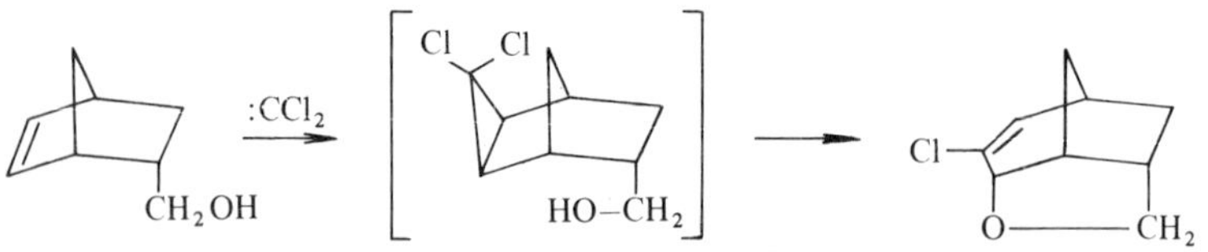

$$(3.5) [9]$$

The reaction of dichlorocarbene with 1,2-glycols is of little synthetic utility, however, it provides evidence for the existence of alkoxychlorocarbene intermediates. Specifically, products were found which could be accounted for if the alkoxychlorocarbene intermediate has a lifetime sufficient to be trapped by an internal nucleophile prior to decomposition.

Reaction of *cis*-1,2-cyclooctanediol with dichlorocarbene yields cyclooctanone, cyclooctene, and the dichlorocarbene adduct of cyclooctene according to equation 3.6 [10]. The formation of cyclooctanone is expected from a pinacol rearrangement of the α-hydroxy carbonium ion formed by decomposition of the α-hydroxyalkoxychlorocarbene adduct. On the other hand, reaction of the alkoxychlorocarbene with an adjacent alkoxide anion yields a cyclic dialkoxycarbene which decomposes with loss of carbon dioxide and formation of cyclooctene. Addition of dichlorocarbene to cyclooctene yields 2,2-dichlorobicyclo[8.1.0]-nonane.

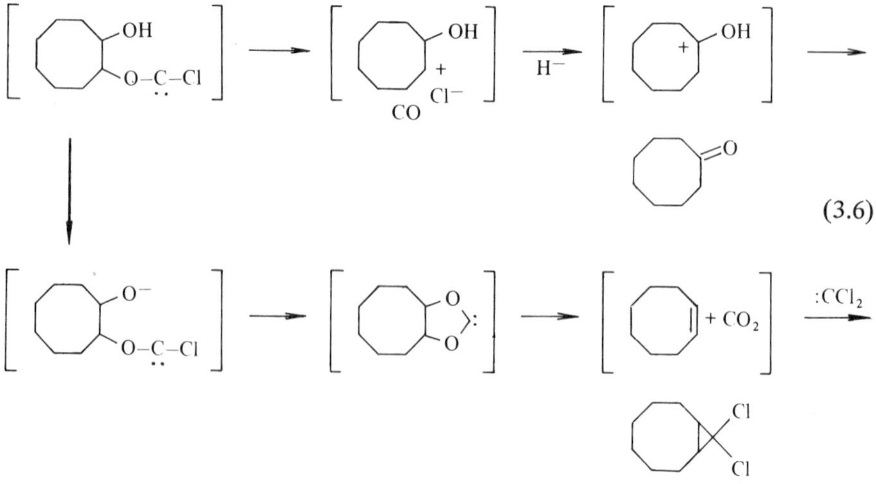

$$(3.6)$$

A similar reaction sequence accounts for the formation of 1,1-dichloro-2,2,3,3-tetramethylcyclopropane in 15% yield from the reaction of pinacol with dichlorocarbene. Likewise, reaction of *meso*-dihydrobenzoin with dichlorocarbene gave a 15% yield of *cis*-stilbene [10] (Eq. 3.7).

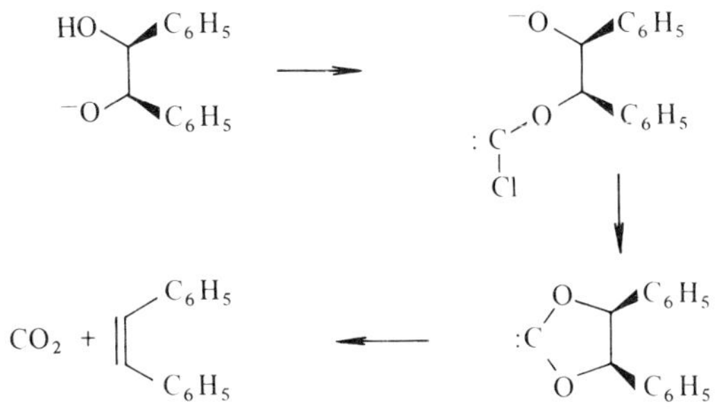

(3.7)

3.4 Carbene Addition to Imines

The facile addition of dichlorocarbenes to olefins under phase transfer conditions has also been observed with imines. The carbon-nitrogen double bond reacts with dichlorocarbene under conditions similar to those required for isolated alkenes. Thus, dichlorocarbene addition to C,N-diaryl substituted Schiff's bases afford good yields of 1,3-diaryl-2,2-dichloroaziridines (Eq. 3.8). Hydrolysis of the C,N-diarylaziridines (IV) to the corresponding aryl-α-chloroacetanilides (V) is also reported (Eq. 3.9) and examples of both processes are recorded in Table 3.3.

Table 3.3. Carbene addition to imines

R^1	R^2	IV % Yield	References	V % Yield	References
C_6H_5	C_6H_5	79	[12]	–	–
C_6H_5	C_6H_5	88	[11]	–	–
4-$CH_3OC_6H_4$	C_6H_5	90	[11]	48	[12]
4-ClC_6H_4	C_6H_5	78	[11]	–	–
4-$NO_2C_6H_4$	C_6H_5	52	[12]	–	–
4-$NO_2C_6H_4$	C_6H_5	67	[11]	–	–
C_6H_5	4-$CH_3C_6H_4$	85	[11]	–	–
C_6H_5	4-ClC_6H_4	83	[11]	–	–
C_6H_5	4-$NO_2C_6H_4$	69	[11]	–	–
4-$NO_2C_6H_4$	4-$CH_3OC_6H_4$	59	[12]	–	–
4-$NO_2C_6H_4$	$C_6H_5CH_2$	–	–	67	[12]
4-$NO_2C_6H_4$	c-C_6H_{11}	–	–	74	[12]
i-C_3H_7	t-C_4H_9	–	–	10	[12]

$$R^1-CH=N-R^2 \xrightarrow{:CCl_2} R^1-\overset{\overset{\displaystyle CCl_2}{\displaystyle \diagup\diagdown}}{CH-N}-R^2 \quad IV \qquad (3.8)$$

$$R^1-\overset{\overset{\displaystyle CCl_2}{\displaystyle \diagup\diagdown}}{CH-N}-R^2 \xrightarrow{H_2O} R^1-CHCl-CO-NH-R^2 \quad V \qquad (3.9)$$

3.5 Addition to Primary Amines: Synthesis of Isonitriles

The Hofmann carbylamine synthesis of isonitriles by the reaction of chloroform and base with primary amines was first reported over 100 years ago, but the reaction was not preparatively useful. For this reason, the two step process involving formylation of the amine followed by dehydration was the method of choice. The phase transfer carbylamine reaction is inexpensive and convenient and the yields are such that it is competitive with the dehydration technique in many cases [13–15].

The mechanism of the reaction (see Eq. 3.10) is believed to involve initial coordination of the electrophilic carbene to the nitrogen lone pair. The zwitterion thus formed undergoes proton transfer from nitrogen to carbon yielding an N-substituted aminodichloromethane. Base catalyzed beta-elimination of the elements of HCl followed by alpha-elimination of another mole of HCl yields the desired isonitrile.

$$R-NH_2 + :CCl_2 \rightarrow R-\overset{+}{N}H_2-\overset{-}{C}Cl_2 \rightarrow$$

$$R-NH-CHCl_2 \xrightarrow{HO^-} R-N=CHCl \xrightarrow{HO^-} R-N\!\!\equiv\!\!C: \qquad (3.10)$$

A series of isonitriles has been prepared by the Hofmann carbylamine reaction and several examples are recorded in Table 3.4.

Methyl and ethyl are more conveniently prepared using dibromocarbene (from bromoform) and the yields reported are therefore for the bromoform case.

Table 3.4. Phase transfer carbylamine synthesis of isonitriles

$R-NH_2 \cdot \xrightarrow{:CX_2} RNC + 2HX$

R	% Yield	References
CH_3	24	[14]
C_2H_5	47	[14]
n-C_4H_9	60	[13, 14]
t-C_4H_9	66	[15]
$C_6H_5CH_2$	55	[13, 14]
n-$C_{12}H_{25}$	41	[13, 14]
c-C_6H_{11}	48	[13, 14]
C_6H_5	57	[13, 14]

3.6 Reaction With Hydrazine, Secondary, and Tertiary Amines

The reaction of hydrazine with chloroform and base to yield diazomethane was reported by Staudinger some sixty years ago but remained of limited utility. The application of 18-crown-6 mediated phase transfer catalysis to this reaction has made it a preparatively useful method. By this method, hydrazine hydrate can be conveniently converted to diazomethane in almost 50% yield (see Eq. 3.11) [16].

$$H_2N-NH_2 \xrightarrow[\text{18-crown-6}]{\text{CHCl}_3,\text{ KOH}} CH_2=N=N \quad (48\%) \quad (3.11)$$

Secondary amines react readily with dichlorocarbene to yield secondary formamides in good yield [17, 18]. Apparently, the secondary amine coordinates with dichlorocarbene, and after proton transfer a dialkylaminodichloromethane is produced. The latter undergoes basic hydrolysis under phase transfer conditions which evidently stops before appreciable cleavage of the formamide occurs. The reaction is formulated in equation 3.12 and several examples are tabulated in Table 3.5.

$$R^1R^2N-H + :CCl_2 \rightarrow [R^1R^2\overset{+}{N}H-\overset{-}{C}Cl_2] \rightarrow R^1R^2N-CHCl_2 \xrightarrow{H_2O}$$

$$R^1R^2N-CHO \quad (3.12)$$

Table 3.5. Preparation of secondary formamides from secondary amines

R^1	R^2	% Yield	References
C_2H_5	C_2H_5	78	[17]
		62	[18]
n-C_3H_7	n-C_3H_7	45	[17]
i-C_3H_7	i-C_3H_7	47	[18]
n-C_4H_9	n-C_4H_9	83	[17]
i-C_4H_9	i-C_4H_9	66	[17]
		70	[18]
n-C_5H_{11}	n-C_5H_{11}	90	[18]
−(CH$_2$)$_4$−		39	[17]
−(CH$_2$)$_5$−		54	[17]
		79	[18]
CH_3	c-C_6H_{11}	62	[17]
c-C_6H_{11}	c-C_6H_{11}	42	[17]
		28	[18]
CH_3	C_6H_5	76	[17]
		78	[18]
CH_3CH_2	C_6H_5	88	[18]
$CH_2=CH-CH_2$	$CH_2=CH-CH_2$	61	[17]
		60	[18]
C_6H_5	C_6H_5	68	[18]

Tertiary amines are generally inert to dichlorocarbene under phase transfer catalytic conditions. In fact, tertiary amines have been utilized as catalysts in the dichlorocyclopropanation reaction of olefins (see Sects. 1.8, 2.2) [19]. The inertness of tertiary amines is probably a function of its reluctance to transfer an alkyl group from cationic nitrogen to anionic carbon after the initial coordination step has occurred.

One example of the reaction of dichlorocarbene with a tertiary amine under phase transfer conditions has been reported [20]. 5,7-Diphenyl-1,3-diazaadamantan-6-one is transformed into 1,5-diphenyl-N,N'-diformylbispidin-9-one in 23% yield. This unusual reaction may be rationalized as follows. Coordination of dichlorocarbene with a nitrogen lone pair yields a zwitterion (VI). Neighboring nitrogen assists fragmentation and the iminium zwitterion VII results. Protonation and hydrolysis of VII followed by dichlorocarbene reaction with the resulting secondary amine yields the bis-formamide VIII. The sequence is formulated in equation 3.13.

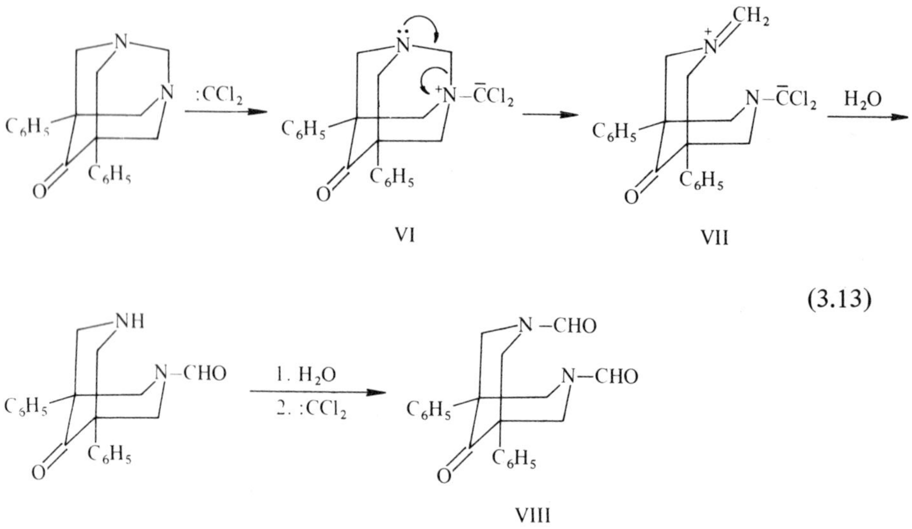

(3.13)

3.7 Dehydration With Dichlorocarbene

The reaction of dichlorocarbene with primary amides, amidines, thioamides, and aldoximes all yield the corresponding nitriles. N,N-Disubstituted ureas likewise yield the corresponding N,N-disubstituted amine nitrile. This reaction amounts to a dehydration in the case of primary amides, aldoximes, and N,N-disubstituted ureas; the elements of hydrogen sulfide are lost from thioamides, and HCN from amidines.

An economical mechanistic proposal assumes initial coordination of the carbene to the amide carbonyl. Intramolecular transfer of a proton from the positively charged nitrogen to the negatively charged carbon of the zwitterion constitutes the second step. The final step is base catalyzed removal of the imine proton with concurrent fragmentation producing the nitrile and formate after hydrolysis. The sequence is illustrated in equation 3.14.

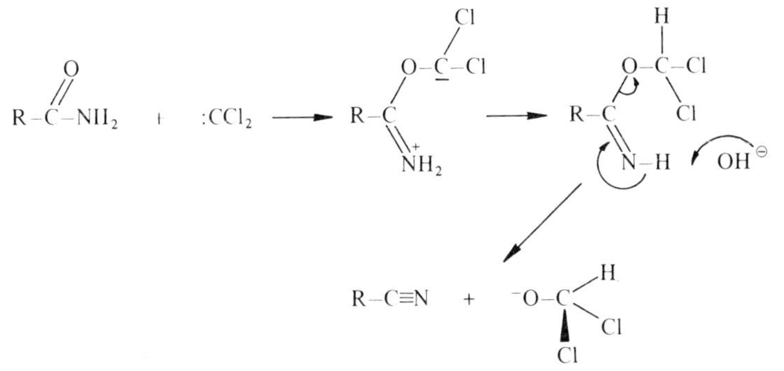

$$R-\overset{O}{\overset{\|}{C}}-NH_2 + :CCl_2 \longrightarrow R-\overset{O-C(Cl)_2}{\underset{NH_2}{C}} \longrightarrow R-C(\text{O-CHCl}_2)=N-H \xrightarrow{OH^-} R-C\equiv N + {}^-O-CHCl_2 \quad (3.14)$$

An alternative to this proposal is that dichlorocarbene reacts with the hydroxyl of the imino tautomer to yield a zwitterionic intermediate. Base catalyzed deprotonation then yields the nitrile according to equation 3.15.

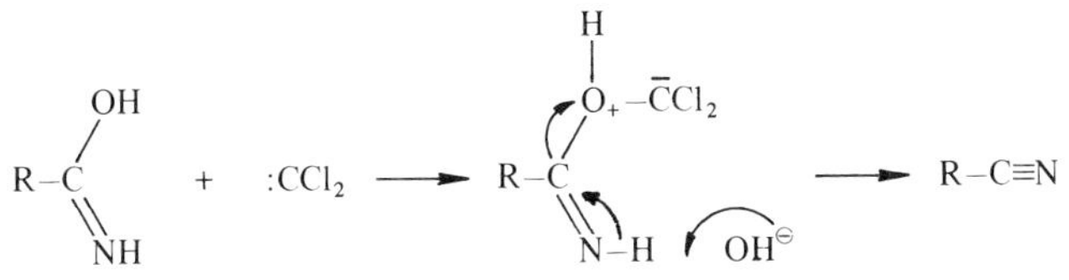

$$R-\overset{OH}{\underset{NH}{C}} + :CCl_2 \longrightarrow R-C(\overset{+}{O}-\bar{C}Cl_2)=N-H \xrightarrow{OH^-} R-C\equiv N \quad (3.15)$$

Table 3.6. Dehydration with dichlorocarbene

Substrate	Product	% Yield	Ref.
$C_6H_5\overset{O}{\overset{\|}{C}}-NH_2$	$C_6H_5-C\equiv N$	84 / 85	[21] / [22]
$C_6H_5-\overset{S}{\overset{\|}{C}}-NH_2$	$C_6H_5-C\equiv N$	67	[21]
$C_6H_5-\overset{NH}{\overset{\|}{C}}-NH_2$	$C_6H_5-C\equiv N$	92	[23]
$C_6H_5-CH=N-OH$	$C_6H_5-C\equiv N$	51	[21]
$4\text{-}Cl-C_6H_4-\overset{NH}{\overset{\|}{C}}-NH_2$	$4\text{-}Cl-C_6H_4-C\equiv N$	85	[23]
$4\text{-}Cl-C_6H_4-\overset{O}{\overset{\|}{C}}-NH_2$	$4\text{-}Cl-C_6H_4-C\equiv N$	79	[22]

Table 3.6 (continued)

Substrate	Product	% Yield	Ref.
4-CH$_3$O–C$_6$H$_4$–C(=NH)–NH$_2$	4-CH$_3$O–C$_6$H$_4$–C≡N	78	[23]
3-CH$_3$O–C$_6$H$_4$–C(=O)–NH$_2$	3-CH$_3$O–C$_6$H$_4$–C≡N	72	[22]
4-NO$_2$–C$_6$H$_4$–C(=O)–NH$_2$	4-NO$_2$–C$_6$H$_4$–C≡N	40	[22]
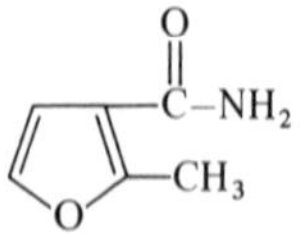 3-(C(=O)NH$_2$)-2-CH$_3$-furan	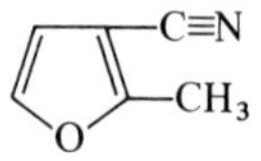 3-(C≡N)-2-CH$_3$-furan	80	[22]
C$_6$H$_5$–CH=CH–C(=O)–NH$_2$	C$_6$H$_5$–CH=CH–C≡N	52	[21]
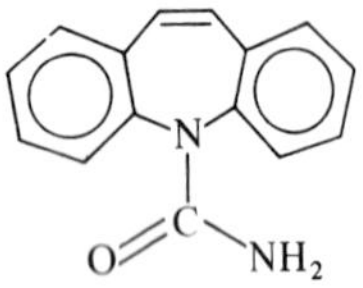 5H-dibenz[b,f]azepine-5-carboxamide	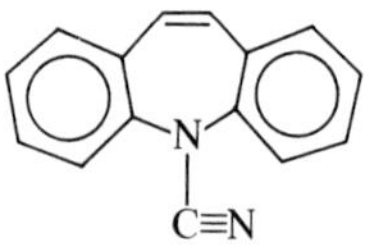 5H-dibenz[b,f]azepine-5-carbonitrile	82	[21]
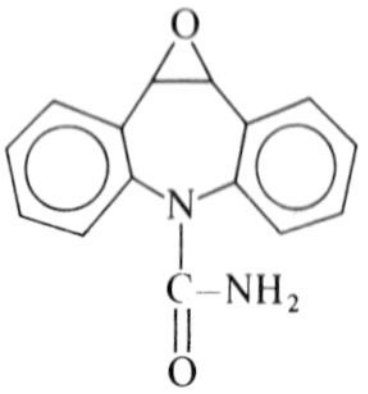 10,11-epoxy-5H-dibenz[b,f]azepine-5-carboxamide	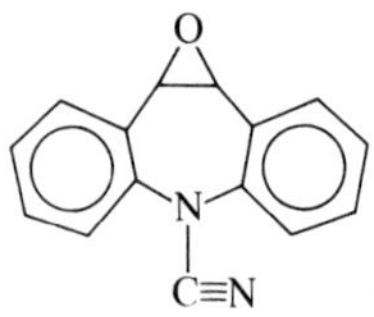 10,11-epoxy-5H-dibenz[b,f]azepine-5-carbonitrile	58	[21]
CH$_3$–C(=NH)–NH$_2$	CH$_3$–C≡N	61	[23]
CH$_3$–CH$_2$–C(=O)–NH$_2$	CH$_3$–CH$_2$–C≡N	45 7	[21] [22]
CH$_3$–CH$_2$–C(=NH)–NH$_2$	CH$_3$–CH$_2$–C≡N	84	[23]
n-C$_3$H$_7$–C(=O)–NH$_2$	n-C$_3$H$_7$–C≡N	55	[22]
i-C$_3$H$_7$–C(=NH)–NH$_2$	i-C$_3$H$_7$–C≡N	90	[23]
i-C$_3$H$_7$–C(=O)–NH$_2$	i-C$_3$H$_7$–C≡N	94	[22]

Table 3.6 (continued)

Substrate	Product	% Yield	Ref.
$t\text{-}C_4H_9\text{-}\overset{O}{\underset{\|}{C}}\text{-}NH_2$	$t\text{-}C_4H_9\text{-}C\equiv N$	89	[22]
$CH_3(CH_2)_4\text{-}\overset{O}{\underset{\|}{C}}\text{-}NH_2$	$CH_3\text{-}(CH_2)_4\text{-}C\equiv N$	95	[22]
$c\text{-}C_6H_{11}\text{-}\overset{O}{\underset{\|}{C}}\text{-}NH_2$	$c\text{-}C_6H_{11}\text{-}C\equiv N$	95	[22]
$C_6H_5\text{-}CH_2\text{-}\overset{O}{\underset{\|}{C}}\text{-}NH_2$	$C_6H_5\text{-}CH_2\text{-}C\equiv N$	75	[21]
$H_2N\text{-}\overset{O}{\underset{\|}{C}}(CH_2)_4\text{-}\overset{O}{\underset{\|}{C}}\text{-}NH_2$	$NC(CH_2)_4\text{-}C\equiv N$	12	[21]

3.8 Miscellaneous Reactions of Dichlorocarbene

Reaction of dichlorocarbene with substituted benzaldehydes affords the corresponding mandelic acids. It might be supposed that this reaction does not involve dichlorocarbene but rather its precursor, the trichloromethide ion. Nucleophilic addition of the trichloromethyl anion to benzaldehyde should yield phenyl trichloromethyl carbinol. In control experiments, however, phenyl trichloromethyl carbinol was not hydrolyzed to mandelic acid under the reaction conditions. It was suggested that the reaction involves dichlorocarbene addition to the benzaldehyde carbonyl to form 2,2-dichloro-3-phenyloxirane followed by rearrangement to α-chlorophenylacetyl chloride which then is hydrolyzed as shown in equation 3.16. Mandelic, 4-methylmandelic, and 4-methoxymandelic acids were produced in 75%, 80%, and 80% yields, respectively [24].

$$Ar\text{-}CHO \xrightarrow{:CCl_2} Ar\text{-}\overset{O}{\overset{/\backslash}{CH\text{-}CCl_2}} \rightarrow Ar\text{-}CHCl\text{-}CO\text{-}Cl$$

$$\xrightarrow{Cl_3C^-} Ar\text{-}\underset{HO}{\overset{H}{\underset{\|}{C}}}\text{-}CCl_3 \not\rightarrow Ar\text{-}\underset{OH}{\overset{H}{\underset{\|}{C}}}\text{-}CO_2H \tag{3.16}$$

Thus far, there is only one dichlorocarbene reaction with a thioether reported [25]. Dichlorocarbene reacts with the allylic sulfide to form an ylid which undergoes a facile [2, 3] sigmatropic rearrangement. The product obtained after hydrolysis and chromatography is a mixture of β,γ-unsaturated-S-phenyl esters according to equation 3.17.

Reactions of Dichlorocarbene With Non-Olefinic Substrates

(3.17)

53%

Chloroform yields both the trichloromethyl anion and dichlorocarbene as reactive intermediates under basic phase transfer conditions. The trichloromethyl anion reacts with phenylmercuric chloride under these conditions to yield phenyl(trichloromethyl)mercury (72%). The product is unstable, however, to the 50% aqueous sodium hydroxide solution usually used in phase transfer catalysis. When 10–15% aqueous sodium hydroxide solution was used, while maintaining the ionic strength by addition of potassium fluoride, the product survived. Reasonable yields of the mercury compound were thus obtained and the reaction was successfully extended to bromodichloromethane [yielding 64% of phenyl(bromodichloromethyl)mercury] and bromoform [yielding phenyl(tribromomethyl)mercury, 54%]. The transformation is illustrated in equation 3.18 [26].

$$C_6H_5-Hg-Cl \xrightarrow[\text{NaOH/KF}]{CHX_3/Q^+X^-} C_6H_5-Hg-CX_3 \qquad (3.18)$$

References

1. Tabushi, I., Yoshida, Z., Takahashi, N.: J. Am. Chem. Soc. *92*, 6670 (1970).
2. Tabushi, I., Aoyama, Y., Takahashi, N., Gund, T. M., Schleyer, P. V. R.: Tetrahedron Let. *1973*, 107.
3. Dehmlow, E. V., Tetrahedron *27*, 4071 (1971).
4. Greibrokk, T.: Tetrahedron Let. *1972*, 1665.
5. Makosza, M., Fedorynski, M.: Rocz. Chem. *46*, 311 (1972).
6. Tabushi, I., Yoshida, Z., Takahashi, N.: J. Am. Chem. Soc. *93*, 1820 (1971).
7. Hiyama, T., Tsukanaka, M., Nozaki, H.: J. Am. Chem. Soc. *96*, 3713 (1974).
8. Makosza, M., Jerzak, B., Fedorynski, M.: Rocz. Chem. *49*, 1783 (1975).
9. Sasaki, T., Eguchi, S., Kiriyama, T.: J. Org. Chem. *38*, 2230 (1973).
10. Stromquist, P., Radcliffe, M., Weber, W. P.: Tetrahedron Let. *1973*, 4523.
11. Graefe, J.: Z. Chem. *14*, 469 (1974).
12. Makosza, M., Kacprowicz, A.: Rocz. Chem. *48*, 2129 (1974).
13. Weber, W. P., Gokel, G. W.: Tetrahedron Let. *1972*, 1637.

14. Weber, W. P., Gokel, G. W., Ugi, I. K.: Angew. Chem., Int. Ed. *11*, 530 (1972).
15. Gokel, G. W., Widera, R. P., Weber, W. P.: Organic Syntheses *55*, 96 (1975).
16. Sepp, D. T., Scherer, K. V., Weber, W. P.: Tetrahedron Let. *1974*, 2983.
17. Graefe, J., Fröhlich, I., Mühlstädt, M.: Z. Chem. *14*, 434 (1974).
18. Makosza, M., Kacprowicz, A.: Rocz. Chem. *49*, 1627 (1975).
19. Isagawa, K., Kimura, Y., Kwon, S.: J. Org. Chem. *39*, 3171 (1974).
20. Sasaki, T., Eguchi, S., Kiriyama, T., Sakito, Y.: J. Org. Chem. *38*, 1648 (1973).
21. Saraie, T., Ishiguro, I., Kawashima, K., Morita, K.: Tetrahedron Let. *1973*, 2121.
22. Höfle, G.: Z. Naturforsch. *28b*, 831 (1973).
23. Graefe, J.: Z. Chem. *15*, 301 (1975).
24. Merz, A.: Synthesis *1974*, 724.
25. Andrews, G., Evans, D. A.: Tetrahedron Let. *1972*, 5121.
26. Fedorynski, M., Makosza, M.: J. Organometal. Chem. *51*, 89 (1973).

4. Dibromocarbene and Other Carbenes

4.1 Introduction

The success of phase transfer cyclopropanations with dichlorocarbene naturally stimulated a search for related examples. Numerous trihalomethanes besides chloroform were subjected to phase transfer conditions in the expectation that a wide variety of dihalocyclopropanes would result. Notable among these attempts is the phase transfer generation of dibromocarbene which is discussed in Sects. 4.2–4.5. In addition to the miscellany of dihalocarbenes, the principal successes have been in the formation of sulfur containing (Sect. 4.6) and unsaturated carbenes (Sect. 4.7).

The mechanism of carbene formation has been discussed in Sect. 2.2 and phase transfer catalysis has been discussed in general terms in Sects. 1.4 and 1.7–1.9. The principles presented at earlier stages in this book generally apply to the carbenes discussed in this chapter, so detailed mechanistic discussions will be limited to those cases which are either peculiar or non-obvious.

4.2 Dibromocarbene Addition to Simple Olefins

Early attempts to generate dibromocarbenes were conducted under conditions virtually identical to those used for the preparation of dichlorocarbene. Unfortunately, both the time and excess of bromoform required for the cyclopropanation of olefins were unacceptably high [1]. The high kinetic acidity of bromoform should have made these reactions straightforward, but dibromocarbene apparently undergoes hydrolysis with greater facilty than does dichlorocarbene [2]. The high cost of bromoform and a desire to decrease reaction times have led to several improvements in methodology for dibromocarbene generation. Addition of, for example, small amounts of lower alcohols (e.g., ethanol and n-butanol) leads to increased reaction rates and increases in yield of the dibromocarbene-olefin adducts of 10–30% [3]. It has been suggested that alcohol specifically solvates the tribromomethyl anion, retarding its decomposition.

An alternate interpretation of the effect of alcohol is as follows: like chloroform, bromoform is deprotonated at the interface and, because the tribromomethyl anion undergoes hydrolysis more readily than the trichloromethyl anion, much of it is destroyed. A small amount of alcohol present allows alkoxide anion to be generated and this is more soluble in organic media when paired with a quaternary ammonium ion than is the hydroxide ion. It may be therefore, that the dibromocarbene is generated in the bulk organic phase and not lost to rapid hydrolysis only when bromoform is deprotonated in the bulk organic phase. If a soluble base were required for the reaction, a trialkylamine should be a good catalyst in line with Makosza's suggestions (see Sects. 1.7 and 2.2) [4]. In fact, tributylamine is found to effectively catalyze the reaction of bromoform and base with simple olefins. Likewise, β-hydroxyethyltrialkylammonium salts are efficient catalysts for this reaction. Moreover, the latter require no additional alcohol as co-catalyst [5]. Dibromocarbene has also been generated in the presence of catalytic amounts of dibenzo-18-crown-6 [6] which is effective but appears to offer no special advantage over quaternary salts.

Dibromocarbene has been successfully generated under phase transfer conditions and added to a variety of olefinic substrates. These include isolated double bond systems, styrenes, conjugated dienes, allenes, cyclopropanated olefins, vinyl ethers, allylic halides, and enynes. It is interesting to note that with the latter class of compounds, dibromocarbene addition to double bonds appears to be favored over addition to triple bonds. The simple addition of dibromocarbene according to equation 4.1 to a number of substrates is recorded in Table 4.1.

$$CHBr_3 + R_2C=CR_2 \xrightarrow[\text{catalyst}]{\text{Base, solvent}} R_2C\underset{\diagdown}{\overset{CBr_2}{\diagup}}CR_2 \qquad (4.1)$$

Table 4.1. Simple addition of dibromocarbene to carbon-carbon double bonds

Number of carbons	Alkene	% Yield	Catalyst	References
3	$CH_2=CH-CH_2-Br$	51	$C_6H_5-CH_2-\overset{\overset{CH_3}{\mid_+}}{\underset{\underset{CH_3}{\mid}}{N}}-CH_2-CH_2-OH$	[5]
5	$(CH_3)_2C=CH(CH_3)$ (cis 2-methyl-2-butene structure)	73	BTEAC	[1]
		81	BTEAC/ROH	[3]
		64	BTEAC	[2]
5	$(CH_3)_2C=CH-CH_2Cl$	62	BTEAC/ROH	[3]
6	1-hexene	61	TBA	[4]

Table 4.1 (continued)

Number of carbons	Alkene	% Yield	Catalyst	References
6	 cyclohexene	72	BTEAC	[1]
		46	BTEAC	[2]
		57	BTEAC/ROH	[3]
		76	TBA	[4]
		74	DB-18-C-6	[6]
		71	$C_6H_5-CH_2-\overset{+}{\underset{CH_3}{N}}(CH_3)-CH_2-CH_2-OH$	[5]
6	$n\text{-}C_4H_9O-CH=CH_2$	50	BTEAC/ROH	[3]
7	1-heptene	67	TBA	[4]
7	$(CH_3)_2C=C=C(CH_3)_2$	10 diadduct	BTEAC	[1]
7	$(CH_3)_2C=CH-CH_2-CO_2CH_3$	89	BTEAC/ROH	[3]
8	1-octene	75	TBA	[4]
8	Cyclooctene	70	BTEAC	[1]
8	1,5-cyclooctadiene	27 diadduct	BTEAC	[7]
8	Styrene	13	BTEAC	[7]
		66	BTEAC	[1]
		78	BTEAC/ROH	[3]
		49	BTEAC	[2]
		88	TBA	[4]
		77	$C_6H_5-CH_2-\overset{+}{\underset{CH_3}{N}}(CH_3)-CH_2-CH_2-OH$	[5]
8	$CH_2=\underset{CH_3}{C}-C\equiv C-\underset{CH_3}{C}=CH_2$	34 diadduct to C=C bonds	BTEAC	[8]
8	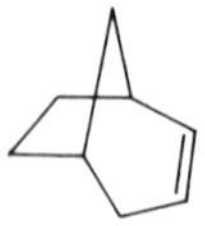 norbornene	94	BTEAC	[9]
8	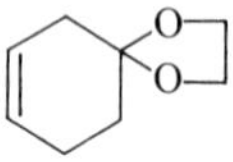 cyclohexene dioxolane	–	–	[10]
9	α-methylstyrene	80	BTEAC/ROH	[3]
		48	BTEAC	[2]

Table 4.1 (continued)

Number of carbons	Alkene	% Yield	Catalyst	References
9	E-propenyl benzene	23	BTEAC	[7]
10	o-divinyl benzene	50 diadduct	BTEAC	[1]
10	 (cyclopropylidene)-C$_6$H$_5$	50	BTEAC/ROH	[11]
11	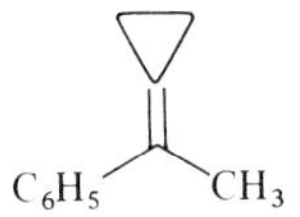 (cyclopropylidene)(C$_6$H$_5$)(CH$_3$)	22	BTEAC/ROH	[11]
14	E-stilbene	32 / 35	BTEAC / DB-18-C-6	[7] / [6]
14	1,1-diphenylethylene	34	BTEAC	[7]
16	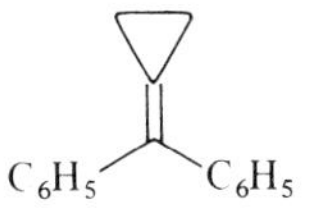 (cyclopropylidene)(C$_6$H$_5$)(C$_6$H$_5$)	60	BTEAC/ROH	[11]

The addition of dibromocarbene to prochiral olefins, when conducted in the presence of a chiral β-hydroxyethyltrialkylammonium salt leads to products which possess small enantiomeric excesses. Thus styrene and β-methylstyrene undergo chiral dibromocyclopropanation in 73% and 60% yields respectively, but their optical rotations (at the sodium D-line) are less than one degree [5].

4.3 Dibromocarbene Addition to Strained Alkenes

Addition of dibromocarbene to strained carbon-carbon double bonds can lead to rearranged products. These products result from ionization of the initial 1,1-dibromocyclopropane adduct with ring opening to yield an allylic carbonium ion/bromide anion pair. Alkyl group shifts followed by ion pair collapse can then lead to rearranged products (see Sect. 2.6). Three examples of this phenomenon are illustrated as equations 4.2–4.4 [9, 12, 13]. For related examples, see equations 2.17–2.21.

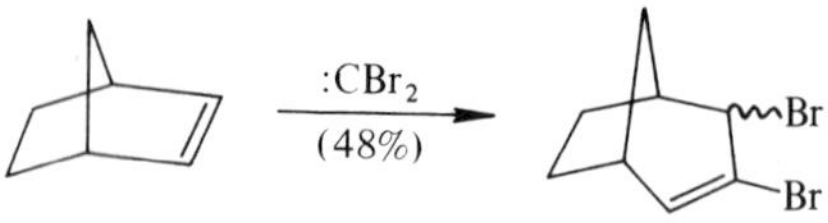

4.4 Dibromocarbene Addition to Indoles

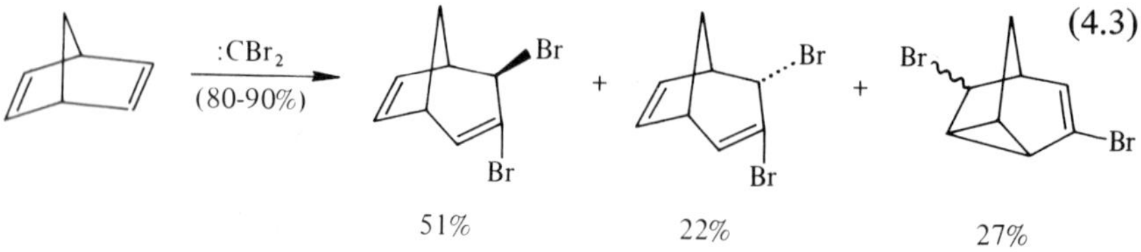

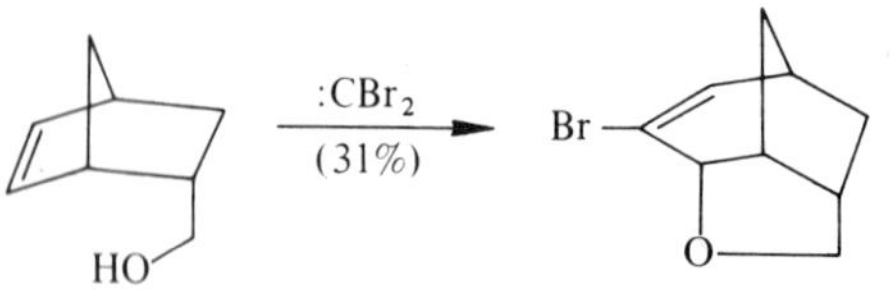

The addition of dibromocarbene to indoles leads to 3-bromoquinolines (Eqs. 4.5 and 4.6) [32]. Both quaternary ammonium salts and crown ethers are effective PT catalysts.

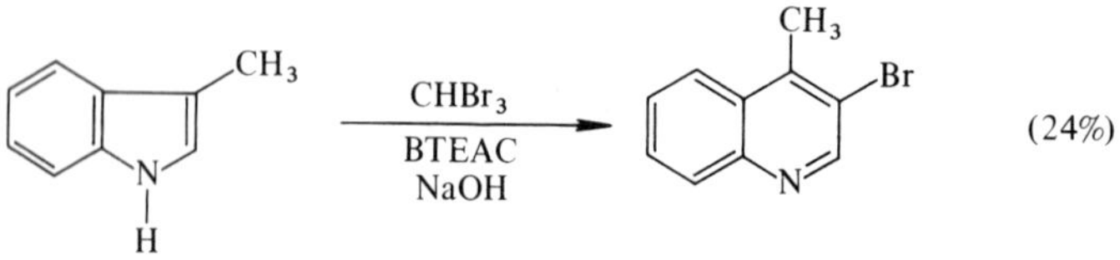

4.5 Dibromocarbene Addition to Michael Acceptors

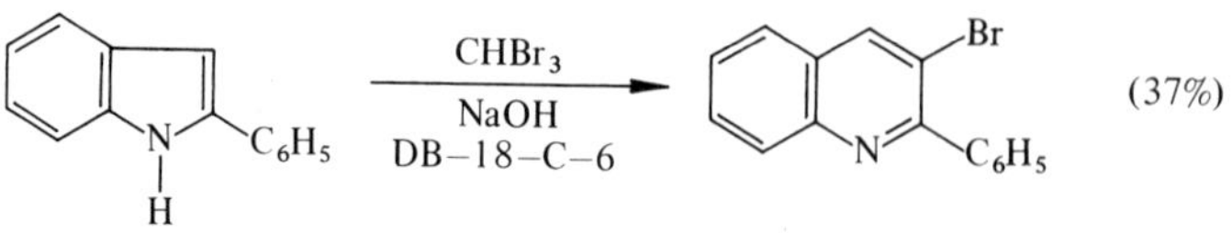

Numerous examples of dibromocarbene addition to the carbon-carbon double bond of α,β-unsaturated esters and ketones have been reported. It is of interest that with these substrates, simple quaternary alkylammonium salts (like BTEAC) are efficient phase transfer catalysts [14, 15]. Neither excess bromoform nor added alcohol are required to obtain good yields of the 1,1-dibromoadducts. It is possible in this case, that the reaction involves Michael addition of the tribromomethyl anion to the β-carbon atom of the activated double bond, followed by an intramolecular displacement of bromide resulting in ring closure, rather than a direct addition of dibromocarbene to the carbon-carbon double bond. The two possibilities are formulated in equation 4.7.

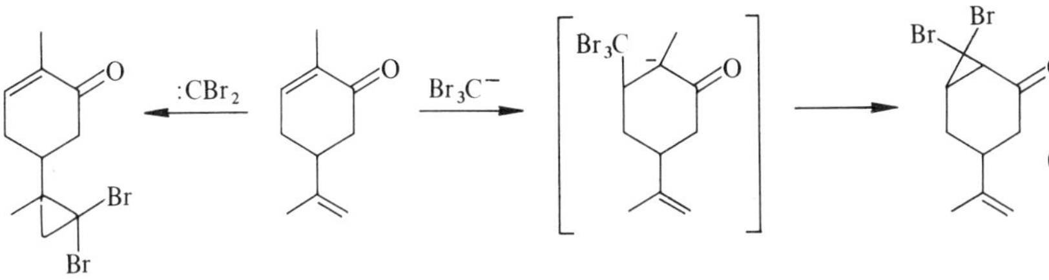

$$(4.7)$$

The observation that dibromocarbene adds to the α,β-unsaturated double bond is consistent with this proposal [14]. Dibromocarbene is electrophilic and is expected to react with the more electron rich (isolated) carbon-carbon double bond rather than with the α,β-unsaturated carbonyl system. The reaction is successful only for those α,β-unsaturated carbonyl compounds bearing a substituent on the α-carbon. This limitation has been circumvented by protecting the ketone function prior to cyclopropanation [14]. A number of examples are recorded in Table 4.2.

Table 4.2. Dibromocyclopanation of α, β-unsaturated carbonyl compounds

Substrate	% Yield	Catalyst	References
	85	BTEAC	[14]
	79	BTEAC	[15]
	25	BTEAC	[14]
	21	BTEAC	[15]
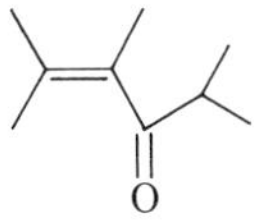	30	BTEAC	[15]
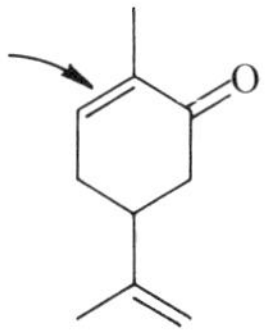	87	BTEAC	[15]

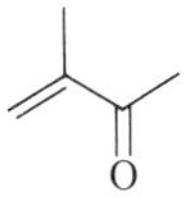

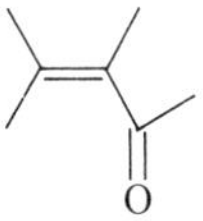

Table 4.2 (continued)

Substrate	% Yield	Catalyst	References
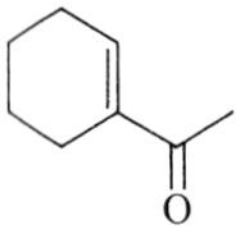 cyclohexenyl methyl ketone	20	BTEAC	[15]
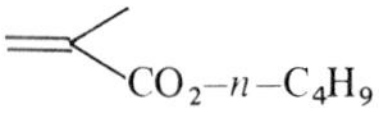 CH$_2$=C(CO$_2$-n-C$_4$H$_9$)-	76	BTEAC/ROH	[3]
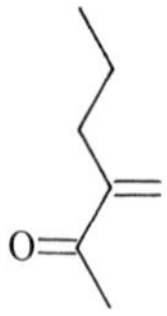 methylene ketone with propyl	50	BTEAC	[14]
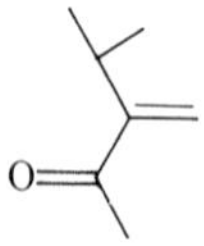 methylene ketone with isopropyl	90	BTEAC	[14]
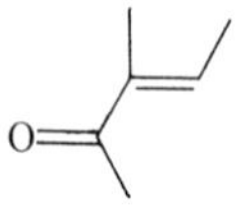 methyl vinyl ketone derivative	70	BTEAC	[14]
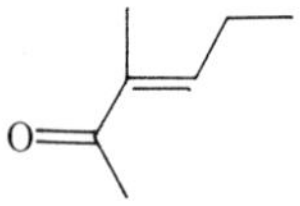 methyl vinyl ketone derivative	80	BTEAC	[14]
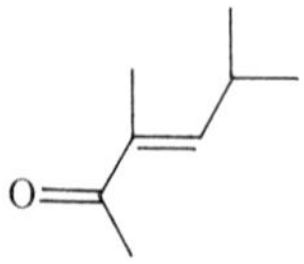 methyl vinyl ketone derivative	30	BTEAC	[14]
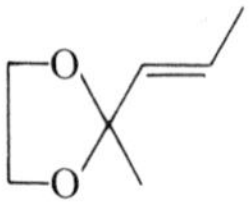 dioxolane with vinyl	30	BTEAC	[14]
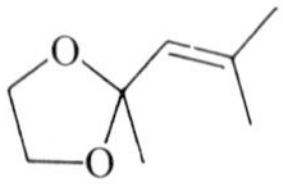 dioxolane with isopropenyl	60	BTEAC	[14]

4.6 Other Reactions of Dibromocarbene

Dibromocarbene has been utilized in place of dichlorocarbene in the phase transfer Hofmann carbylamine reaction (see Sect. 3.5). Methyl and ethyl isocyanide are low boiling and separation of these products from residual chloroform is difficult. The purification can be facilitated by use of bromoform in these two cases (Eq. 4.8) [16].

$$R-NH_2 \xrightarrow{:CBr_2} R-NC + 2HBr \qquad (4.8)$$

Reaction of dibromocarbene with several adamantyl alcohols has been reported to give the corresponding bromides in low yield [17]. Three such reactions are illustrated in equations 4.9–4.11 [17]. The reaction is analogous to the more thoroughly studied reaction of alcohols with dichlorocarbene (see Sect. 3.3).

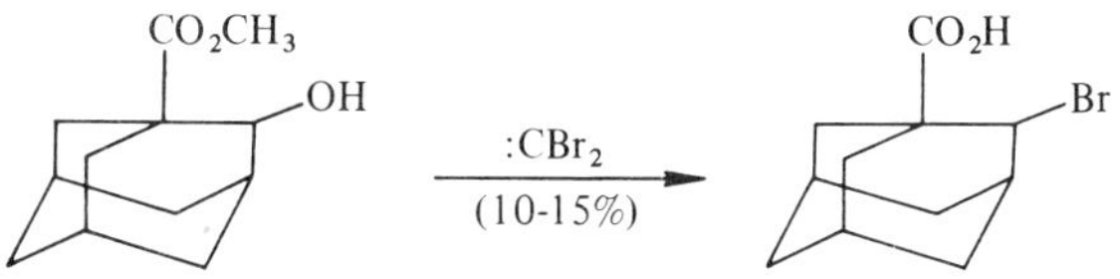

(4.9)

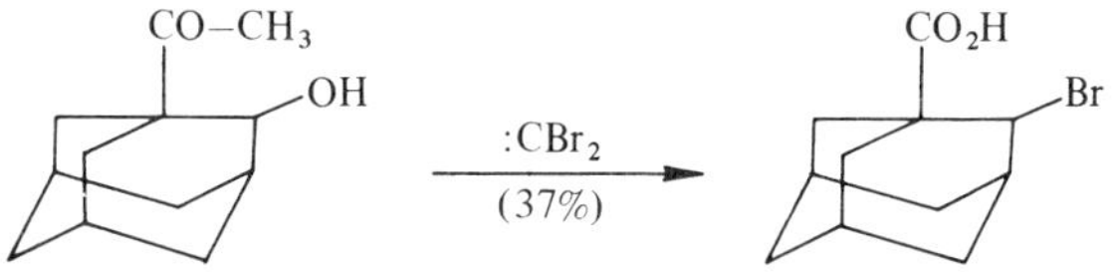

(4.10)

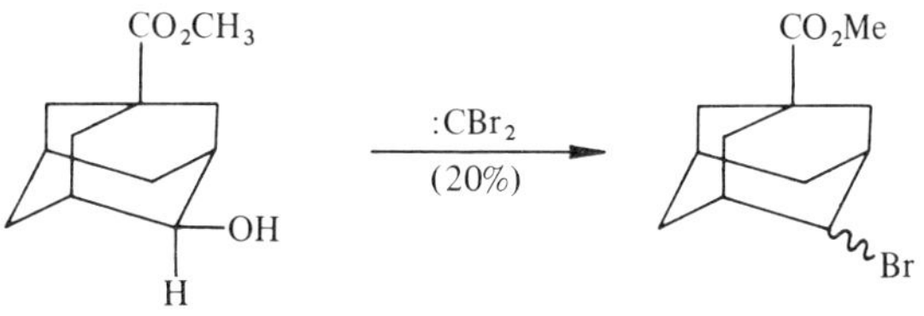

(4.11)

4.7 Other Halocarbenes

In addition to the phase transfer catalyzed generation of dichlorocarbene (Chaps. 2 and 3) and dibromocarbene (Sect. 4.2–4.6), most of the other possible dihalocarbenes have been formed by this method. Fluorodichloromethane (Freon 21) reacts with 50% aqueous base under phase transfer catalysis to give fluorochlorocarbene (:CFCl) [18, 19]. Fluorodibromo- [19] and fluorodiiodomethane [20] each react to form a fluorohalocarbene. Chlorodiiodomethane and triiodomethane (iodoform) have likewise been subjected to phase transfer conditions and have yielded the expected carbenes [21]. On the other hand, chlorodifluoromethane (Freon 22) did not yield the desired difluorocarbene and it appears that no attempt to prepare either chlorobromo-

or iodobromocarbene has yet been reported. The reactions of these carbenes are straightforward and occur according to the principles set forth in Sects. 2.4 and 4.2. A number of examples of halocarbene additions to simple olefins is reported in Table 4.3.

Table 4.3. Reactions of halomethanes with olefins

Number of carbons	Olefin	Carbene	% Yield	References
3	$CH_2=CH-CH_2OH$	:CFCl	15	[18]
4	$(CH_3)_2C=CH_2$	:CFBr	40–50	[19]
	$HOCH_2(CH_3)C=CH_2$	:CFCl	24	[18]
5	$(CH_3)_2C=CHCH_3$	:CFCl	43	[18]
	$C_2H_5(CH_3)C=CH_2$	:CFI	–	[20]
		:CCII	65	[21]
		:CI$_2$	4	[21]
6	$(CH_3)_2C=C(CH_3)_2$	:CFBr	40–50	[19]
		:CFCl	45–60	[19]
	$(C_2H_5)_2C=CH_2$	:CFCl	71	[18]
	$t\text{-}C_4H_9-CH=CH_2$	:CFI	–	[20]
		:CCII	56	[21]
	Cyclohexene	:CFCl	45–60	[19]
		:CFI	20	[20]
		:CCII	53	[21]
8	$C_6H_5CH=CH_2$	:CFBr	40–50	[19]
		:CFI	60	[20]
		:CFCl	45–60	[19]
		:CCII	49	[21]
		:CI$_2$	21	[21]
	$4\text{-}Cl\text{-}C_6H_4\text{-}CH=CH_2$	:CCII	–	[21]
		:CI$_2$	59	[21]
9	$C_6H_5(CH_3)C=CH_2$	:CFI	17	[20]
14	$(C_6H_5)_2C=CH_2$	:CI$_2$	20	[21]
		:CCII	59	[21]
	$E\text{-}C_6H_5CH=CH-C_6H_5$	:CI$_2$	2	[21]

4.8 Phenylthio- and Phenylthio(chloro)carbene

Sulfur substituted carbenes have been generated by the phase transfer method with reasonable success. Treatment of either phenyl(chloromethyl) sulfide or phenyl-(dichloromethyl) sulfide with 50% aqueous sodium hydroxide solution in the presence of both an olefin and a phase transfer catalyst (usually BTEAC) results in the cyclopropanation of the olefin according to equation 4.12. Several examples of the cyclopropanation of olefins with phenylthio- and phenylthio(chloro)carbene are recorded in Table 4.4 [22, 23].

$$R_2C=CR_2 + C_6H_5-S-CHClY \xrightarrow[\text{BTEAC}]{\text{50\% NaOH}} \begin{array}{c} Y-C-C_6H_5 \\ / \backslash \\ R_2C-CR_2 \end{array} \quad (Y = H, Cl) \quad (4.12)$$

Table 4.4. Cyclopropanation of olefins with phenylthio and phenylthio(chloro)carbene

Number of carbons	Olefin	Carbene	% Yield	References
4	Z-CH$_3$CH=CH–CH$_3$	:CHSC$_6$H$_5$	65	[23]
	E-CH$_3$CH=CHCH$_3$	:CHSC$_6$H$_5$	79	[23]
	CH$_3$CH$_2$O–CH=CH$_2$	:CHSC$_6$H$_5$	60	[23]
5	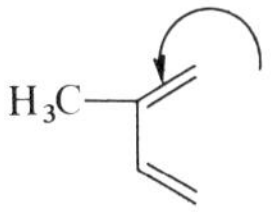 H$_3$C—(isoprene)	:CClSC$_6$H$_5$	51	[22]
6	CH$_3$(CH$_2$)$_3$–O–CH=CH$_2$	:CClSC$_6$H$_5$	58	[22]
	CH$_3$(CH$_2$)$_3$–S–CH=CH$_2$	:CClSC$_6$H$_5$	49	[22]
	Cyclohexene	:CHSC$_6$H$_5$	67	[23]
	Cyclohexene	:CClSC$_6$H$_5$	44	[22]
8	C$_6$H$_5$CH=CH$_2$	:CHSC$_6$H$_5$	70	[23]
	C$_6$H$_5$CH=CH$_2$	:CClSC$_6$H$_5$	63	[22]

Methylthio(chloro)carbene has been generated under phase transfer conditions from dichloromethyl methyl sulfide. This carbene cyclopropanates tetramethylethylene in 43% yield, accompanied by a 24% yield of 1,2-dichloro-1,2-*bis*-methylthioethylene [24].

4.9 Unsaturated Carbenes

A number of unsaturated carbenes have been generated under phase transfer conditions. The most extensive work has been reported for dimethylvinylidene carbene. Dimethylvinylidene carbene has been generated by base catalyzed γ-elimination of HCl from 3-chloro-3-methyl-1-butyne and by α-elimination of HBr from 1-bromo-3-methyl-1,2-butadiene (see Eq. 4.13). Both quaternary alkylammonium salts and various crown ethers have proved to be effective catalysts for this reaction. Yields of dimethylvinylidenecyclopropane products obtained under phase transfer catalytic

Dibromocarbene and Other Carbenes

conditions are higher than those obtained by previous methods, all of which required rigorous exclusion of moisture. With simple olefins, both types of phase transfer catalysts gave approximately equal yields of dimethylvinylidenecyclopropanes. With olefins having hydrophilic functional groups (OH, $-CO_2R$), crown ethers, however, gave improved yields of the dimethylvinylidenecyclopropane products. Examples of dimethylvinylidene carbene additions to olefinic substrates are recorded in Table 4.5.

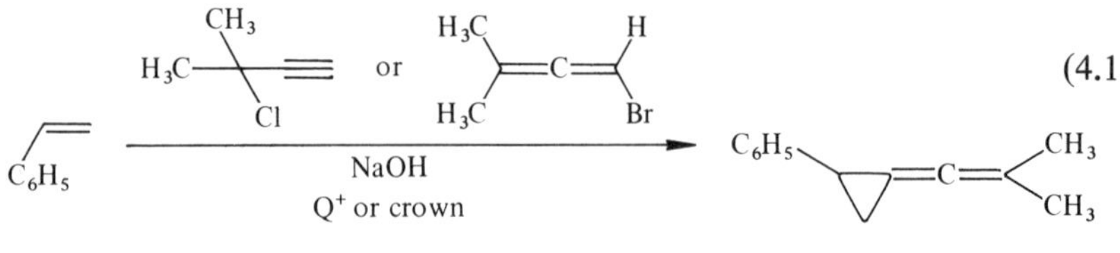

$$\begin{array}{c} H_3C-\overset{CH_3}{\underset{Cl}{C}}\!\!-\!\!\!\equiv \quad \text{or} \quad \overset{H_3C}{\underset{H_3C}{}}\!\!\!>\!\!\!=\!\!C\!\!=\!\!\overset{H}{\underset{Br}{}} \\ \underset{C_6H_5}{\overset{=}{}} \quad \xrightarrow[\text{Q}^+ \text{ or crown}]{\text{NaOH}} \quad \underset{}{\overset{C_6H_5}{\triangle}}\!\!=\!\!C\!\!=\!\!\overset{CH_3}{\underset{CH_3}{}} \end{array} \quad (4.13)$$

Table 4.5. Addition of dimethylvinylidene carbene to olefins

Number of carbons	Alkene	Catalyst	% Yield	Carbene precursor[a]	References
a) Simple alkenes					
5	$(CH_3)_2C=CHCH_3$	TCMAC	84	B	[27]
		DC-18-C-6	37	A	[26]
	Cyclopentene	TCMAC	27	A	[28]
6	$CH_3(CH_2)_3-CH=CH_2$	TCMAC	25	B	[27]
	$(CH_3)_2C=C(CH_3)_2$	TCMAC	68	B	[27]
	$(CH_3)_2C=C(CH_3)_2$	TCMAC	87	A	[28]
	Cyclohexene	TCMAC	61	B	[27]
		TCMAC	37	A	[28]
7	Norbornene	DC-18-C-6	19	A	[26]
	Norbornadiene	DC-18-C-6	18 mono	A	[26]
8	Camphene	BTEAC	6	A	[25]
	1,5-cyclooctadiene	BTEAC	19 mono	A	[25]
10	α-pinene	BTEAC	20	A	[25]
	β-pinene	BTEAC	31	A	[25]
	(Δ^3-carene) 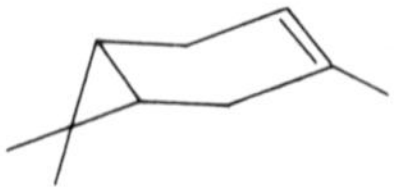	BTEAC	20	A	[25]

Table 4.5 (continued)

Number of carbons	Alkene	Catalyst	% Yield	Carbene precursor[a]	References
b) Styrenes					
8	$C_6H_5-CH=CH_2$	DC-18-C-6	69	A	[26]
		TCMAC	81	A	[28]
9	$C_6H_5C(CH_3)=CH_2$	BTEAC	52	A	[25]
	$C_6H_5CH=CH-CH_3$	BTEAC	36	A	[25]
	Indene	BTEAC	16	A	[25]
c) Conjugated olefins					
4	1,3-butadiene	DC-18-C-6	58 mono	A	[26]
5	Isoprene	BTEAC	23 mono, head 3 mono, tail	A	[25]
		DC-18-C-6	38	A	[26]
8	$(CH_3)_2C=CH-CH=C(CH_3)_2$	BTEAC	28 mono	A	[25]
		DC-18-C-6	59 mono	A	[26]
8	$CH_2=C(CH_3)-C{\equiv}C-C(CH_3)=CH_2$	BTEAC	38 mono	A	[25]
d) Unsaturated alcohols					
4	$CH_2=C(CH_3)-CH_2-OH$	BTEAC	14	A	[25]
5	$(CH_3)_2C=CH-CH_2OH$	TCMAC	16	A	[28]
7	$(CH_3)_2C=CH-CH_2OAc$	BTEAC	15 Alcohol isolated	A	[25]
		DC-18-C-6	46	A	[26]
9	$C_6H_5-CH=CH-CH_2OH$	BTEAC	21	A	[25]
e) Unsaturated ethers					
5	Dihydropyran	DB-18-C-6	38	A	[26]
6	$(CH_3)_2CH-CH_2O-CH=CH_2$	DC-18-C-6	35	A	[26]
		TCMAC	48	A	[28]
9	$C_2H_5O-CH(CH_3)-OCH_2CH=C(CH_3)_2$	TCMAC	82	A	[28]
10	$CH_3(CH_2)_3C(C_2H_5)-CH_2-OCH=CH_2$	DC-18-C-6	65	A	[26]

[a] Carbene precursors: A = 3-chloro-3-methylbutyne; B = 1-bromo-3-methyl-1,2-butadiene.

The addition of dimethylvinylidene carbene to three types of non-olefinic substrates has also been examined. Dimethylvinylidene carbene reacts with azobenzene to yield 1-phenyl-2-isobutenylbenzimidazole (10%) (Eq. 4.14). This product may arise from a rearrangement of the initial carbene adduct of the azo linkage [29].

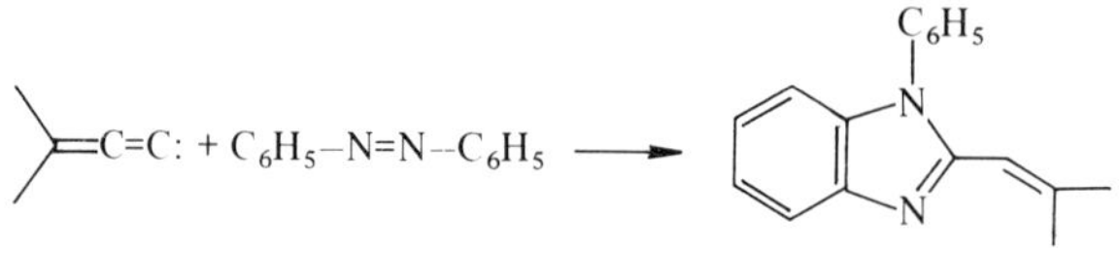

$$\diagup\!\!\!\!\diagdown C=C: + C_6H_5-N=N-C_6H_5 \longrightarrow \text{[benzimidazole product]} \qquad (4.14)$$

Reaction of dimethylvinylidene carbene with 3-methyl-2-butenyl-methylsulfide leads to 4-methylthio-3,3,6-trimethylhept-5-ene-1-yne (36%). The reaction sequence, illustrated in equation 4.15, has been proposed to account for the product obtained. Apparently, the unsaturated carbene favors reaction with sulfur rather than the carbon-carbon double bond [28].

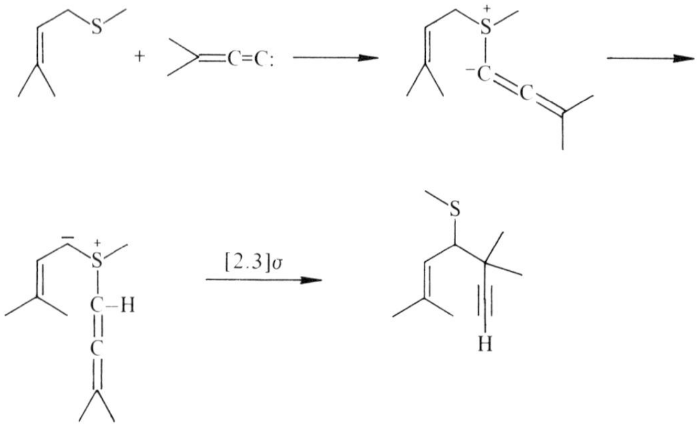

$$(4.15)$$

An attempt to insert dimethylvinylidene carbene into the tertiary benzylic C–H bond of isopropylbenzene was unsuccessful [28].

Pentamethylenevinylidene carbene has been generated by base catalyzed γ-elimination of HCl from 1-chloro-1-ethynylcyclohexane using dicyclohexyl-18-crown-6 or 18-crown-6 as phase transfer catalyst. This unsaturated carbene has been trapped successfully by styrene (40%, see Eq. 4.16), 2,5-dimethyl-2,4-hexadiene (36%), and isoprene (32%) [26].

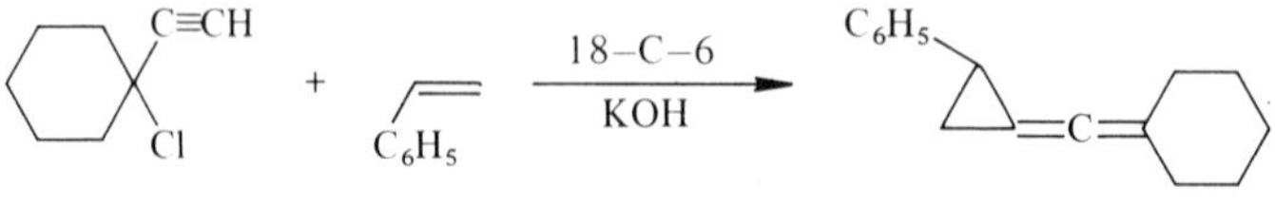

$$\text{[1-chloro-1-ethynylcyclohexane]} + \text{[styrene]} \xrightarrow[\text{KOH}]{18-C-6} \text{[cyclopropane product]} \qquad (4.16)$$

Dicyclopropylvinylidene has been generated by treatment of 2-(N-nitrosoacetylamino)-1,1-dicyclopropylethanol with 50% NaOH in the presence of a catalytic amount of tricaprylmethylammonium chloride (TCMAC) and trapped *in situ* by cyclohexene to give 7-(dicyclopropylmethylene)-bicyclo[4.1.0]-heptane in 64% yield [30]. Likewise, 2-cyclopropylpropylidene has been generated from 2-(N-nitrosoacetylamino)-2-cyclopropyl-2-propanol and trapped by reaction with cyclohexene to yield 7-(cyclopropylethylidene)-bicyclo-[4.1.0]-heptane (52%) [30]. Cyclohexylidene carbene has been generated from 1-(N-nitrosoacetylaminomethyl)cyclohexanol and trapped *in situ* by a number of olefins (see Table 4.6 and Eq. 4.17).

(4.17)

Table 4.6. Reactions of cyclohexylidene carbenes with olefins

Olefin	% Yield	Catalyst	References
Cyclohexene	78	TCMAC	[31]
Cycloheptene	61	TCMAC	[31]
Cyclooctene	60	TCMAC	[31]
$CH_2=CH-O-C_6H_5$	60	TCMAC	[31]
$CH_2=CH-O-C_2H_5$	83	TCMAC	[31]
$CH_2=CH-O-t-C_4H_9$	60	TCMAC	[31]

References

1. Skattebøl, L., Abskharoun, G. A., Greibrokk, T.: Tetrahedron Let. *1973*, 1367.
2. Fedoryński, M., Makosza, M.: Bull. Acad. Polon. Sci. *19*, 105 (1971).
3. Makosza, M., Fedoryński, M.: Synthetic Comm. *3*, 305 (1973).
4. Makosza, M., Kacprowicz, A., Fedoryński, M.: Tetrahedron Let. *1975*, 2119.
5. Hiyama, T., Sawada, H., Tsukanaka, M., Nozaki, H.: Tetrahedron Let. *1975*, 3013.
6. Kostikov, R. R., Molchanov, A. P.: Zhur. Organich. Khim. *11*, 1767 (1975).
7. Dehmlow, E. V., Schönefeld, J.: Ann. *744*, 42 (1971).
8. Dehmlow, S. S., Dehmlow, E. V.: Ann. *1973*, 1753.
9. Kraus, W., Klein, G., Rothenwöhrer, W., Sadio, H.: Synthesis *1972*, 485.
10. Allan, A. R., Baird, M. S.: J. Chem. Soc., Chem. Commun. *1975*, 172.
11. Dunkelblum, E., Singer, B.: Synthesis *1975*, 323.
12. Jefford, C. W., de los Heros, V., Burger, U.: Tetrahedron Let. *1976*, 703.
13. Sasaki, T., Eguchi, S., Kiriyama, T.: J. Org. Chem. *38*, 2230 (1973).
14. Barlet, M. R.: C. R. Acad. Sci. Paris *278C*, 621 (1974).
15. Sydnes, L., Skattebøl, L.: Tetrahedron Let. *1975*, 4603.
16. Weber, W. P., Gokel, G. W., Ugi, I. K.: Angew. Chem. Int. Ed. *11*, 530 (1972).
17. Tabushi, I., Aoyama, Y.: J. Org. Chem. *38*, 3447 (1973).
18. Chau, L. V., Schlosser, M.: Synthesis *1973*, 112.
19. Weyerstahl, P., Blüme, G., Müller, C.: Tetrahedron Let. *1971*, 3869.

20. Weyerstahl, P., Mathias, R., Blume, G.: Tetrahedron Let. *1973,* 611.
21. Mathias, R., Weyerstahl, P.: Angew. Chemie Int. Ed. *13,* 132 (1974).
22. Makosza, M., Bialecka, E.: Tetrahedron Let. *1971,* 4517.
23. Boche, G., Schneider, D. R.: Tetrahedron Let. *1975,* 4247.
24. Moss, R. A., Pilkiewicz, F. G.: Synthesis *1973,* 209.
25. Sasaki, T., Eguchi, S., Ogawa, T.: J. Org. Chem. *39,* 1927 (1974).
26. Sasaki, T., Eguchi, S., Ohno, M., Nakata, F.: J. Org. Chem. *41,* 2408 (1976).
27. Patrick, T.: Tetrahedron Let. *1974,* 1407.
28. Julia, S., Michelot, D., Linstrumelle, G.: C. R. Acad. Sc. Paris *278C,* 1523 (1974).
29. Sasaki, T., Eguchi, S., Ogawa, T.: Heterocycles *3,* 193 (1975).
30. Newman, M. S., Gromelski, S. J.: J. Org. Chem. *37,* 3220 (1972).
31. Newman, M. S., Din, Z. U.: J. Org. Chem. *38,* 547 (1973).
32. Kwon, S., Nishimura, Y., Ikeda, M., Tamura, Y.: Synthesis *1976,* 249.

5. Synthesis of Ethers

5.1 Introduction

One of the most useful applications of phase transfer catalysis in nucleophilic substitution has been in the Williamson ether synthesis. The reaction of an alkoxide anion with an alkyl halide or sulfonate to give either symmetrical or unsymmetrical ethers (depending on reactants) shows significant improvement in convenience, reaction rate, and yield when conducted under phase transfer catalytic conditions.

5.2 Mixed Ethers: The Mechanism

Freedman and Dubois [1] have found that a variety of ethers can be prepared by the reaction of either primary or secondary alcohols with inexpensive primary alkyl chlorides in the presence of excess concentrated aqueous sodium hydroxide solution and tetrabutylammonium hydrogen sulfate as catalyst. The formation of *n*-butyl *n*-hexyl ether is shown in equation 5.1. Typically, the S_N2 displacement reaction was not successful with secondary alkyl halides even under phase transfer conditions.

$$n\text{-}C_4H_9OH + n\text{-}C_6H_{13}Cl \xrightarrow[\text{Bu}_4\text{NHSO}_4]{\text{aq. NaOH}} n\text{-}C_4H_9\text{-}O\text{-}n\text{-}C_6H_{13} \qquad (5.1)$$

Nevertheless, this procedure is significantly easier than the traditional Williamson ether synthesis which requires prior generation of the alkoxide salt, usually by reaction of the alcohol with a strong base such as sodium hydride, sodamide, or sodium metal.

In the two phase Williamson ether synthesis, the base used is concentrated aqueous sodium hydroxide. Ordinarily, 50% aqueous sodium hydroxide is used, but even more concentrated solutions seem to be more effective. The alcohol in solution is deprotonated by hydroxide either in the aqueous phase or at the interface and then solubilized in the organic phase by ion pairing with the quaternary ammonium ion.

Synthesis of Ethers

The success of the reaction depends on the preferential extraction of tetrabutylammonium alkoxide rather than the corresponding hydroxide salt. The preference of the quaternary ammonium ion for alkoxide over hydroxide can be attributed to two factors. First, the soft (larger, more polarizable) quat prefers to pair with the soft (larger, more polarizable) alkoxide rather than the hard hydroxide ion. Second, hydroxide is more effectively solvated by the aqueous medium than is the alkoxide and prefers to remain in that phase. Under phase transfer conditions, secondary alcohols react more slowly than do primary alcohols, probably due to the lower acidity of secondary alcohols or to steric factors.

After deprotonation of alcohol (Eq. 5.2) and ion pairing with the quaternary ion (Eq. 5.3), reaction occurs between it and the alkyl halide (Eq. 5.5). For each mole of ether formed, a mole of quaternary ammonium halide is also formed. Exchange of the nucleofuge for a molecule of nucleophile followed by phase transfer completes the catalytic cycle which is shown in equations 5.2–5.6.

$$ROH + NaOH \rightleftharpoons NaOR + H_2O \tag{5.2}$$

$$NaOR\ (aq) + QCl\ (aq) \rightleftharpoons QOR\ (aq) + NaCl\ (aq) \tag{5.3}$$

$$QOR\ (aq) \rightleftharpoons QOR\ (org) \tag{5.4}$$

$$QOR\ (org) + R'Cl\ (org) \rightleftharpoons R-O-R'\ (org) + QCl\ (org) \tag{5.5}$$

$$QCl\ (org) \rightleftharpoons QCl\ (aq) \tag{5.6}$$

Alkyl iodides are less satisfactory in this reaction than are the chlorides for two reasons. First, the iodides are generally more expensive than the corresponding chlorides. Second, iodide ion tends to poison the phase transfer catalysts. Iodide ion preferentially pairs with quaternary alkylammonium ion in nonpolar solution. As a result, exchange to form the quaternary ammonium alkoxide which would

Table 5.1. Preparation of unsymmetrical ethers [1]

Alcohol	Halide	Mole % TBAB	% Yield
$CH_2=CHCH_2-OH$	$C_6H_5CH_2Cl$	5[a]	72
$n\text{-}C_4H_9-OH$	$C_6H_5CH_2Cl$	5	92
$HO-(CH_2)_4-OH$	CH_3CH_2Cl	6	93
$CH_3OCH_2CH(CH_3)-OH$	$n\text{-}C_4H_9Cl$	23	97
$n\text{-}C_4H_9OCH_2CH_2-OH$	$n\text{-}C_4H_9Cl$	5	82
$n\text{-}C_4H_9OCH_2CH_2-OH$	$c\text{-}C_6H_{11}Cl$	6	[b]
$CH_3(CH_2)_7-OH$	$n\text{-}C_4H_9Cl$	5	95

[a] Benzyltriethylammonium chloride as catalyst.
[b] 50% of cyclohexene only.

lead to product is hampered and the reaction is slowed or halted. In the case examined by Freedman and Dubois [1], the ether formation using 5 mole-% of tetrabutylammonium iodide is only two thirds as fast as the reaction utilizing an equivalent amount of tetrabutylammonium hydrogen sulfate.

A kinetic study conducted in tetrahydrofuran solvent revealed that the reaction is first order in catalyst, first order in alkyl halide and first order in alcohol. The low energy of activation (13.9 ± 0.5 kcal/mole) and the entropy (-26.5 ± 1.6 eu) are consistent with the bimolecular nature of the process. Finally, the reaction rate was insensitive to stirring rate beyond that needed to maintain effective mass transfer. The preparation of a variety of unsymmetrical ethers by this method is recorded in Table 5.1.

5.3 Rate Enhancement in the Williamson Reaction

The rate acceleration observed in phase transfer Williamson ether syntheses probably arises from several sources [2]. A major component to the energy barrier for bimolecular nucleophilic substitution reactions is removal of solvent molecules from the nucleophile. Desolvation of the nucleophile permits close approach by the nucleophile to the carbon of the alkylating agent facilitating reaction. Desolvation of a nucleophilic anion is energetically costly particularly in protic solvents which can hydrogen bond to the negatively charged anion. In this regard, Williamson type reactions have often been carried out by treating a sodium alkoxide dissolved in the corresponding alcohol with a primary alkyl bromide or iodide. In phase transfer catalysis, the anion is transferred from an aqueous phase into a nonpolar organic phase by ion-pairing with a large lipophilic cation. While a few water molecules unquestionably remain bound to the anion after transfer to the organic phase [3], the anion is significantly less solvated in the nonpolar organic phase than in the aqueous phase. Desolvation is not the only effect which must be considered. The reactivity of the anion in nonpolar solution also changes with the size of the cation with which it is paired. For small distances, the larger the cation the greater the reactivity of the anion. Thus the relative rates of reaction of tetrabutylammonium phenoxide and potassium phenoxide with n-butyl bromide to yield n-butyl phenyl ether have been compared in a number of solvents [2]. In dioxane, the tetrabutylammonium phenoxide ion pair reacts some 3×10^4 faster than the potassium phenoxide ion pair [2]. On the other hand, little difference in relative reaction rate was observed between the two salts in the ionizing solvent dimethylformamide.

An explanation of the cation size and solvent effect can be found in the energy required to separate the two charges: $E = (q_1 q_2)/(Dr^2)$. This energy is related to the distance (r) between the ions and the dielectric constant of the medium (D). The separation of the charge centers (r) in the tetrabutylammonium phenoxide ion pair is larger than in the potassium phenoxide ion pair. Tetrabutylammonium phenoxide is therefore more energetic and the nucleophilic anion is more reactive. As a consequence, alkoxide anions exhibit enhanced reactivity under phase transfer conditions.

5.4 Methylation

Similar results have been reported using a phase transfer catalyzed alkylation where dimethyl sulfate functioned as electrophile [4]. In this case, the reaction rate was sensitive to changes in the stirring rate suggesting that the reaction may be non-catalytic and occurring in the polar solvent (dimethyl sulfate).

Table 5.2. Phase transfer catalyzed formation of methyl ethers [4]

$$ROH + (CH_3)_2SO_4 \xrightarrow{NaOH/TBAI} ROCH_3 + Na^+(CH_3OSO_3)^- + H_2O$$

Alcohol	% Yield
n-$C_5H_{11}OH$	$\simeq 90$
$(CH_3)_3CCH_2OH$	70
n-$C_3H_7CH(OH)CH_3$	60
n-$C_6H_{13}OH$	$\simeq 90$
n-$C_7H_{15}OH$	$\simeq 90$
$C_6H_5CH_2OH$	92
n-$C_8H_{17}OH$	$\simeq 90$
$C_6H_5CH_2CH_2OH$	$\simeq 90$
$C_6H_5CH(OH)CH_3$	93
$C_6H_5CH(OH)CCl_3$	86
$C_6H_5CH=CHCH_2OH$	$\simeq 90$
$C_6H_5CH=CHCH_2OH$	85
$(HC\equiv C)_2C(OH)$-n-C_4H_9	95
$(HC\equiv C)_2C(OH)$-c-C_6H_{11}	$\simeq 90$
$(HC\equiv C)_2C(OH)C_6H_5$	97
$(C_6H_5)_3COH$	80

5.5 Phenyl Ethers

The phase transfer method has also been used in the preparation of a variety of phenolic ethers [5]. It is interesting that alkylation of phenoxide anions generally yield products of both C and O-alkylation with one favored to a greater or lesser extent depending on solvent. Sodium 2-naphthoxide, for example, exclusively C-benzylates in such hydrogen bonding solvents as ethanol [6], whereas O-alkylation is favored in tetrahydrofuran. As the size of the cation associated with 2-naphthoxide is increased (for example from Li^+ to R_4N^+), O-alkylation tends to be favored. Consistent with these observations is the report that 2-naphthol O-alkylates under phase transfer conditions [5]. The successful O-alkylation of the isomeric nitrophenols and salicylaldehyde are also noteworthy. These observations are consistent with the view that reaction occurs in the organic rather than the aqueous phase. The benzyltributylammonium bromide catalyzed two-phase alkylation of phenols in dichloromethane yields exclusively products of O-alkylation as recorded in Table 5.3.

A number of sterically hindered phenols have been alkylated with dimethyl sulfate under phase transfer conditions. When methyl iodide was substituted for dimethyl sulfate, it was found that a stoichiometric amount of catalyst was required to achieve complete reaction [5]. The benzyltributylammonium bromide catalyzed alkylation of several sterically hindered phenols are recorded in Table 5.4.

Table 5.3. Preparation of alkyl aryl ethers ArOR [5]

ArOH	RX	% Yield
C_6H_5OH	CH_3I	95
	$CH_2=CHCH_2Br$	77
	CH_2—$CHCH_2Cl$ (epoxide)	77
	n-C_4H_9Br	85
	c-$C_5H_{11}Br$	73
	$C_6H_5CH_2Cl$	86
	$C_6H_5CHBrCH_3$	91
	$BrCH_2CO_2C_2H_5$	86
2-$NO_2C_6H_4OH$	CH_3I	81
3-$NO_2C_6H_4OH$	CH_3I	79
4-$NO_2C_6H_4OH$	CH_3I	83
2-$OHCC_6H_4OH$	$[CH_3O]_2SO_2$	78
	$[C_2H_5O]_2SO_2$	83
4-$CH_3OC_6H_4OH$	$[CH_3O]_2SO_2$	91
2-$HO_2CC_6H_4OH$	$[CH_3O]_2SO_2$	78[a]
	$[C_2H_5O]_2SO_2$	85[a]
4-t-$C_4H_9C_6H_4OH$	CH_3I	96
	$CH_2=CHCH_2Br$	77
	$C_6H_5CH_2Br$	83
1-$C_{10}H_7OH$	CH_2–$CHCH_2Cl$ (epoxide)	42[b]
2-$C_{10}H_7OH$	CH_3I	92
	$[CH_3O]_2SO_2$	92
	$[C_2H_5O]_2SO_2$	81
	CH_2–CH_2CH_2Cl (epoxide)	41
2-C_6H_5–C_6H_4OH	$[CH_3O]_2SO_2$	81

[a] Refers to formation of ester 2-RO_2C–C_6H_4OR.
[b] Accompanied by 20% di-1-naphthoxymethane.

Synthesis of Ethers

Table 5.4. Methylation of sterically hindered phenols with dimethyl sulfate

ArOH	% Yield
2,6-dimethylphenol[a]	92
2,6-dimethoxyphenol	91
2,6-diisopropylphenol	84
2,6-dimethyl-4-t-butylphenol	79
2,6-di-t-butylphenol	83
2,6-di-t-butyl-4-methylphenol	84
2,6-di-t-butyl-4-methoxyphenol	85
2,4,6-tri-t-butylphenol	93

[a] Methyl iodide used.

There is also a single case reported in which a phenyl methyl ether was prepared from the corresponding aryl chloride. Potassium methoxide, activated by dicyclohexyl-18-crown-6 displaces a chloride ion from 1,2-dichlorobenzene to give 2-chloroanisole in fair yield. None of the 3-isomer is isolated, indicating that the reaction is a nucleophilic substitution rather than a benzyne reaction [22].

5.6 Methoxymethyl Ethers of Phenol

The protection of phenolic hydroxyls as the methoxymethyl ethers can also be achieved effectively under phase transfer conditions in which methoxymethyl chloride is the electrophile. The solid potassium phenoxide salt is solubilized by 18-crown-6 in dry acetonitrile which also contains the alkylating agent [7]. The protection of two phenolic ketones in good yield are formulated in equations 5.7 and 5.8.

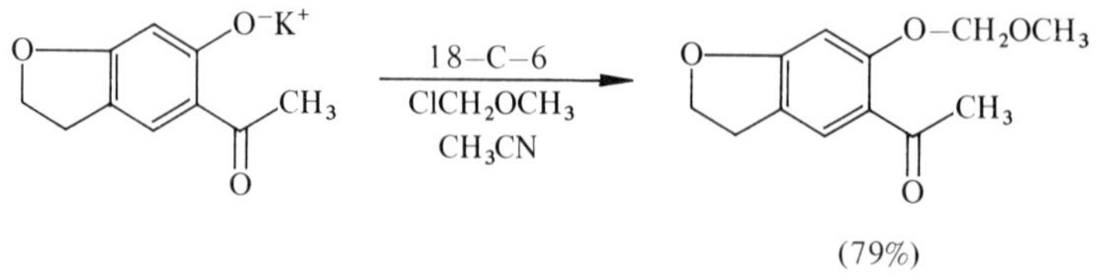

$$\text{ArO}^-\text{K}^+ \xrightarrow[\text{CH}_3\text{CN}]{\text{18-C-6, ClCH}_2\text{OCH}_3} \text{ArO-CH}_2\text{OCH}_3 \quad (79\%) \quad (5.7)$$

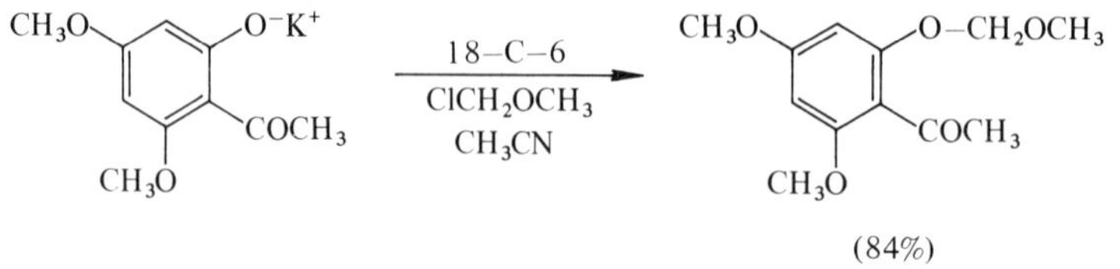

$$\text{ArO}^-\text{K}^+ \xrightarrow[\text{CH}_3\text{CN}]{\text{18-C-6, ClCH}_2\text{OCH}_3} \text{ArO-CH}_2\text{OCH}_3 \quad (84\%) \quad (5.8)$$

5.7 Diethers From Dihalomethanes

Catechols likewise undergo O-alkylation under phase transfer conditions, yielding methylenedioxy derivatives on reaction with methylene bromide [8]. This reaction constitutes a useful route to this commonly occurring oxygen heterocycle as well as providing a facile protection method for 1,2-dihydroxyarenes. Dibromomethane rather than the more reactive diiodomethane was used in this reaction because the latter is a source of iodide ion which poisons the phase transfer catalyst by selectively ion-pairing with the quaternary ammonium cation. Phenoxide ion can apparently compete successfully with bromide but not iodide in the formation of an extractable ion pair under these conditions.

Table 5.5. Protection of 1,2-dihydroxybenzenes [8]

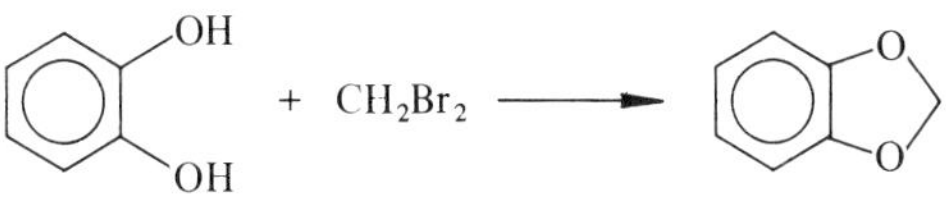

Reactant	Product	% Yield
Catechol	Benzo-1,3-dioxole	76
2,3-dihydroxybenzaldehyde	Piperonal	80
4-methylcatechol	4-methylbenzo-1,3-dioxole	86
2,3-dihydroxynaphthalene	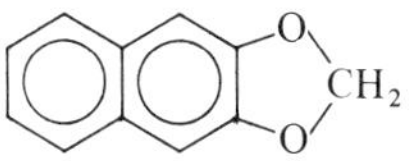 (naphtho-1,3-dioxole)	82

It seems likely that dichloromethane could be used in this procedure instead of dibromomethane because after the first chlorine has been substituted, the second substitution should be extremely fast. Cyclization to form a five-membered ring is a favorable process and the alkylating reagent at this stage will be an alkoxymethyl halide. Substitution of the first chlorine in dichloromethane is not improbable, especially in light of the known reaction of both alkoxide and phenoxide anions with it under phase transfer conditions to yield symmetrical dialkyl or diaryl acetals of form-

Table 5.6. Acetals formed from dichloromethane

Reactant	Product	% Yield	References
n-C_4H_9OH	(n-$C_4H_9O)_2CH_2$	65	[9]
c-$C_6H_{11}OH$	(c-$C_6H_{11}O)_2CH_2$	79	[9]
$C_6H_5CH_2OH$	$(C_6H_5CH_2O)_2CH_2$	77	[9]
1-$C_{10}H_7OH$	$(1$-$C_{10}H_7O)_2CH_2$	20	[5]

Synthesis of Ethers

aldehyde [9]. Note also (Table 5.3) that the alkylation of 1-naphthol with epichlorohydrin in dichloromethane solution yielded (20%) dinaphthoxymethane as a by-product [5]. Likewise, the major product resulting from the attempted alkylation of 4-t-butylphenol with benzyl chloride in dichloromethane (phase transfer conditions) was not the anticipated aryl benzyl ether but di-4-t-butylphenoxymethane [5].

5.8 The Koenigs-Knorr Reaction

Although secondary alkyl halides are generally unsatisfactory in the phase transfer catalyzed Williamson ether synthesis, the more reactive α-haloethers are as useful as, and more reactive than, simple alkyl halides. Thus the Koenigs-Knorr reaction has been successfully executed under phase transfer conditions [10]. α-2-Bromo-3,4,5,6-

Table 5.7. PT catalyzed Koenigs-Knorr reactions

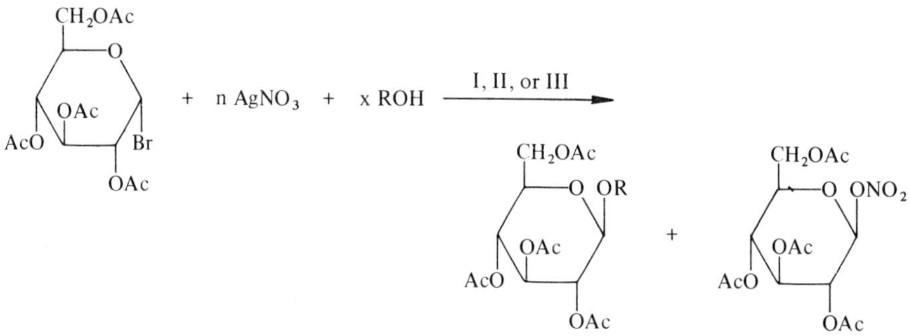

ROH	I A : B	II A : B	III A : B
CH_3OH	100 : 0	100 : 0	98 : 2
i-C_3H_7OH	100 : 0	100 : 0	89 : 11
t-C_4H_9OH	100 : 0	95 : 5	64 : 36
Cyclohexanol	100 : 0	98 : 2	73 : 27
Diglyme	–	0 : 100	0 : 100

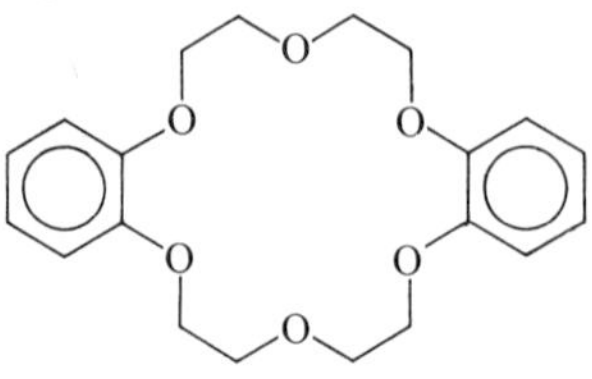

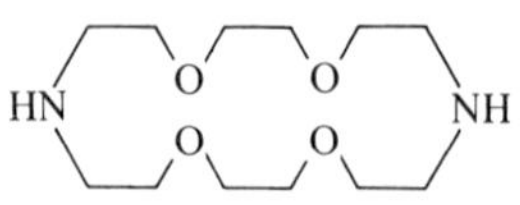

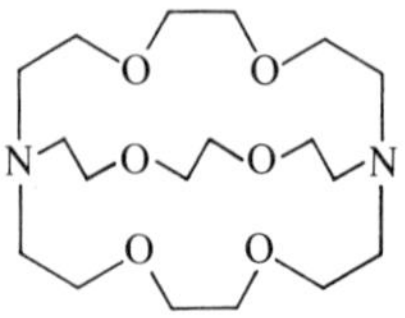

80

tetraacetoxyglucose reacts with various alcohols in the presence of silver nitrate and a catalytic amount of dibenzo-18-crown-6 [11, 12], 1,10-diaza-18-crown-6 [12], or [2,2,2]-cryptate [12] to yield the product of nucleophilic displacement (with inversion): β-alkoxy-3,4,5,6-tetraacetoxyglucoside. With secondary and tertiary alcohols, formation of the nitric acid ester is competitive.

5.9 Epoxides

The oxygen containing three-membered ring systems known either as oxiranes or epoxides are special ethers in the sense that they are both readily formed and highly reactive. These substances are commonly formed by one of three methods: intramolecular nucleophilic displacement, methylenation of a carbonyl compound or epoxidation of an olefin. All three synthetic approaches are represented in the phase transfer literature.

Carbon acids such as chloroacetonitrile and chloromethyl p-tolyl sulfone are readily deprotonated by concentrated aqueous sodium hydroxide under phase transfer conditions. In the presence of aldehydes or ketones, the carbanion undergoes condensation to yield an intermediate α-alkoxy chloride. Intramolecular nucleophilic displacement of halide yields either a glycidic nitrile or α, β-epoxysulfone. The overall process is, in fact, a phase transfer variant of the Darzens condensation [13] (see Eq. 5.9). Either benzyltriethylammonium chloride (BTEAC) or dibenzo-18-crown-6 (DB-18-C-6) catalyzes the phase transfer Darzens condensation of chloroacetonitrile [14, 15]. The results are presented in Table 5.8.

$$Cl-CH_2CN + QOH \rightleftharpoons Q^+Cl-\bar{C}HCN + H_2O$$

$$Cl-\bar{C}HCN + \underset{R_2}{\overset{R_1}{>}}C=O \longrightarrow NC-CH-\underset{\underset{R_2}{Cl}}{\overset{\overset{O^-}{|}}{C}}-R_1 \longrightarrow NC\overset{O}{\underset{R_2}{\triangle}}R_1 \qquad (5.9)$$

Table 5.8. Phase transfer synthesis of glycidic nitriles

R_1	R_2	Catalyst	% Yield	References
CH_3	CH_3	BTEAC	60	[14]
$-(CH_2)_4-$		BTEAC	65	[14]
$-(CH_2)_5-$		BTEAC	79	[14]
$-(CH_2)_4-CH-$	CH_3	BTEAC	78	[14]
C_6H_5	H	BTEAC	75	[14]
C_6H_5	H	DB-18-C-6	78	[15]
C_6H_5	CH_3	BTEAC	80	[14]
C_6H_5	C_6H_5	BTEAC	55	[14]

Synthesis of Ethers

Likewise, BTEAC catalyzes the related condensation of chloromethyl p-tolyl sulfone with carbonyl compounds (see Eq. 5.10). The synthesis of several α,β-epoxy-sulfones are recorded in Table 5.9.

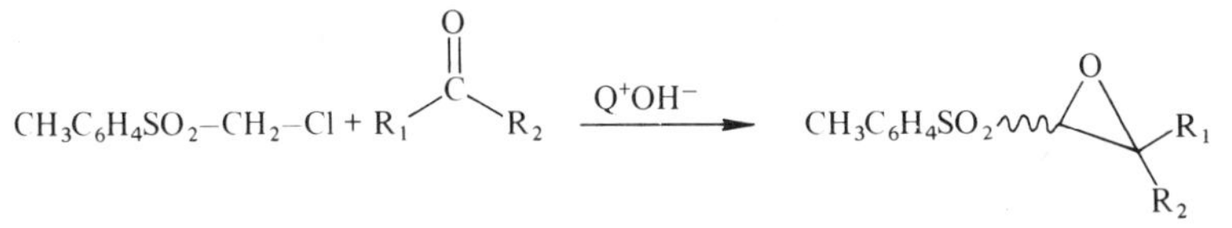

$$CH_3C_6H_4SO_2-CH_2-Cl + R_1\overset{O}{\underset{}{C}}R_2 \xrightarrow{Q^+OH^-} CH_3C_6H_4SO_2\sim\overset{O}{\underset{R_2}{\triangle}}R_1 \qquad (5.10)$$

Table 5.9. PTC synthesis of α,β-epoxysulfones [16]

R_1	R_2	% Yield
CH_3	CH_3	91
$i\text{-}C_3H_7$	H	65
$-(CH_2)_5-$		90
C_6H_5	H	60
C_6H_5	C_6H_5	90

Two groups have reported the successful methylenation of carbonyl compounds with phase transfer generated sulfonium ylids. The methylenation of benzaldehyde with trimethylsulfonium ylid is shown in equation 5.11 and is discussed in detail in Chapter 14 [17–19].

$$\rangle S^\pm \xrightarrow{Q^+OH^-} \rangle S^+-CH_2^- + C_6H_5CHO \rightarrow C_6H_5\overset{O}{\overset{/\backslash}{CH-CH_2}} \qquad (5.11)$$

α,β-Epoxy ketones have been prepared by the Michael addition of either hydrogen peroxide anion or t-butyl hydroperoxide anion (generated under phase transfer conditions) to α,β-unsaturated ketones [20]. Of particular interest is the finding that chiral phase transfer catalysts derived from naturally occurring alkaloids induce optical activity in the epoxidized olefins. This method is the best example yet published of an asymmetric phase transfer reaction (see Table 5.10 for examples).

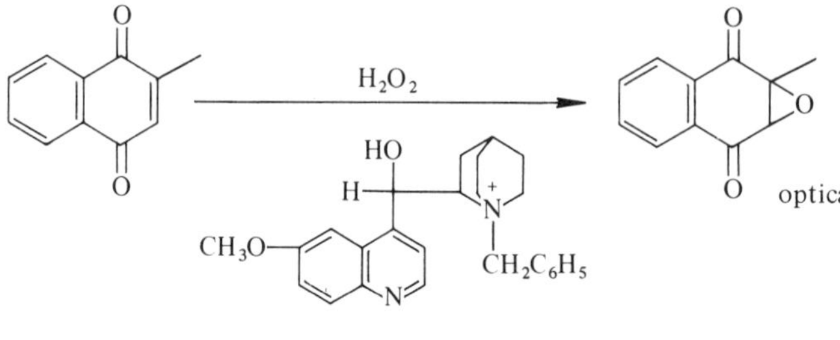

$$(5.12)$$

optically active product

Oxidation of cyclohexene by aqueous 30% hydrogen peroxide catalyzed by MoO_3 or H_2WO_4 and quaternary alkylammonium salts led to reasonable yields (30–40%) of cyclohexene oxide as well as 1,2-cyclohexanediol (23–28%). The quat serves to phase transfer the hydrogen peroxide into the organic phase. This stabilizes the hydrogen peroxide and permits its efficient utilization as a reoxidant [21] (see Sect. 11.5).

Table 5.10. Synthesis of α,β-epoxy ketones by PTC

Alkene	Oxidant	Catalyst	$[\alpha]^{RT}$ of Epoxide	
E–C_6H_5–CH=C(CH_3)NO_2	H_2O_2	X	– 3°	578 mµ
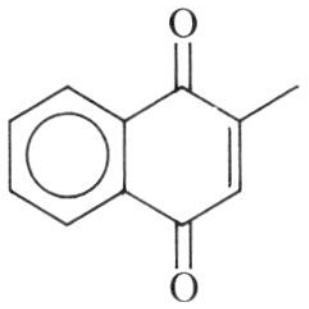 [2-methyl-1,4-naphthoquinone]	H_2O_2	X	–12°	436 mµ
	t-C_4H_9OOH	X	– 8°	436 mµ
C_6H_5–C(O)–CH=CH–C_6H_5	H_2O_2	X	–51°	578 mµ
	t-C_4H_9OOH	X	+ 24°	578 mµ
	H_2O_2	Y	+49°	578 mµ
4-Cl–C_6H_4–CH=CH–C(O)–C_6H_5	H_2O_2	X	–62°	578 mµ
4-O_2N–C_6H_4–CH=CH–C(O)–C_6H_5	H_2O_2	X	–33°	578 mµ
2-CH_3O–C_6H_4–C(O)–CH=CH–C_6H_5	H_2O_2	Y	+ 29°	578 mµ
	H_2O_2	X	–34°	578 mµ[a]
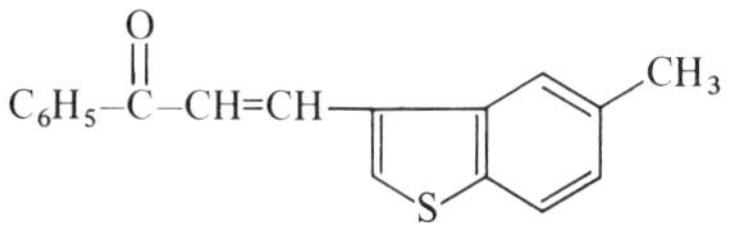 C_6H_5–C(O)–CH=CH–(5-methylbenzothiophene)	H_2O_2	X	–56°	578 mµ
2-$C_{10}H_7$–C(O)–CH=CH–C_6H_5	H_2O_2	X	–31°	578 mµ

X = CH_3O-[quininium-N-benzyl salt with OH]

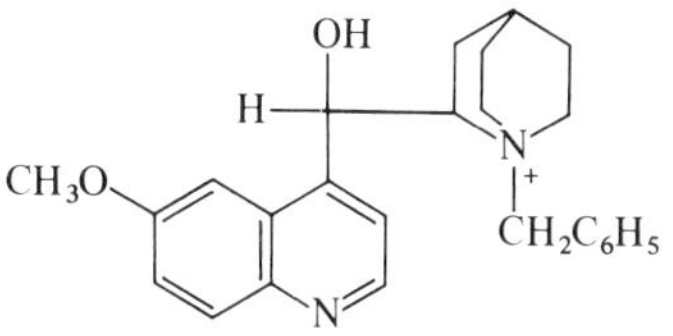

Y = Quinidine-$C_6H_5CH_2Cl$ salt.

[a] 24% enantiomeric excess estimated by means of chemical shift reagents.

References

1. Freedman, H. H., Dubois, R. A.: Tetrahedron Let. *1975*, 3251.
2. Ugelstad, J., Ellingsen, T., Berge, A.: Acta Chem. Scand. *20*, 1593 (1966).
3. Starks, C. M., Owens, R. M.: J. Am. Chem. Soc. *95*, 3613 (1973).
4. Merz, A.: Angew. Chem. Int. Ed. *12*, 846 (1973).
5. McKillop, A., Fiaud, J. C., Hug, R. P.: Tetrahedron *30*, 1379 (1974).
6. Kornblum, N., Seltzer, R., Haberfield, P.: J. Am. Chem. Soc. *85*, 1148 (1963) and references cited therein.
7. Rall, G. J. H., Oberholzer, M. E., Ferreira, D., Roux, D. G.: Tetrahedron Let. *1976*, 1033.
8. Bashall, A. P., Collins, J. F.: ibid *1975*, 3489.
9. Dehmlow, E. V., Schmidt, J.: ibid *1976*, 95.
10. Koenigs, W., Knorr, E.: Chem. Ber. *34*, 957 (1901).
11. Knöchel, A., Rudolph, G., Thiem, J.: Tetrahedron Let. *1974*, 551.
12. Knöchel, A., Rudolph, G.: ibid *1974*, 3739.
13. Newman, M. S., Magerlein, B. J.: Organic Reactions 5, p. 413. New York, N. Y.: J. Wiley 1957.
14. Jończyk, A., Fedoryński, M., Makosza, M.: Tetrahedron Let. *1972*, 2395.
15. Makosza, M., Ludwikow, M.: Synthesis *1974*, 665.
16. Jończyk, A., Bańko, K., Makosza, M.: J. Org. Chem. *40*, 266 (1975).
17. Merz, A., Märkl, G.: Angew. Chem. Int. Ed. *12*, 845 (1973).
18. Yano, Y., Okonogi, T., Sunaga, M., Tagaki, W.: Chem. Commun. *1973*, 527.
19. Hiyama, T., Mishima, T., Sawada, H., Nozaki, H.: J. Am. Chem. Soc. *97*, 1626 (1975).
20. Helder, R., Hummelen, J. C., Laane, R. W. R. M., Wiering, J. S., Wynberg, H.: Tetrahedron Let. *1976*, 1831.
21. Starks, C. M., Napier, D. R.: S. African Patent 71-1495.
22. Sam, D. J., Simmons, H. E.: J. Am. Chem. Soc. *96*, 2252 (1974).

6. Synthesis of Esters

6.1 Introduction

Esters are commonly synthesized from carboxylic acids by reaction of the acid with an excess of alcohol containing a catalytic amount of a mineral acid. In cases where practical considerations dictate it, the acid can be converted to an acyl halide (usually the chloride) and then condensed with the appropriate alcohol. A less commonly used procedure involves direct alkylation of the carboxylate ion with an alkyl halide. Even when this latter procedure (Eq. 6.1) involves a silver carboxylate and alkyl chloride, the reaction is of marginal practical value.

$$R-COO^- Ag^+ + R'-Cl \rightarrow R-COO-R' + AgCl \qquad (6.1)$$

The phase transfer method facilitates the dissolution of carboxylates in nonpolar media. In these solutions, due to relatively poor solvation of anions, carboxylate is an effective nucleophile and reacts readily with alkyl halides. The catalysis of such reactions by amines, ammonium salts and crown ethers and the synthesis of esters by the phase transfer technique is the subject of this chapter.

6.2 Tertiary Amines and Quaternary Ammonium Salts

One of the earliest reactions in which quaternary ammonium catalysis was recognized was in the alkylation of fatty acids with epichlorohydrin [1]. Maerker, Carmichael, and Port reported in 1961 that when stearic acid was boiled with epichlorohydrin containing benzyltrimethylammonium chloride, less than 10% ester was formed. In contrast, when the sodium salt of stearic acid was reacted under similar conditions, high yields of the glycidic ester were obtained (Eq. 6.2). In the absence of catalyst,

$$n\text{-}C_{17}H_{35}COO^- Na^+ + \overset{O}{\overset{\diagup\diagdown}{CH_2-CH}}-CH_2Cl \rightarrow$$
$$n\text{-}C_{17}H_{35}COO-CH_2-\overset{O}{\overset{\diagup\diagdown}{CH}}-CH_2 + NaCl \qquad (6.2)$$

neither reaction was useful. It was found that under these conditions the mechanism of alkylation was as expected (initial attack at the epoxide) and the reaction was indifferent to the presence of small amounts of water. Moreover, it was suggested that the efficacy of the quaternary salt as a catalyst depended on its solubility in epichlorohydrin. Similar results had been reported somewhat earlier in the patent literature [2].

Hennis and coworkers reported that the reaction of carboxylates with alkyl chlorides is catalyzed by tertiary amines, the combination of tertiary amines and sodium iodide or quaternary ammonium compounds [3, 4]. It was found that the ester formation was catalyzed by quaternary ammonium salts and that these were generated *in situ* from the amine and alkyl halide. Sodium iodide in 2-butanone converted the alkyl chloride to the more reactive iodide (the Finkelstein reaction) which in turn alkylated amine. Alkyl iodides added directly to the reaction mixture were even more active co-catalysts but the preformed quaternary ammonium salt was the most effective catalyst. It was suggested that the enhanced solubility of the quaternary ammonium carboxylate or the lack of tight ion pairing in this salt might account for the

Table 6.1. Quaternary ion and amine catalyzed ester formation

Carboxylate salt	Alkyl halide	Catalyst	% Yield	Ref.
CH_3CO_2Na	$C_6H_5CH_2Cl$	NaI/Et_3N	97	[3]
CH_3CO_2K	$CH_3=CHCH_2Br$	TMEDA	90	[5]
	$CH_3CH=CHCH_2Br$	TMEDA	84	[5]
	$n\text{-}C_6H_{13}Br$	TMEDA	90	[5]
	$C_6H_5CH_2Cl$	TMEDA	100	[5]
	$C_6H_5CH_2Br$	TMEDA	90	[5]
	$Br(CH_2)_4CO_2C_2H_5$	TMEDA	100	[5]
	$n\text{-}C_8H_{17}Br$	TOPAC	68	[8]
	$n\text{-}C_6H_{13}CHBrCH_3$	TMEDA	0	[5]
	$Br(CH_2)_3Br$	TMEDA	100	[5]
	$Br(CH_2)_4Br$	TMEDA	100	[5]
	Br–cyclopentene–Br	TOPAC	75 (di)	[8]
$CH_3CH_2CO_2Na$	CH_3Cl	NaI/Et_3N	96	[3]
	$CH_2=CHCH_2Cl$	NaI/Et_3N	80	[3]
	$n\text{-}C_4H_9Cl$	NaI/Et_3N	87	[3]
$C_6H_5CO_2Na$	CH_3Cl	NaI/Et_3N	96	[3]
	$CH_2=CHCH_2Cl$	NaI/Et_3N	93	[3]
	$n\text{-}C_4H_9Cl$	NaI/Et_3N	85	[3]
	$C_6H_5CH_2Cl$	NaI/Et_3N	89	[3]
2-OH-$C_6H_4CO_2Na$	CH_3Cl	NaI/Et_3N	98	[3]
	$CH_2=CHCH_2Cl$	NaI/Et_3N	87	[3]
	$n\text{-}C_4H_9Cl$	NaI/Et_3N	92	[3]
	$C_6H_5CH_2Cl$	NaI/Et_3N	89	[3]

TMEDA = $(CH_3)_2NCH_2CH_2N(CH_3)_2$; TOPAC = $(C_8H_{17})_3N^+(C_3H_7)Cl^-$.

enhanced reactivity. These suggestions accord with the low catalytic activity of tetramethylammonium halides compared to triethylammonium iodide.

Normant and coworkers have recently reported that the reaction of solid potassium acetate with benzyl chloride in acetonitrile is catalyzed by a wide variety of tertiary di- and polyamines (Eq. 6.3) [5]. The catalytic efficiency depends on the cation as-

$$CH_3COO^- K^+ + C_6H_5CH_2Cl \rightarrow CH_3CO_2CH_2C_6H_5 + KCl \tag{6.3}$$

sociated with acetate anion; the efficiency increases with increasing softness (and decreasing lattic energy).

The catalytic activity could be due to the coordination and enhanced solubilization of potassium acetate by the amine or it could be due to *in situ* quaternary ion formation (or both). Although the authors rule out the latter as a mechanism of catalysis, it seems unlikely that there is no component of this mechanism operating. In light of the observation that Wittig-Horner-Emmons reactions require no additional catalyst [6] and that certain α-phosphoryl sulfoxides catalyze basic two-phase alkylations [7], it is clear that more study is required. The alkylation of several carboxylate anions are recorded in Table 6.1.

6.3 Noncatalytic Esterification in the Presence of Ammonium Salts

Several non-catalytic systems have been reported to give good yields of esters. For example, if an equimolar mixture of a carboxylic acid, an alkyl halide and a tertiary amine (most commonly triethylamine) are heated for a few hours at 140 °C, excellent yields of esters are usually obtained. Under these conditions, methylene chloride was found to give methanediol-*bis*-esters [9]. Examples of esters synthesized by this approach are presented in Table 6.2.

Table 6.2. Esters from acids, halides, and triethylamine [9]

Acid	Alkyl halide	% Yield ester[a]
Maleic	*n*-decyl chloride	74
Benzoic acid	*t*-butyl bromide	12
Capric	*n*-dodecyl chloride	82
Isophthalic	*n*-decyl chloride	87
Butyl hydrogen phthalate	*n*-butyl chloride	90
Butyl hydrogen phthalate	*n*-octyl chloride	84
Phthalic	*n*-octyl chloride	98
Butyl hydrogen phthalate	*s*-octyl chloride	89
Butyl hydrogen phthalate	Methylene dichloride	72
Butyl hydrogen phthalate	Ethylene dichloride	91

[a] Yield based on acid.

Synthesis of Esters

The homogeneous reaction of tetrabutylammonium carboxylates (ion pair extraction) with methyl iodide in a two-phase system consisting of aqueous base and dichloromethane has also been reported. In most cases where the carboxylic acid was monobasic, excellent to quantitative yields of ester were observed. It was also noted that the esterification was insensitive to steric hindrance in the acid component under these conditions (see Eq. 6.4) [10].

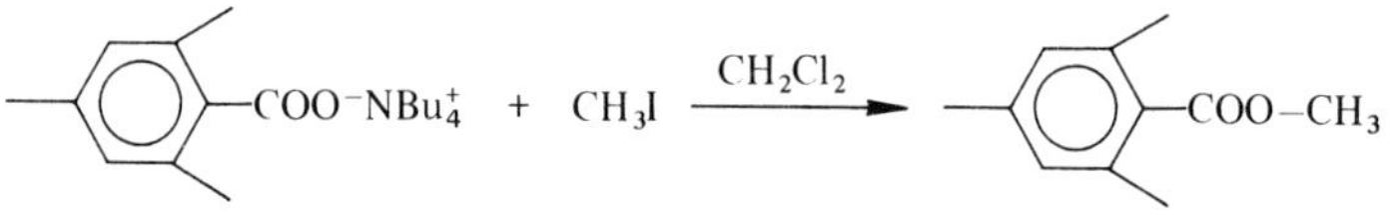

$$\text{Ar–COO}^-\text{NBu}_4^+ + \text{CH}_3\text{I} \xrightarrow{\text{CH}_2\text{Cl}_2} \text{Ar–COO–CH}_3 \tag{6.4}$$

Guided by earlier work [9, 10] a general synthesis of methylene diesters was developed which utilized dichloromethane as solvent and alkylating agent. The reaction (Eq. 6.5) was relatively slow, but ultimate yields were good [11]. The methylene diesters prepared in this fashion are recorded in Table 6.3.

$$2\ \text{R–COOH} + \text{CH}_2\text{Cl}_2 \xrightarrow{\text{TBA}^+\ \text{OH}^-} (\text{R–CO}_2)_2\text{CH}_2 \tag{6.5}$$

The reaction of quaternary ammonium carboxylates with primary alkyl halides in DMF solution at room temperature results in good yields of esters [12]. (See Table 6.4).

6.4 Polycarbonate Formation

An important industrial reaction and one of the earliest applications of phase transfer catalysis by quaternary ammonium salts is the synthesis of polycarbonates [13]. Typically, 2,2-(4,4'-dihydroxydiphenyl)propane(bisphenol A) is dissolved in concentrated aqueous sodium hydroxide and exposed to a dichloromethane solution of phosgene. Salts such as benzyltriethylammonium chloride or tertiary amines catalyze the condensation polymerization [13–18]. The two phase polymerization is formulated in equation 6.6.

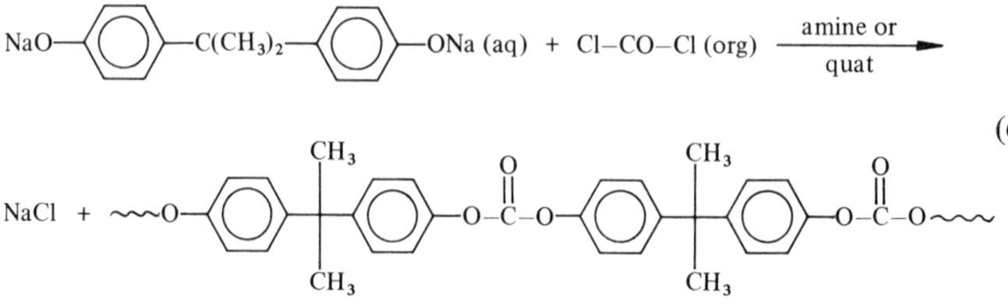

$$\text{NaO–C}_6\text{H}_4\text{–C(CH}_3)_2\text{–C}_6\text{H}_4\text{–ONa (aq)} + \text{Cl–CO–Cl (org)} \xrightarrow{\text{amine or quat}} \tag{6.6}$$

$$\text{NaCl} + \sim\!\!\sim\!\!\text{O–C}_6\text{H}_4\text{–C(CH}_3)_2\text{–C}_6\text{H}_4\text{–O–C(=O)–O–C}_6\text{H}_4\text{–C(CH}_3)_2\text{–C}_6\text{H}_4\text{–O–C(=O)–O}\!\sim\!\!\sim$$

Table 6.3. Diesters from dichloromethane [11]

Acid	% Yield
Butanoic	79
Pivalic	80
Benzoic	88
4-chlorobenzoic	84
4-nitrobenzoic	60
4-methoxybenzoic	85
Mesitoic	87

Table 6.4. Reactions of tetramethyl- and tetraethylammonium carboxylates with alkyl halides to yield esters [12]

Carboxylate	Alkyl halide	% Yield	Ammonium salt[a]
Formate	C_2H_5Br	80	E
Acetate	$NCCH_2Cl$	80	E
Acetate	n-C_4H_9Cl	86	E
Acetate	n-C_4H_9Br	90	E
Acetate	s-C_4H_9Cl	48	E
Acetate	$C_6H_5CH_2Cl$	100	E
Oxalate	$C_6H_5CH_2Cl$	b	E
Acrylate	CH_3I	87	M
Cyanoacetate	CH_3I	90	M
Malonate	CH_3I	b	M
Glycolate	C_2H_5Br	100	M
Ketomalonate	CH_3I	b	M
Tartrate	CH_3I	60 (di)	M
Malate	CH_3I	92 (di)	M
Adipate	n-C_4H_9Br	b	E

[a] E = tetraethyl; M = tetramethyl.
[b] Diester product, no yield reported.

6.5 Crown Catalyzed Esterification

The alkylation of carboxylate ions has also been catalyzed by crown ethers. Ordinarily, the reaction is conducted with the alkyl halide in an organic phase such as acetonitrile which is in contact with the solid carboxylate salt. The combination of the crown and the solvent (at least for the more polar solvents like acetonitrile) aid in the dissolution of the salt. The poorly solvated carboxylate anion is an effective nucleophile in such situations and alkylates readily. The potassium acetate-crown ether ion pair is a reactive nucleophile but appears to be a relatively weak base. The synthesis of several esters by crown ether phase transfer of carboxylate salts is recorded in Table 6.5.

Synthesis of Esters

Table 6.5. Crown catalyzed ester synthesis

Carboxylate	Substrate	Products	% Yield	Ref.
CH_3COOK	1,2-dibromoethane	Ethylene diacetate	90	[19]
	C_2H_5O—△—OBs	C_2H_5O—△—OAc	100	[20]
	$n\text{-}C_6H_{13}Br$	$n\text{-}C_6H_{13}OAc$	100	[19]
CH_3COOK	$n\text{-}C_8H_{17}Br$	$n\text{-}C_8H_{17}OAc$	96	[19]
	$2\text{-}C_8H_{17}Br$	$2\text{-}C_8H_{17}OAc$	≃ 90	[19]
		+ octenes	≃ 10	[19]
	$C_6H_5CH_2Br$	$C_6H_5CH_2OAc$	100	[19]
$KO_2CCH_2CH_2CO_2K$	$4\text{-}NO_2C_6H_4CH_2Br$	$(4\text{-}NO_2C_6H_4CH_2OCOCH_2)_2$	97	[19]
$E\text{-}KO_2CCH=CHCO_2K$	$4\text{-}NO_2C_6H_4CH_2Br$	$(4\text{-}NO_2C_6H_4CH_2OCOCH)_2$	98	[19]

The amount of potassium acetate which dissolves either in benzene or acetonitrile in the presence of 18-crown-6 is slightly lower than the amount of crown present. At concentrations of 0.55 and 1.0 molar crown in benzene, for example, potassium acetate is soluble to the extent of 0.4 and 0.8 molar, respectively. Similarly, in acetonitrile, $0.14\,M$ crown dissolves 0.1 equivalent of salt [19]. In the ester displacement reaction, the crown-potassium acetate ion pair must be more soluble than the crown-potassium bromide ion pair or one would expect catalyst poisoning.

6.6 Crown Catalyzed Phenacyl Ester Synthesis

A particularly useful application of the crown mediated ester synthesis is in the preparation of phenacyl esters [21, 22]. The crown catalyzed reaction of potassium carboxylates with α,4-dibromoacetophenone yields derivatives which are usually crystalline and which can be readily detected by uv spectroscopy [23]. Moreover, the phenacyl ester is a useful protecting group which can be removed under mild conditions [24]. Phenacyl ester formation has also been catalyzed by tertiary amines [5]. Examples of the condensation according to equation 6.7 are given in Table 6.6.

$$R-COO^-K^+ + Br-CH_2-CO-C_6H_4-Br-4 \rightarrow R-COO-CH_2-CO-C_6H_4-Br-4$$
(6.7)

Likewise α-keto and α-aldehydo esters have been prepared by the reaction of potassium phenylacetate with α-bromo ketones or aldehydes by an 18-crown-6 catalyzed displacement reaction in acetonitrile or benzene solution. These compounds could be isolated after short reaction times, or they could be converted on continued heating to 2(5H)furanones, the product of intramolecular aldol condensation and

Table 6.6. Crown mediated phenacyl ester synthesis

Acid	Halide[a]	% Yield	References
Formic	A	93	[22]
Acetic	A	98	[22]
	B	88	[5][b]
Propanoic	A	98	[22]
Pivalic	A	95	[22]
Heptanoic	A	99	[22]
Benzoic	A	93	[22]
2-iodobenzoic	A	92	[22]
2-methylbenzoic	A	90	[22]
Mesitoic	A	98	[22]
4-t-butylbenzoic	A	92	[22]
Malic	B	97	[21]
Malonic	B	94	[21]
Methylmalonic	B	95	[21]
Succinic	B	98	[21]
Fumaric	B	80	[21]
Glutaric	B	97	[21]
Adipic	B	95	[21]

[a] A = 4-$BrC_6H_4COCH_2Br$; B = $C_6H_5COCH_2Br$.

[b] = Catalyst is: CH_3N⌒NCH_3 (piperazine)

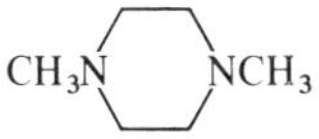

subsequent dehydration. The sequence is formulated in equation 6.8 [25]. The results of several such condensation reactions are presented in Table 6.7.

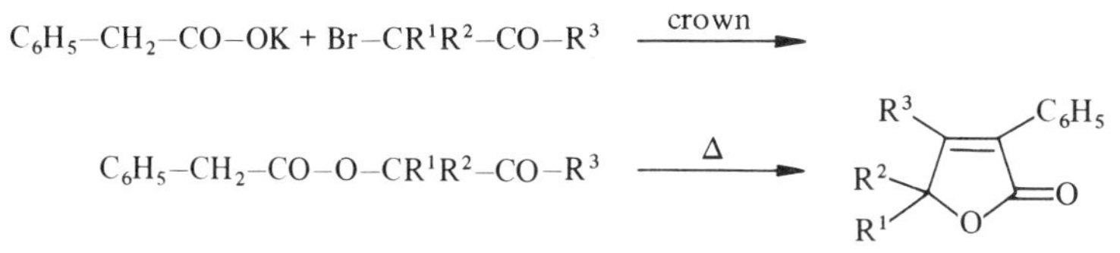

$$C_6H_5-CH_2-CO-OK + Br-CR^1R^2-CO-R^3 \xrightarrow{crown}$$

$$C_6H_5-CH_2-CO-O-CR^1R^2-CO-R^3 \xrightarrow{\Delta} \begin{array}{c} R^3 \diagup\!\!\!\diagdown C_6H_5 \\ R^2 \diagdown\!\!\!\diagup =O \\ R^1 \quad O \end{array} \quad (6.8)$$

Table 6.7. Synthesis of 2(5H)furanones from phenylacetate esters

R^1	R^2	R^3	% Yield
C_6H_5	C_6H_5	H	90
C_6H_5	4-$CH_3OC_6H_4$	H	81
C_6H_5	H	4-$CH_3OC_6H_4$	80
4-$CH_3OC_6H_4$	H	C_6H_5	75
C_6H_5	4-BrC_6H_4	H	62
C_6H_5	4-NCC_6H_4	H	20
C_6H_5	H	C_6H_5	80
CH_3	H	CH_3	90

Phenylacetate ion appears to be an excellent nucleophile in nonpolar media, but a relatively weak base. Substitution of the α-bromoaldehydes seems very much more favorable than an aldol type condensation, even for the unhindered aldehydes [25]. A related example of this nucleophile/base balance has been observed in the case of 4-bromocyclohexenone [8]. Phase transferred acetate ion displaces bromide (73%) rather than dehydrohalogenating the system to phenol (Eq. 6.9) [8].

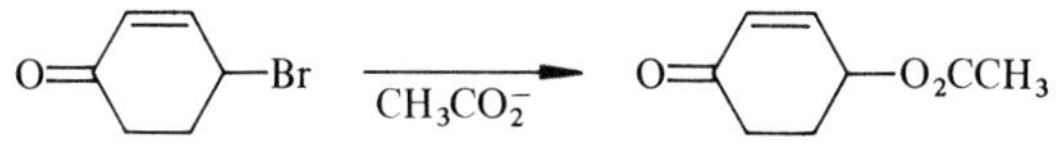

$$O=\underset{CH_3CO_2^-}{\longrightarrow} Br \quad \longrightarrow \quad O=\underset{}{\bigcirc}-O_2CCH_3 \tag{6.9}$$

6.7 Crown Catalyzed Esterification of BOC-Amino Acids to Chloromethylated Resins

The Merrifield solid phase peptide synthesis requires the formation of an esterified resin as the first step. The substitution of an N-protected amino acid for a chloride in chloromethylated polystyrene must be quantitative or products which are one amino acid too short will ultimately be isolated. 18-Crown-6 has been shown to catalyze quantitative ester formation between the potassium salts of boc-amino acids and chloromethylated resin in DMF solution. Dichloromethane was reactive under these conditions [26]. The results of these displacement reactions according to equation 6.10 are presented in Table 6.8.

Table 6.8. Esterification of polystyrene resin [26]

Boc-amino acid	% Yield
Boc-Leu	101
Boc-Pro	96
Boc-Phe	99
Boc-Val	96
Boc-Ser	97
Boc-Ile	93
Boc-ε-Z-Lys	99
Boc-D-Ala	92
Boc-N^{im}-Tos-His	100
Boc-Gly	96
Boc-Ser (OBzl)	92
Boc-Tyr (OBzl)	89

Boc = t-butyloxycarbonyl; Tos = p-toluenesulfonyl; Z = benzyloxycarbonyl; Bzl = benzyl.

Polymer–$C_6H_4CH_2Cl$ + $(CH_3)_3C$–O–$\overset{\overset{O}{\|}}{C}$–NH–CHR–COOK → (6.10)

Polymer–C_6H_4–CH_2–O–CO–CHR–NH–CO–O–$C(CH_3)_3$

Crown complexed potassium acetate has recently been used to initiate the polymerization of acrylic acid [27].

6.8 Cryptate and Resin Catalyzed Esterifications

Not only will crown ethers, quaternary ions, and amines catalyze the esterification reaction, but cryptates are also effective in this application. Potassium carboxylates react in good yield with a variety of alkyl halides in the presence of a catalytic amount of [2.2.2]-cryptate to yield the corresponding esters [28]. The results are summarized in Table 6.9.

Table 6.9. Cryptate catalyzed ester formation [28]

Carboxylate	Bromide	% Yield
CH_3COOK	$C_6H_5CH_2Br$	99
CH_3COOK	$CH_3(CH_2)_6CH_2Br$	85
CH_3CH_2COOK	$BrCH_2CH_2Br$	89 (di)
CH_3CH_2COOK	$C_6H_5CH_2Br$	85
CH_3CH_2COOK	$CH_3(CH_2)_6CH_2Br$	90
$(CH_3)_2CHCOOK$	$C_6H_5CH_2Br$	95
$(CH_3)_2CHCOOK$	$CH_3(CH_2)_6CH_2Br$	99

The alkylation of carboxylates to yield esters has also been catalyzed by quaternary ammonium groups bound to a macroreticular resin. Amberlite IRA-904 was converted to the hydroxyl form and then treated with the acid to form ammonium carboxylate ion pairs. The alkylation was carried out by a batch process in such solvents as ether, hexane, or ethanol, as illustrated in equation 6.11 [29]. The results are summarized in Table 6.10.

$$\text{Resin–}\overset{+}{N}(CH_3)_3 {}^-OH + R^1\text{–COOH} \rightarrow \text{Resin–}\overset{+}{N}(CH_3)_3 {}^-O_2C\text{–}R^1 + H_2O$$

$$\downarrow R^2X \quad (6.11)$$

$$\text{Resin–}\overset{+}{N}(CH_3)_3 X^- + R^1\text{–COO–}R^2$$

Synthesis of Esters

Table 6.10. Ester formation catalyzed by resin bound quats

Acid	Halide	% Yield
C_6H_5COOH	CH_3I	90
	$i\text{-}C_3H_7Br$	60
	$C_2H_5O_2CCH_2Br$	99
	$C_6H_5CH_2Cl$	73
$c\text{-}C_6H_{11}COOH$	CH_3I	76
	$i\text{-}C_3H_7Br$	28
$C_6H_5CH=CHCOOH$	CH_3	97
	$i\text{-}C_3H_7Br$	59
$n\text{-}C_{11}H_{23}COOH$	CH_3I	93
	$i\text{-}C_3H_7Br$	52

6.9 Synthesis of Sulfonate and Phosphate Esters by PTC

The formation of both sulfonate esters and phosphate esters have been achieved under phase transfer conditions. Good yields of sulfonate esters were obtained when 3-hydroxy steroids (aromatic A ring) were reacted with dialkylaminosulfonyl chlorides in a two phase system consisting of benzene and aqueous base. A typical acylation using pyrrolidine sulfonyl chloride in the presence of catalytic benzyl-triethylammonium chloride is formulated in equation 6.12 [30].

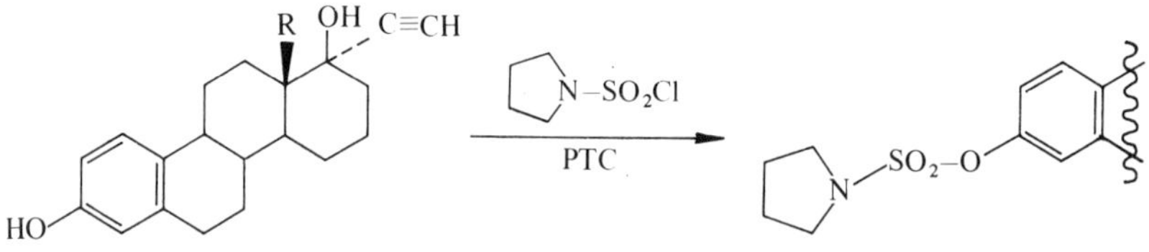

(6.12)

Phenols have been phosphorylated under phase transfer conditions in the presence of a nucleophilic catalyst [31, 32]. The reaction of 4-nitrophenol with dimethoxythiophosphoryl chloride is ordinarily slow and leads to a mixture of the desired methyl parathion and hydrolysis products. Addition of N-methylimidazole enhanced the rate but the best results were obtained when both the imidazole and a quaternary ammonium salt (TBAB) were used at the same time. The co-catalysis was accounted for in terms of nucleophilic activation of the acylating agent by imidazole and solubilization of the phenoxide by ion pairing with the quaternary ion. The overall transformation is formulated in equation 6.13.

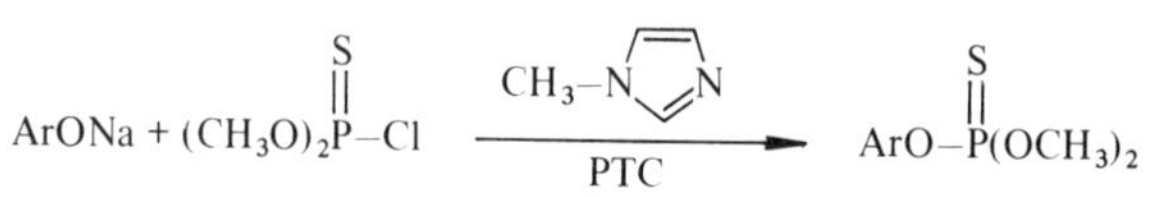

$$ArONa + (CH_3O)_2\overset{S}{\overset{\|}{P}}-Cl \xrightarrow[PTC]{CH_3-N\diagup\hspace{-0.5em}\diagdown N} ArO-\overset{S}{\overset{\|}{P}}(OCH_3)_2 \qquad (6.13)$$

References

1. Maerker, G., Carmichael, J. F., Port, W. S.: J. Org. Chem. *26*, 2681 (1961).
2. Mueller, A. C.: U.S. Patent 2,772, 296, November 27, 1956.
3. Hennis, H. E., Easterly, J. P., Collins, L. R., Thompson, L. R.: Ind. Eng. Chem. Prod. Res. Dev. *6*, 193 (1967).
4. Hennis, H. E., Thompson, L. R., Long, J. P.: Ind. Eng. Chem. Prod. Res. Dev. *7*, 96 (1968).
5. Normant, H., Cuvigny, T., Savignac, P.: Synthesis *1975*, 805.
6. Mikolajczyk, M., Grzejszcak, S., Midura, W., Zatorski, A.: Synthesis *1975*, 393.
7. Mikolajczyk, M., Grzejszcak, S., Zatorski, A., Montanari, F., Cinquini, M.: Tetrahedron Let. *1975*, 3757.
8. Toru, T., Kurozumi, S., Tanaka, T., Miura, S., Kobayashi, M., Ishimoto, S.: Synthesis *1974*, 867.
9. Mills, R. H., Farrar, M. W., Weinkauff, O. J.: Chemistry and Industry *1962*, 2144.
10. Brandström, A.: "Preparative Ion Pair Extraction", pp. 109–111, 155–156. Läkemedel, Sweden: Apotekarsocieteten/Hässle 1974.
11. Holmberg, K., Hansen, B.: Tetrahedron Let. *1975*, 2303.
12. Wagenknecht, J. H., Baizer, M. M., Chrumz, J. L.: Syn. Commun. *2*, 215 (1975).
13. Schnell, H.: Angew. Chem. *68*, 633 (1956).
14. British Patent 808,490, April 6, 1956, Bayer.
15. German Patent F 17,167, March 26, 1955, Bayer.
16. Oxenrider, B. C., Park, F., Hetterly, R. M.: U.S. Patent 3,297,634, January 10, 1967, Allied.
17. Schnell, H.: Chemistry and Physics of Polycarbonates, pp. 37–41, 93–94. New York, N. Y.: Interscience Publishers (1964).
18. Curtius, U., Bockum, K., Böllert, V., Fritz, H., Fritz, G., Nentwig, J.: U.S. Patent 3,326,958, June 20, 1967, Bayer.
19. Liotta, C. L., Harris, H. P., McDermott, M., Gonzalez, T., Smith, K.: Tetrahedron Let. *1974*, 2417.
20. Maryanoff, C. A., Ogura, F., Mislow, K.: ibid *1975*, 4095.
21. Grushka, E., Durst, H. D., Kikta, E. J.: J. Chromatogr. *112*, 673 (1975).
22. Durst, H. D.: Tetrahedron Let. *1974*, 2421.
23. Durst, H. D., Milano, M., Kikta, E. J., Connelly, S. A., Grushka, E.: Anal. Chem. *47*, 1797 (1975).
24. Hendrickson, J. B., Kendall, C.: Tetrahedron Let. *1970*, 343.
25. Padwa, A., Dehm, D.: J. Org. Chem. *40*, 3139 (1975).
26. Roeske, R. W., Gesellchen, P. D.: Tetrahedron Let. *1976*, 3369.
27. Yamada, B., Yasuda, Y., Matsushita, T., Otsu, T.: Polymer Letters *14*, 277 (1976).
28. Akabori, A., Ohtomi, M.: Bull. Chem. Soc. Jap. *48*, 2991 (1975).
29. Cainelli, G., Maneschalchi, F.: Synthesis *1975*, 723.
30. Schwarz, S., Weber, G.: Z. Chem. *15*, 270 (1975).
31. Ridgway, R. W., Greenside, H. S., Freedman, H. H.: J. Am. Chem. Soc. *98*, 1979 (1976).
32. Freedman, H. H.: U.S. Patent 3,972,887, August 3, 1976, Dow.

7. Reactions of Cyanide Ion

7.1 Introduction

Phase transfer catalysis found one of its earliest applications in cyanide displacement reactions. Starks studied this reaction in some detail and it was due in large part to his work that the concepts of phase transfer were so clearly understood at an early stage [1, 2]. There was, of course, other early interest in cyanide ion and these efforts should not be overlooked. Brandström had solubilized cyanide ion by his ion pair extraction technique [3]; Solodar had facilitated the benzoin condensation by utilizing tetrabutylammonium cyanide [4]; and Durst [5] and Liotta [6] both used crown ethers to phase transfer cyanide ion. A good deal of other work has now been carried out on phase transferred cyanide ion and is the subject of this chapter.

7.2 The Mechanism and General Features of the Cyanide Displacement Reaction

Because of Starks' early discussion of this reaction, it is one of the standard examples of phase transfer catalysis. The simple nucleophilic displacement (S_N2) of a nucleofuge by cyanide ion (Eq. 7.1) is well understood and the mechanism is discussed in Sect. 1.4 in general terms and in comparison with the mechanism of cyclopropanation in Sect. 2.2. Because of these earlier discussions, to which the reader is referred, only the salient features of the mechanism will be set forth here.

$$R-X + NaCN \xrightarrow{Q^+X^-} R-CN + NaX \qquad (7.1)$$

The catalytic cycle accounting for the essential features of the cyanide displacement reaction is illustrated in equation 7.2. Note that this is the cycle shown in Sect. 1.4, except that the anions are identified. In essence, the process involves an equilibrium between sodium cyanide and quaternary ammonium cyanide in the aqueous phase, followed by a phase transfer equilibrium occurring across the phase boundary. Once the quaternary ammonium cyanide is present in the organic phase, nucleophilic displacement can take place leading to product and regeneration of the phase transfer catalyst. There are several specific features of the phase transfer cyanide displacement

$$Q^+CN^- + R-Cl \rightarrow R-CN + Q^+Cl^- \quad \text{(organic phase)}$$

~~~~~~~~~~~~~~~~~~~~~~~~~~~~~~~~~~ (phase boundary) \qquad (7.2)

$$Q^+CN^- + NaCl \rightleftarrows NaCN + Q^+Cl^- \quad \text{(aqueous phase)}$$

reaction to which attention should be called [1, 2].

1. The nucleophilic displacement reaction is presumed to occur in the organic phase and this bimolecular step is assumed to be rate controlling.

2. As expected for a bimolecular process, the reaction is faster with primary than with secondary halides. Likewise, n-octyl methanesulfonate was found to undergo cyanide displacement at a greater rate than did the corresponding bromide. Moreover, elimination competes with substitution in the case of secondary halides; elimination was the exclusive process observed in the reaction of cyanide ion with cyclohexyl halides.

3. The reaction is a true phase transfer process, the reaction rate being independent of stirring rate beyond a minimal mixing requirement (*ca.* 250 rpm) [2, 7].

4. The reaction appears not to be micellar inasmuch as poor surfactants (such as tetradodecylammonium salts) are effective catalysts.

The phase transfer cyanide reaction appears to be superior even to the reaction conducted in dipolar aprotic media. s-Chlorooctane, for example, yields 85–90% substitution products and only 10–15% elimination products under phase transfer conditions [1], whereas the homogeneous reaction in dimethyl sulfoxide resulted in only 70% of s-cyanooctane [8]. The phase transfer reaction is applicable for chloride, bromide, or methanesulfonate leaving groups but is less satisfactory when the nucleofuge is either iodide or p-toluenesulfonate (tosylate). This is due to the fact that the large, lipophilic and polarizable quaternary cations tend to ion pair irreversibly with iodide and tosylate. This difficulty can often be overcome by renewing the aqueous reservoir of nucleophile.

Crown ethers have also been utilized as phase transfer catalysts in solid-liquid phase transfer cyanide displacements. These reactions are generally carried out in methylene chloride or acetonitrile solution with 18-crown-6 as catalyst and solid potassium cyanide as nucleophile source [5, 6]. Small amounts of water are found not to affect the course of the reaction [5], suggesting some hydration of cyanide ion under these conditions. This is not surprising inasmuch as Starks reported that in the liquid-liquid phase transfer process, four to five molecules of water apparently accompanied each nucleophile into nonpolar solution [2]. It seems likely that if water were or could be rigorously excluded, (i.e., "naked anions" obtained), the reactivity of cyanide would be even higher. Despite the apparent similarity of the solid-liquid and liquid-liquid phase transfer processes, it should be noted that qualitative differences in the relative reactivity of primary alkyl halides (R–Cl vs. R–Br) have been observed for the crown and quaternary ion cases [2, 6]. Specifically, Starks found that for the reaction of cyanide ion with n-octyl halides, methanesulfonate was a better leaving group than bromide which in turn was better than chloride. Liotta, on the other hand, found that benzyl chloride was more susceptible to nucleophilic displacement than benzyl bromide under crown-phase transfer conditions in acetonitrile solution [6].

Tertiary amines are also known to effect the phase transfer addition of cyanide ion to primary, allylic, and benzylic halides [9]. The reported effect of amine structure on catalytic efficiency closely parallels that reported by Hennis for ester formation in a two-phase system (see Sect. 1.7). Both the nitrogen of the amine and the carbon bearing halide of the alkyl bromide must be sterically accessible for the reaction to succeed. Thus, *n*-hexylamine is effective in concert with *n*-butyl bromide but the combinations of either *s*-butyl bromide and *n*-hexylamine or *n*-butyl bromide and cyclohexylamine are not. Tertiary amines are generally more effective than secondary or primary amines. In addition, the yields of primary nitriles decrease dramatically with the size of the primary alkyl bromide from quantitative with *n*-butyl to only 6% with *n*-decyl bromide when *n*-hexylamine is used as phase transfer catalyst. On the other hand, tributylamine was equally useful as a catalyst for the quantitative conversion of either 1-bromohexane or 1-bromodecane to the corresponding nitriles [9]. In general, these observations accord with those of Hennis and coworkers indicating that this reaction is an example of *in situ* formation of and catalysis by quaternary ammonium salts [10].

Finally, it should be noted that the reaction of cyanide ion with 1-bromooctane to yield 1-octyl cyanide has been catalyzed by a quaternary ammonium catalyst bound to a polystyrene resin. The resin constituted one of three phases, the other two being aqueous cyanide reservoir and toluene containing the substrate. The kinetics observed in this three-phase system are strikingly similar to those observed by Starks [11].

7.3 The Formation of Alkyl Cyanides

Alkyl cyanides have been prepared by three major variations of the phase transfer method. These include the liquid-liquid phase transfer method (including catalysis by quats and amines), the crown catalyzed solid-liquid method and the tri-phase catalytic method. Each of these methods has been discussed with reference to the mechanism in the preceding section and it remains largely to exemplify these approaches. It should be noted, however, that numerous stoichiometric syntheses of aliphatic nitriles are available, including those involving dipolar aprotic solvents [12].

Stoichiometric amounts of tetraethylammonium cyanide react with aliphatic bromides in dichloromethane, acetonitrile or DMSO to give reasonable yields of the corresponding nitriles [13]. These reactions are clearly related to, but not actually examples of, phase transfer catalysis. It is interesting, however, that under these homogeneous conditions, tetraethylammonium cyanide reacts in acetonitrile with neopentyl bromide to give the corresponding nitrile (see Eq. 7.3). Bimolecular displacements on such sterically hindered substrates are usually quite difficult to effect.

$$(C_2H_5)_4N^+CN^- + (CH_3)_3CCH_2-Br \xrightarrow{CH_3CN} (CH_3)_3CCH_2-CN + (C_2H_5)_4N^+Br^-$$

(7.3)

The preparation of a number of aliphatic nitriles is tabulated below (Table 7.1). Several examples which are noncatalytic are included in this table for the benefit of the reader interested more in the product than in the mode of synthesis.

Table 7.1. Synthesis of aliphatic nitriles

Substrate	Catalyst[a]	% Yield	Solvent[b]	Ref.
a) Primary substrates				
C_2H_5Br	TEAC-*st*	65	A	[13]
n-C_4H_9Br	TEAC-*st*	74	A	[13]
	HA	100	0	[9]
i-C_4H_9Br	HA	2	0	[9]
	TEAC-*st*	75	C	[13]
n-$C_5H_{11}Br$	TEAC-*st*	80	A	[13]
t-$C_4H_9CH_2Br$	TEAC-*st*	54	B	[13]
n-$C_6H_{13}Br$	HA	66	0	[9]
	TBA	100	0	[9]
	18-C-6	100	B	[6]
n-$C_6H_{13}Cl$	18-C-6	90	B	[6]
n-$C_7H_{15}Br$	HA	39	0	[9]
	TBA	100	0	[9]
n-$C_8H_{17}Br$	HDTBP	95	0	[1]
	HA	32	0	[9]
	DMEBr	87	0	[15]
n-$C_8H_{17}Cl$	HDTBP	95	0	[1]
	cryptate	93	0	[14]
	HDTBP	94	0	[14]
n-$C_8H_{17}OSO_2CH_3$	HDTBP	95	0	[1]
n-$C_8H_{17}I$	HDTBP	5–30	0	[1]
n-$C_8H_{17}OTs$	HDTBP	5–30	0	[1]
n-$C_{10}H_{21}Br$	TEAC-*st*	91	A	[13]
n-$C_{12}H_{25}Br$	HA	6	0	[9]
	TBA	99	0	[9]
$Br(CH_2)_2Br$	TEAC-*st*	70	A	[13]
$Br(CH_2)_3Br$	TEAC-*st*	86	A	[13]
	18-C-6	97	B	[6]
	18-C-6	100	C_6H_6	[6]

Table 7.1 (continued)

Substrate	Catalyst[a]	% Yield	Solvent[b]	Ref.
$Br(CH_2)_3Cl$	TEAC-*st*	85 mono	A	[13]
	18-C-6	100 di	B	[6]
$Cl(CH_2)_3Cl$	18-C-6	96	B	[6]
$Br(CH_2)_4Br$	TEAC-*st*	90	A	[13]
	18-C-6	100	B	[6]
$Cl(CH_2)_4Cl$	18-C-6	95	B	[6]
b) Secondary substrates				
$i\text{-}C_3H_7Br$	TEAC-*st*	72	C	[13]
$s\text{-}C_4H_9Br$	TEAC-*st*	80	A	[13]
	18-C-6	70	B	[6]
	HA	0	0	[9]
$c\text{-}C_5H_9Br$	TEAC-*st*	68	B	[13]
$c\text{-}C_6H_{11}Br$	TBA	0	0	[9]
	18-C-6	elimination only		[6]
$c\text{-}C_6H_{11}Cl$	18-C-6	elimination only		[6]
$s\text{-}C_8H_{17}Br$	18-C-6	57 17-elimination	B	[6]
$s\text{-}C_8H_{17}Cl$	HDTBP	85–90	0	[1]
	18-C-6	78 3-elimination	B	[6]
c) Allylic and benzylic substrates				
$CH_2=CHCH_2Br$	TEAC-*st*	50	A	[13]
	TBA	79	0	[9]
$CH_3CH=CHCH_2Br$	TEAC-*st*	55	A	[13]
$C_6H_5CH_2Br$	18-C-6	100	B	[6]
	TEAC-*st*	74	A	[13]
	TBA	100	0	[9]
$C_6H_5CH_2Cl$	18-C-6	94	B	[6]
	18-C-6	90–95	A or B	[5]
$4\text{-}NO_2C_6H_4CH_2Cl$	18-C-6	85–95	B	[5]
$4\text{-}ClC_6H_4CH_2Cl$	18-C-6	85–95	B	[5]
$3,4\text{-}(CH_3O)_2C_6H_3CH_2Cl$	18-C-6	85–95	B	[5]
$2\text{-}BrCH_2C_6H_4CH_2Br$	TEAC-*st*	85-di	A	[13]

Table 7.1 (continued)

Substrate	Catalyst[a]	% Yield	Solvent[b]	Ref.
d) Miscellaneous substrates				
$C_2H_5OCOCH_2Br$	TEAC-*st*	60	A	[13]
$C_2H_5OCOC(CH_3)_2Br$	TEAC-*st*	80	B	[13]
$(C_2H_5O)_2CHCH_2Br$	TEAC-*st*	85	B	[13]
CH_3OCH_2Cl	TEAC-*st*	30	A	[13]
$(CH_3)_3SiCl$	18-C-6	40–50	A	[5]

[a] Catalysts:

$Et_4N^+CN^-$-*stoichiometric*	TEAC-*st*
n-hexylamine	HA
Tributylamine	TBA
N-dodecyl-N-methylephedrinium bromide	DMEBr
Hexadecyltributylphosphonium bromide	HDTBP
18-crown-6	18-C-6
none	0
(structure with $n-C_{14}H_{29}$)	cryptate

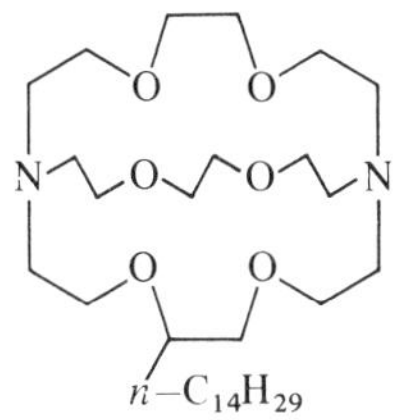

[b] Solvents:

CH_2Cl_2	A
CH_3CN	B
DMSO	C
none	0

7.4 Formation of Acyl Nitriles

Nucleophilic displacement may be effected at sp^2 as well as at saturated carbon, although in the former case, the reaction is probably an addition-elimination sequence. Benzoyl chloride, for example, reacts with sodium cyanide under liquid-liquid phase transfer conditions to yield benzoyl cyanide according to equation 7.4 [16]. Several examples of this reaction are given in Table 7.2. The preparation of aroyl cyanides

$$Ar-CO-Cl + NaCN \xrightarrow[CH_2Cl_2]{Q^+Cl^-} Ar-CO-CN + NaCl \qquad (7.4)$$

Table 7.2. Preparation of aroyl cyanides [16]

Acyl halide	Yield of aroyl cyanide	Yield of dimer
C_6H_5COCl	60	30
4-$CH_3C_6H_4COCl$	72	–
4-$CH_3OC_6H_4COCl$	60	–
4-ClC_6H_4COCl	22	46

has hitherto involved reaction of the chloride with mercuric [17], silver [18, 19], or cuprous [20] cyanide. Although more convenient than the metal catalyzed reaction, the yields obtained by the phase transfer method are limited by dimer formation which occurs as a side reaction. It appears that this dimer arises from benzoylation of a dicyano-alkoxide intermediate according to equation 7.5. Attempts to extend this reaction to aliphatic acyl chlorides have thus far been unsuccessful.

$$Ar-CO-CN \xrightarrow{^-CN} Ar-\underset{|}{\overset{O^-}{C}}(CN)_2 \xrightarrow{Ar-CO-Cl} Ar-C(CN)_2-O-CO-Ar \quad (7.5)$$

7.5 Synthesis of Cyanoformates

A wide variety of chloroformates have been converted into the corresponding cyanoformates by reaction with potassium cyanide under crown ether catalyzed phase transfer conditions [21]. Dimer formation was not a problem in these reactions as it was with acyl chlorides (see Sect. 7.4). Unfortunately, this reaction fails to convert either phosgene or *t*-butyl chloroformate into the corresponding nitrile. Of interest is the observation that in the absence of a small amount of water, this reaction is extremely sluggish. It is likely that the water is needed to assist in degrading the potassium cyanide crystal lattice, although its role in hydration of the anion may also be important. The transformation of chloroformates into cyanoformates is formulated in equation 7.6 and examples are given in Table 7.3.

$$R-O-CO-Cl + KCN \xrightarrow{\text{18-Crown-6}} R-O-CO-CN + KCl \quad (7.6)$$

Table 7.3. Preparation of cyanoformates according to equation 7.6 [21]

R	% Yield	R	% Yield
Methyl	76	Isopropyl	62
Ethyl	72	2-octyl	88
n-butyl	90	Cyclohexyl	90
Isobutyl	94	Benzyl	65
2,2,2-trichloroethyl	88	Phenyl	82

7.6 Cyanohydrin Formation

The formation of aldehyde cyanohydrin esters has been achieved under phase transfer catalytic conditions in which the two liquid phases are aqueous sodium or potassium cyanide and a methylene chloride solution of an aromatic aldehyde and an acid chloride. Both quaternary ammonium salts and 18-crown-6 are effective catalysts in this reaction. The formation of aldehyde cyanohydrin esters according to equation 7.7 probably occurs by a mechanism similar to that proposed for the formation of benzoyl cyanide dimers (see Eq. 7.5) [3, 22]. Accordingly, cyanide anion adds to the aldehyde carbonyl group to yield an alkoxide anion which in turn is acylated by the acid chloride.

$$\text{Ar-CHO} + \text{NC}^- \rightarrow \text{Ar-CH(O}^-\text{)-CN} \xrightarrow{\text{C}_6\text{H}_5\text{-CO-Cl}} \text{Ar-CH(O-CO-C}_6\text{H}_5\text{)-CN} \quad (7.7)$$

Attempts to substitute alkyl chloroformates for acid chlorides in this reaction gave cyanoformates rather than the anticipated alkyl cyanohydrin carbonates (Eq. 7.8) [22]. It seems likely that this fact reflects a reactivity difference among the carbonyl groups in the order: chloroformates > aromatic aldehydes > aromatic acyl chlorides.

$$\text{C}_6\text{H}_5\text{-C(=O)-H} + \text{R-O-C(=O)-Cl} \nrightarrow \begin{array}{c} \text{C}_6\text{H}_5\text{-C(O-C(=O)-OR)(CN)-H} \\ \text{C}_6\text{H}_5\text{-C(=O)-H} + \text{RO-C(=O)-CN} \end{array} \quad (7.8)$$

Allylic ethers of aldehyde cyanohydrins have also been prepared under phase transfer conditions [23]. The organic phase, consisting of aldehyde cyanohydrin, an allylic halide (preferably the bromide) and dichloromethane is brought into contact with aqueous sodium hydroxide and tricaprylmethylammonium chloride. Anion exchange in the aqueous phase leads to the tricaprylmethylammonium/hydroxide ion pair, which reacts with the cyanohydrin to form the corresponding alkoxide. Reaction of the alkoxide with the allylic halide leads to the allylic ether of the cyanohydrin (Eq. 7.9). The formation of small amounts of allylic nitriles is apparently due to the side

$$\text{R-CH(OH)-CN} + \text{Q}^+\text{OH}^- \rightarrow \left[\begin{array}{c} \text{R-C(O}^-\text{)(CN)-H} \\ \updownarrow \\ \text{R-CHO} + \text{CN}^- \end{array} \right] \xrightarrow{\substack{R_1\\R_2}\diagdown\text{X}} \begin{array}{c} \text{R-CH(O-CH}_2\text{-CR}_2\text{=CR}_1\text{)-CN} \\ R_1R_2\text{C=CH-CN} \end{array} \quad (7.9)$$

reaction of the halide with a small equilibrium concentration of cyanide ion. Several examples are recorded in Table 7.4.

Table 7.4. Preparation of cyanohydrin ethers according to equation 7.9

R	R_1	R_2	% Yield
$(CH_3)_2CHCH_2$	H	H	46
$(CH_3)_2CHCH_2$	H	CH_3	45
$(CH_3)_2CHCH_2$	CH_3	CH_3	46
$CH_3CH(OCH_3)CH_2$	CH_3	CH_3	27
CH_3	CH_3	CH_3	32
$(CH_3)_2CHCH_2$	CH_3	$(CH_3)_2C=CH(CH_2)_2$	37
CH_3	CH_3	$(CH_3)_2C=CH(CH_2)_2$	30

7.7 The Benzoin Condensation

In addition to the stoichiometric reactions of cyanide discussed above, phase transferred cyanide ion has been used as a catalyst in several organic reactions. Notable among these is the benzoin condensation which was facilitated by the presence of a catalytic amount of tetrabutylammonium cyanide. Thus, addition of a small amount of tetrabutylammonium cyanide to a 50% aqueous methanol solution at room temperature was found to effect the conversion of benzaldehyde to benzoin (Eq. 7.10) [4] whereas the latter was not formed in its absence. Since Solodar's initial finding [4], it has been shown that crown ethers [24] and Amberlite IRA 400 and 410 ion

$$2C_6H_5-CHO + Bu_4N^+CN^- \text{ (cat)} \rightarrow C_6H_5-CHOH-CO-C_6H_5 \qquad (7.10)$$

Table 7.5. Catalytic benzoin condensations

Aldehyde	% Yield	Catalyst	Temperature	Ref.
C_6H_5CHO	70	TBA cyanide	Room temperature	[4]
	78	18-C-6 or DB-18-C-6	60 °C	[24]
	51	Amberlite cyanide	85–90 °C	[25]
$4\text{-}CH_3C_6H_4CHO$	99	18-C-6 or DB-18-C-6	60 °C	[24]
$4\text{-}CH_3OC_6H_4CHO$	11	Amberlite cyanide	85–90 °C	[25]
Furfural	66	18-C-6 or DB-18-C-6	60 °C	[24]
	64	Amberlite cyanide	85–90 °C	[25]

exchange resins [25] in concert with cyanide ion also catalyze the benzoin condensation. In the latter case, the amount of catalyst required was substantial [25]. The results of several benzoin condensation reactions are recorded in Table 7.5.

Cyanide is not the only ion known to catalyze the benzoin reaction. This condensation is also catalyzed by N-alkylthiazolium bromide salts. The micelle forming N-laurylthiazolium bromide is a far better catalyst than the N-butyl substituted salt, indicating that this reaction may be micellar rather than of the phase transfer type. This catalyst is effective in the acyloin reaction (see Eq. 7.11) of aliphatic aldehydes whereas cyanide ion is not (see Table 7.6) [26].

$$2R-CHO \xrightarrow{catalyst} R-CHOH-CO-R \qquad (7.11)$$

Table 7.6. Preparation of acyloins [26][a]

Aldehyde	% Yield
Propanal	16
Butanal	39
Hexanal	67
Octanal	63
Benzaldehyde	95
Furfural	80

[a] 0.07 equivalent of catalyst used.

Table 7.7. Cyanosilylation of quinones [27, 29]

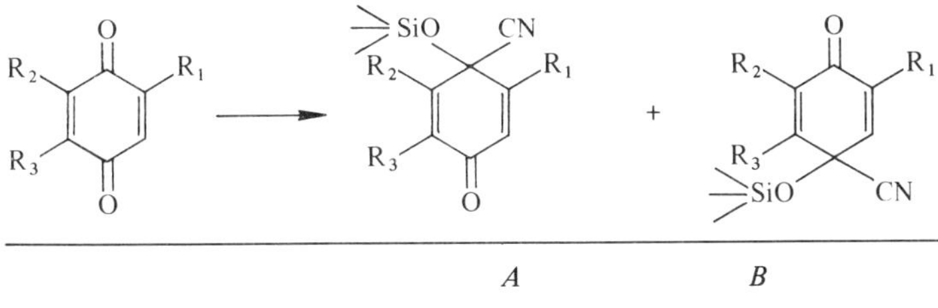

R_1	R_2	R_3	Ratio A:B	% Yield
H	H	H		80
CH_3	H	H	89:11	92
OCH_3	H	H	100:0	80
OCH_3	H	CH_3	100:0	90
OCH_3	OCH_3	H	100:0	65
CH_3	CH_3	H	94:6	100
t-C_4H_9	t-C_4H_9	H	0:100	98
H	$-C_4H_4-$			75
CH_3	$-C_4H_4-$		91:9	96
OCH_3	$-C_4H_4-$		100:0	82

7.8 Hydrocyanation, Cyanosilylation, and Other Reactions

There is currently a single example of olefin hydrocyanation under phase transfer catalytic conditions in the literature. Methacrylonitrile in acetonitrile reacts with potassium cyanide and acetone cyanohydrin in the presence of catalytic 18-crown-6 to yield 1,2-dicyanopropane (92%) according to equation 7.12. The cyanohydrin functions as both a proton donor and a source of cyanide ion [6].

$$CH_2=C(CH_3)CN \xrightarrow[K^+CN^-]{crown} NC-CH_2-\bar{C}(CH_3)CN \xrightarrow{(CH_3)_2C(CN)OH}$$

$$NC-CH_2CH(CH_3)CN \qquad (7.12)$$

The protection of carbonyl functions by reaction with trimethylsilyl cyanide to form the cyanohydrin trimethylsilyl ethers is subject to catalysis by potassium cyanide in the presence of 18-crown-6, tetrabutylammonium cyanide or resin bound tetraalkylammonium cyanide [27, 29]. The nucleophilic cyanide ion is believed to attack the carbonyl yielding a cyanohydrin anion. This anion is silylated by trimethylsilyl cyanide with generation of an equivalent of cyanide anion. The sequence is shown in equation 7.13. Accordingly, 3-pentanone and 4-t-butylcyclohexanone are cyanosil-

$$R^1R^2C=O + NC^- \rightarrow R^1R^2C(CN)-O^- \xrightarrow{(CH_3)_3Si-CN}$$

$$R^1R^2C(CN)-OSi(CH_3)_3 + NC^- \qquad (7.13)$$

ylated in 89% and 99% yields respectively using catalytic amounts of potassium cyanide and 18-crown-6 [28]. The KCN/crown system also catalyzes the $trans$-cyanosilylation reaction of pentanal and 2-octanone with acetonecyanohydrin trimethylsilyl ether to yield the protected cyanohydrins in 76% and 85% yields respectively [28]. The $trans$-cyanosilylation reaction is illustrated in equation 7.14.

$$R^1R^2C=O + (CH_3)_2C(CN)-OSi(CH_3)_3 \xrightarrow[\substack{KCN/18-C-6 \\ -acetone}]{120°} R^1R^2C(CN)-OSi(CH_3)_3 \qquad (7.14)$$

The mono-cyanosilylation of quinones is recorded in Table 7.7. The aldol condensation of ethyl α-trimethylsilyldiazoacetate with aldehydes (Eq. 7.15) was found to be catalyzed by the 18-crown-6 complex of potassium cyanide [30]. A mechanism involving the diazoester enolate anion has been proposed. Several examples of this condensation have been recorded in Table 7.8.

$$R-CHO + (CH_3)_3Si-C(=N_2)-COOC_2H_5 \xrightarrow[crown]{KCN}$$

$$R-CH(OSiMe_3)C(=N_2)-COOC_2H_5 \qquad (7.15)$$

Table 7.8. Aldol condensations of ethyl α-trimethylsilyldiazoacetate [30]

Aldehyde	% Yield of adduct
$CH_3(CH_2)_4CHO$	86
$(CH_3)_2CHCHO$	93
C_6H_5CHO	86
$4\text{-}ClC_6H_4CHO$	93
$4\text{-}CH_3OC_6H_4CHO$	83
$CH_3CH=C(CH_3)CHO$	44

Photochemical cyanation of aromatic hydrocarbons in acetonitrile solution is a higher yield process when the potassium cyanide complex of 18-crown-6 is the cyanide ion source [31] compared to similar reactions in mixed organic aqueous solvent systems [32] (see Eq. 7.16). A ten-fold excess of 18-crown-6/KCN over the aromatic hydrocarbon (present in 10^{-4} M) was used. The yield improvements were attributed to increased activity of cyanide due to diminished hydration of the ion. Biphenyl, naphthalene, phenanthrene, and anthracene were photocyanated in 50%, 15%, 25% and 20% yields respectively; the latter being an equimolar mixture of mono and dicyanation products [31].

$$\text{Ar–H} + \text{KCN/crown} \xrightarrow[\text{MeCN}]{h\nu} \text{Ar–CN} \qquad (7.16)$$

Phase transfer agents have been added to organic solutions in order to solubilize salts which comprise supporting electrolytes in electrolytic syntheses [33]. Both naphthalene and anisole were anodically cyanated in dichloromethane according to equation 7.17. The reaction was found to be clean, and high yield, although in both cases, mixtures of positional isomers were isolated. The fact that the reaction was not current efficient was compensated by the inexpensive nature of the supporting electrolyte.

$$\text{Ar–H} + \text{NC}^- \xrightarrow{-2\epsilon} \text{Ar–CN} + \text{H}^+ \qquad (7.17)$$

References

1. Starks, C. M.: J. Am. Chem. Soc. *93*, 195 (1971).
2. Starks, C. M., Owens, R. M.: J. Am. Chem. Soc. *95*, 3613 (1973).
3. Brandström, A.: "Preparative Ion Pair Extraction", pp. 108–109. Hässle, Sweden 1974: Apotekarsocieteten.
4. Solodar, J.: Tetrahedron Let. *1971*, 287.
5. Zubrick, J. W., Dunbar, B. I., Durst, H. D.: ibid *1975*, 71.
6. Cook, F. L., Bowers, C. W., Liotta, C. L.: J. Org. Chem. *39*, 3416 (1974).

7. Menger, F. M.: J. Am. Chem. Soc. *92*, 5965 (1970).
8. Smiley, R. A., Arnold, C.: J. Org. Chem. *25*, 257 (1960).
9. Reeves, W. P., White, M. R.: Synthetic Commun. *6*, 193 (1976).
10. Hennis, H. E., Thompson, L. R., Long, J. P.: Ind. Eng. Chem. Prod. Res. Dev. *7*, 96 (1968).
11. Regen, S. L.: J. Am. Chem. Soc. *98*, 6270 (1976).
12. Friedman, L., Shechter, H.: J. Org. Chem. *25*, 877 (1960).
13. Simchen, G., Kobler, H.: Synthesis *1975*, 605.
14. Cinquini, M., Montanari, F., Tundo, P.: J. Chem. Soc., Chem. Commun. *1975*, 393.
15. Colonna, S., Fornasier, R.: Synthesis *1975*, 531.
16. Koenig, K. E., Weber, W. P.: Tetrahedron Let. *1974*, 2275.
17. Wöhler, F., Liebig, J.: Ann. *3*, 249 (1832).
18. Habner, H.: Ann. *120*, 330 (1861).
19. Nef, J. U.: Ann. *287*, 307 (1875).
20. Oakwood, T. S., Weisberger, C. A.: "Org. Syn. Coll. Vol. III", p. 112. New York, N. Y.: J. Wiley 1955.
21. Childs, M. E., Weber, W. P.: J. Org. Chem. *41*, 3486 (1976).
22. Unpublished Results of K. E. Koenig, M. E. Childs, and W. P. Weber.
23. Cazes, B., Julia, S.: Tetrahedron Let. *1974*, 2077.
24. Akabori, S., Ohtomi, M., Arai, K.: Bull. Chem. Soc. Japan *49*, 746 (1976).
25. Durr, G.: Compt. Rend. *242*, 1630 (1956).
26. Tagaki, W., Hara, H.: J. Chem. Soc., Chem. Commun. *891* (1973).
27. Evans, D. A., Truesdale, L. K., Carroll, G. L.: J. Chem. Soc., Chem. Commun. *1973*, 55.
28. Evans, D. A., Truesdale, L. K.: Tetrahedron Let. *1973*, 4929.
29. Evans, D. A., Hoffman, J. M., Truesdale, L. K.: J. Am. Chem. Soc. *95*, 5822 (1973).
30. Evans, D. A., Truesdale, L. K., Grimm, K. G.: J. Org. Chem. *41*, 3355 (1976).
31. Beugelmans, R., LeGoff, M.-T., Pusset, J., Roussi, G.: J. Chem. Soc., Chem. Commun. *1976*, 377.
32. Vink, J. A., Verheijdt, P. L., Cornelisse, J., Havinga, E.: Tetrahedron *28*, 5081 (1972).
33. Eberson, L., Helgee, B.: Chem. Scripta *5*, 47 (1974).

8. Reactions of Superoxide Ions

8.1 Introduction

Superoxide salts like potassium and sodium superoxide are virtually insoluble in nonpolar media, and although they are somewhat more soluble in hydroxylic media, they are not stable in these solvents [1]. Potassium superoxide is only sparingly soluble in the dipolar aprotic solvent DMSO [2]. As a result of these solubility problems and the inaccessibility of electrochemically generated superoxide [3] in most laboratories, the chemistry of this potentially interesting nucleophile was little studied until recently.

It has been shown in the past few years that 18-crown-6 and dicyclohexyl-18-crown-6 will complex and solubilize potassium superoxide in such solvents as dimethylsulfoxide, benzene, tetrahydrofuran, dimethylformamide, dimethoxyethane, and even diethyl ether [4]. The interest in the biological role of superoxide and its ready availability in solution has made it the focus of considerable recent attention, some of which is recorded in this chapter.

8.2 Reactions at Saturated Carbon

The dicyclohexyl-18-crown-6 catalyzed reaction of potassium superoxide in benzene solution yields dialkyl peroxides from primary and secondary alkyl bromides, mesylates, and tosylates according to equation 8.1 [5]. The reaction appears to be an $S_N 2$ bimolecular displacement process; $(-)-R-2$-bromooctane undergoes substitution

$$2\ R-X + KO_2 \xrightarrow{\text{crown}} R-O-O-R \qquad (8.1)$$

with inversion of configuration (> 94%) affording the (+)–S,S–di–2–octyl peroxide. The stereochemistry of this transformation was proved by the reaction cycle shown in equation 8.2.

$$\text{(S) } n\text{-}C_6H_{13}\text{-}\underset{\underset{CH_3}{|}}{CH}\text{-}OH \xrightarrow{PBr_3} \text{(R) } n\text{-}C_6H_{13}\text{-}\underset{\underset{CH_3}{|}}{CH}\text{-}Br \quad [\alpha]_D^{25} -42.5$$

$[\alpha]_D^{25} + 8.6$

$$\Bigg\downarrow KO_2/DC\text{-}18\text{-}C\text{-}6 \qquad (8.2)$$

$$\text{(S) } n\text{-}C_6H_{13}\text{-}\underset{\underset{CH_3}{|}}{CH}\text{-}OH \xleftarrow{LiAlH_4} \text{(S,S) } (n\text{-}C_6H_{13}\text{-}\underset{\underset{CH_3}{|}}{CH}\text{-}O)_2 \quad [\alpha]_D^{25} + 39.9$$

$[\alpha]_D^{25} + 7.7$

The superoxide displacement reaction can, and frequently does, yield alcohols and olefins as by-products. The results of several superoxide displacement reactions which yield peroxides, alcohols, and/or olefins are recorded in Table 8.1.

Table 8.1. Superoxide displacements at saturated carbon [5]

Substrate (RX)	Yield of products			Catalyst
	Peroxide	Alcohol	Olefins	
$n\text{-}C_5H_{11}Br$	53	–	–	DC-18-C-6
$n\text{-}C_6H_{13}Br$	54	–	–	DC-18-C-6
$n\text{-}C_7H_{15}Br$	56	–	–	DC-18-C-6
$n\text{-}C_{16}H_{33}Br$	44	21	–	DC-18-C-6
$n\text{-}C_{18}H_{37}Br$	77	21	–	DC-18-C-6
$n\text{-}C_{18}H_{37}Br$	61	18	–	18-C-6
$c\text{-}C_6H_{11}Br$	–	–	67	DC-18-C-6
$c\text{-}C_5H_9Br$	42	–	24	DC-18-C-6
$C_6H_{13}CH(CH_3)Br$	55	–	37	DC-18-C-6
$n\text{-}C_{18}H_{37}OTs$	50	42	–	DC-18-C-6
$n\text{-}C_{18}H_{37}OMs$	46	40	–	DC-18-C-6
$C_6H_{13}CH(CH_3)OTs$	52	13	16	DC-18-C-6
$C_6H_{13}CH(CH_3)OMs$	44	19	14	DC-18-C-6

If the crown catalyzed reaction of potassium superoxide with alkyl halide is carried out in dimethylsulfoxide as solvent, the product is the corresponding alcohol [6–9]. The formation of alcohol rather than dialkyl peroxide has been shown to result from reaction of the alkyl hydroperoxide anion with dimethylsulfoxide to form alkoxide and dimethylsulfone (Eqs. 8.3–8.5) [9]. The alkyl hydroperoxide anion is presumably formed by reduction of the initial alkyl hydroperoxide radical by the superoxide anion [8, 9].

$$RX + O_2^{\cdot -} \rightarrow R\text{-}O\text{-}O\cdot + X^- \qquad (8.3)$$

$$R\text{-}O\text{-}O\cdot + O_2^{\cdot -} \rightarrow R\text{-}O\text{-}O^- + O_2 \qquad (8.4)$$

$$R-O-O^- + CH_3SOCH_3 \rightarrow RO^- + CH_3SO_2CH_3 \qquad (8.5)$$

The first reaction between alkyl halide and superoxide anion in DMSO is a nucleophilic displacement process (Eq. 8.3), since reaction with optically active secondary halides, tosylates, and mesylates leads to alcohols with inversion of configuration at the active center. The small relative rate difference observed between primary and secondary alkyl halides is difficult to reconcile with the usual $S_N 2$ reactivity pattern [6, 9]. Examples of the transformation of alkyl halides into alcohols in DMSO solution are presented in Table 8.2.

Table 8.2. Reactions of KO_2 in DMSO at saturated carbon

Number of carbons	Substrate	% Yield of products			Ref.
		Alcohol	Olefin	Aldehyde or ketone	
4	t-C_4H_9Br	(some)	(some)	–	[7]
6	$CH_3(CH_2)_2C(CH_3)_2Br$	20 inv	30	–	[6]
6	E-1,2-dibromocyclohexane	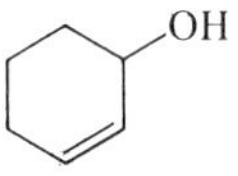 (cyclohexenol) 100	–	–	[7]
6	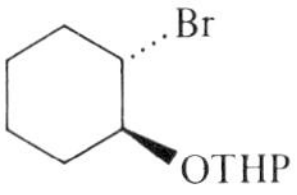 (cyclohexyl with Br, OTHP)	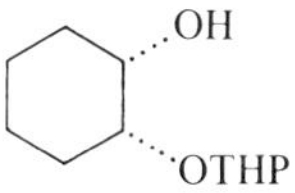 (cyclohexyl with OH, OTHP) 29	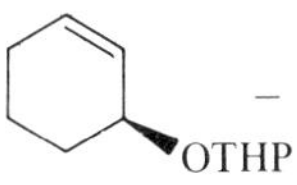 (cyclohexenyl OTHP) 67	–	[7]
7	$C_6H_5CH_2Cl$	41	–	6	[6]
7	$C_6H_5CH_2Br$	75	–	–	[7]
8	$Cl(CH_2)_7CH_3$	34	1	5	[6]
8	$CH_3CHCl(CH_2)_5CH_3$	36	12	1	[6]
8	$Br(CH_2)_7CH_3$	63	1	12	[6]
8	$I(CH_2)_7CH_3$	46	3	11	[6]
8	$CH_3CHBr(CH_2)_5CH_3$	51	34	1	[6]
8	$CH_3CHI(CH_2)_5CH_3$	48	48	1	[6]
8	$TsO(CH_2)_7CH_3$	75	1	1	[6]
8	$CH_3CH(OTs)(CH_2)_5CH_3$	75	23	1	[6]
10	$Br(CH_2)_9CH_3$	80	–	–	[7]
10	$(CH_3)_2C=CH(CH_2)_2\overset{CH_3}{C}=CH-CH_2Br$	75	–	–	[7]
10	Z-4-t-$C_4H_9C_6H_{10}OTs$	95 inv	–	–	[7]
10	E-4-t-$C_4H_9C_6H_{10}OMs$	96 inv	–	–	[7]

8.3 Additions to Carbonyl Groups

Superoxide ion reacts with esters in the presence of 18-crown-6 affording the corresponding acids after aqueous work-up [7, 8, 10]. This reaction was first reported by Corey who found that cleavage occurred as well as displacement in ester mesylates and tosylates [7, 8]. Two examples of this reaction are shown in equations 8.6 and 8.7.

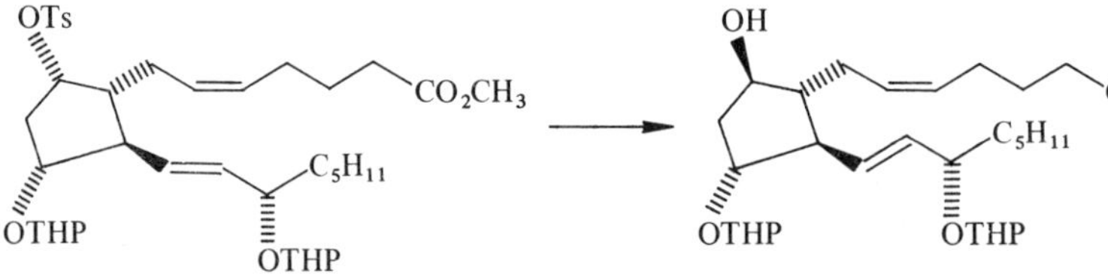

(8.6) [8]

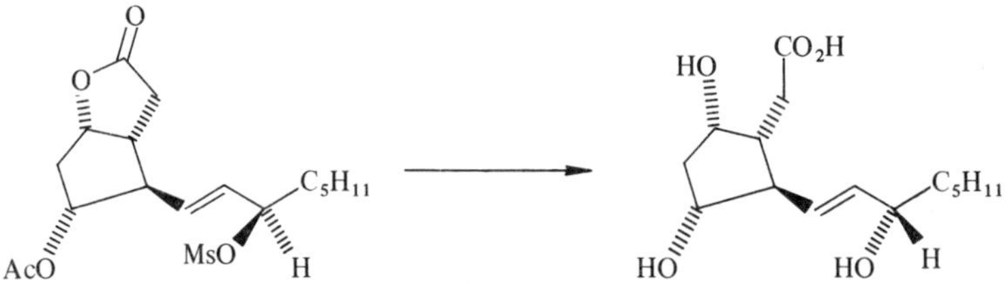

(8.7) [7]

The observation that R-2-octyl acetate yields R-2-octanol with 99% retention rules out the possibility that the reaction occurs by alkyl-oxygen fission [10]. This mechanism is consistent with the normal mode of ester hydrolysis which also occurs by acyl-oxygen fission. Neither amides nor nitriles were affected by superoxide under phase transfer conditions [10]. The data on ester cleavage by superoxide ion is summarized in Table 8.3.

Table 8.3. Superoxide cleavage of esters [10]

Ester	Acid produced	% Yield
$CH_3CO_2CH(CH_3)C_6H_{13}$	Acetic	89
$(CH_3)_3CCO_2CH_3$	Pivalic	81
$C_6H_5CO_2C_2H_5$	Benzoic	88
$n\text{-}C_7H_{15}CO_2CH_3$	Octanoic	98
$n\text{-}C_7H_{15}CO_2CH(CH_3)_2$	Octanoic	98
$n\text{-}C_7H_{15}CO_2C(CH_3)_3$	Octanoic	96
$n\text{-}C_7H_{15}CO_2CH_2C_6H_5$	Octanoic	100
	Benzoic	55
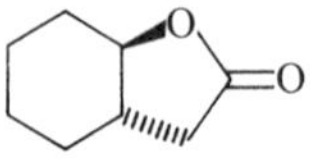	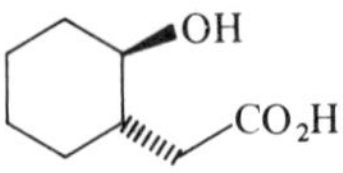	85
$n\text{-}C_9H_{19}CO_2C_6H_5$	Decanoic	84

Additions to Carbonyl Groups

Potassium superoxide has been found to undergo nucleophilic acyl substitution with acid chlorides to yield diacyl peroxides in benzene solvent (Eq. 8.8). This reaction does not appear to require any catalyst [11], indicating that it probably occurs at the solid-liquid interface (see Table 8.4).

$$2R\text{-}\underset{\underset{O}{\|}}{C}\text{-}Cl + 2KO_2 \rightarrow R\text{-}\underset{\underset{O}{\|}}{C}\text{-}O\text{-}O\text{-}\underset{\underset{O}{\|}}{C}\text{-}R + 2KCl + O_2 \qquad (8.8)$$

Table 8.4. Preparation of diaryl peroxides [11]

$$R\text{-}\underset{\underset{O}{\|}}{C}\text{-}Cl \rightarrow R\text{-}\underset{\underset{O}{\|}}{C}\text{-}O\text{-}O\text{-}\underset{\underset{O}{\|}}{C}\text{-}R$$

R	% Yield
C_6H_5-	61
$4\text{-}CH_3OC_6H_4-$	62
$4\text{-}ClC_6H_4-$	69
$2\text{-}ClC_6H_4-$	57
$2\text{-}CH_3C_6H_4-$	53
$CH_3(CH_2)_{12}-$	74
$CH_3(CH_2)_{14}-$	50

Table 8.5. Cleavage reactions of superoxide ion [12]

Substrate	Product	% Yield
Benzil	Benzoic acid	87
Camphoroquinone	Camphoric acid	87
1,2-cyclohexanedione	Adipic acid	53
2-ketoglutaric acid	Succinic acid	42
2-ketophenylacetic acid	Benzoic acid	93
Ethyl-2-ketophenylacetate	Benzoic acid	93
Benzoin	Benzoic acid	98
2-hydroxycyclohexanone	Adipic acid	69
Mandelic acid	Benzoic acid	94
2-hydroxystearic acid	Heptadecanoic acid	77
1-cyclohexylmandelic acid	–	0
1-hydroxycycloheptanecarboxylic acid	–	0
Ethyl mandelate	Benzoic acid	93
2-chlorocyclohexanone	Adipic acid	60
2-chlorocyclooctanone	Octanedioic acid	62
3-Bromocamphor	Camphoric acid	54
Phenacyl chloride	Benzoic acid	72
2-bromo-2-phenylacetic acid	Benzoic acid	90
2-bromooctanoic acid	Heptanoic acid	58
Methyl-2-bromo-2-cyclohexylacetate	Cyclohexanecarboxylic acid	54

The crown catalyzed reaction of potassium superoxide in dry benzene with α-keto-, α-hydroxy-, and α-haloketones, esters and carboxylic acids yields carboxylic acids as formulated in equation 8.9 [12]. The observation that 1-hydroxy-1-carboxy-cycloheptane and α-cyclohexylmandelic acid fail to react under conditions which convert related substrates to acids suggests that a hydrogen alpha to the reactive site is a requisite (see Table 8.5).

$$R-CHX-CO-R' + KO_2 \xrightarrow{crown} R-COOH + R'-COOH \tag{8.9}$$

A variety of chalcones are cleaved by potassium superoxide [13] according to equation 8.10 to yield a mixture of two, and sometimes three, carboxylic acids. Labeling experiments indicate that the superoxide transfers an electron to the enone followed by reaction of the resulting radical with molecular oxygen. An overall mechanism has been postulated for this reaction [13]. The results of a number of chalcone cleavage reactions are presented in Table 8.6.

$$R^1-C_6H_4-CO-CH=CH-C_6H_4-R \xrightarrow{O_2^{\cdot-}} R^1C_6H_4COOH \quad (I)$$
$$+ RC_6H_4COOH + RC_6H_4CH_2COOH \tag{8.10}$$
$$(II) \qquad\qquad (III)$$

Table 8.6. Superoxide cleavage of chalcones [13]

R_1	R	I	II	III
H	H		90	5
4-CH_3	H	62	32	6
4-CH_3O	H	72	29	5
4-Cl	H	73	35	5
4-Br	H	52	33	5
4-CH_3	4-Br	64	24	–
4-Br	4-CH_3	59	43	6
H	4-Br	62	40	–
H	4-Cl	57	39	3
H	2-NO_2	53	35	–
H	3-Br	64	36	–

8.4 Reactions With Aryl Halides

Potassium superoxide in the presence of dicyclohexyl-18-crown-6 has been used to carry out aromatic nucleophilic substitution reactions on halobenzenes activated by electron withdrawing nitro groups [14] as formulated in equation 8.11 (Table 8.7).

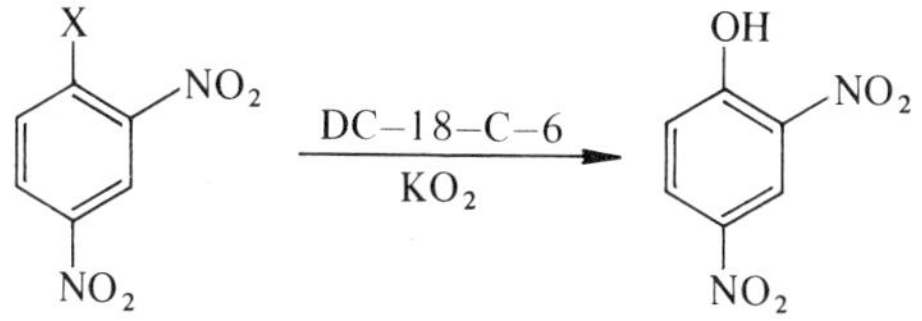

$$\underset{NO_2}{\underset{|}{\text{2,4-}(NO_2)_2C_6H_3X}} \xrightarrow[KO_2]{DC-18-C-6} \underset{NO_2}{\underset{|}{\text{2,4-}(NO_2)_2C_6H_3OH}} \tag{8.11}$$

Table 8.7. Aromatic substitution reactions of superoxide ion

Substrate	Catalyst	% Yield	
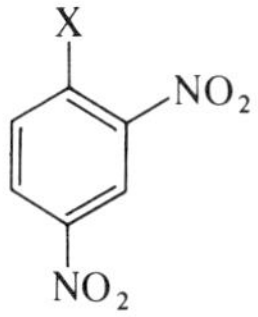 2,4-dinitro-C$_6$H$_3$X, X = F	DC-18-C-6	95	
X = Cl	DC-18-C-6	95	
X = Br	DC-18-C-6	95	
X = I	DC-18-C-6	95	
2-nitrobromobenzene	DC-18-C-6	55	
4-nitrobromobenzene	DC-18-C-6	80	
1,2-dinitrobenzene	DC-18-C-6	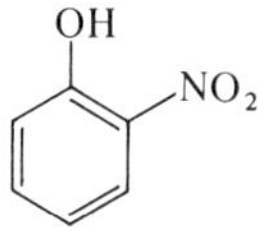 53 (2-nitrophenol)	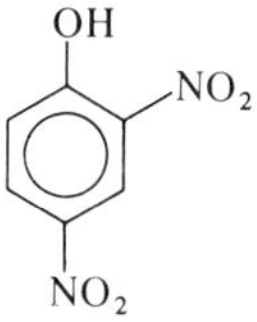 4 (2,4-dinitrophenol)
1,3-dinitrobenzene	DC-18-C-6	35 (2,4-dinitrophenol)	31 (3-nitrophenol)
1,4-dinitrobenzene	DC-18-C-6	90 (4-nitrophenol)	

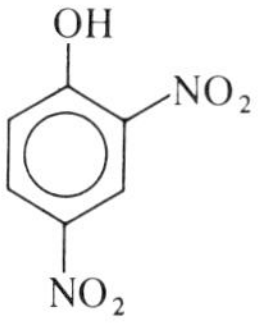

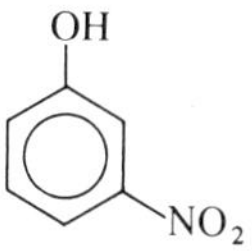

Aromatic nucleophilic substitution by superoxide occurs by a mechanism different from that encountered in aliphatic nucleophilic substitution reactions of this ion. Thus, reaction of ^{18}O enriched potassium superoxide with 1-bromo-2,4-dinitrobenzene catalyzed by dicyclohexyl-18-crown-6 in benzene saturated with unlabeled oxygen results in 2,4-dinitrophenol almost devoid of label. The loss of ^{18}O label in this reaction rules out a direct displacement mechanism. This result is consistent with electron transfer from superoxide anion to the arene to form an intermediate aromatic anion radical which reacts with oxygen (from all sources) to yield phenol. This mechanism is formulated in equation 8.12. Examples of this reaction are presented in Table 8.7.

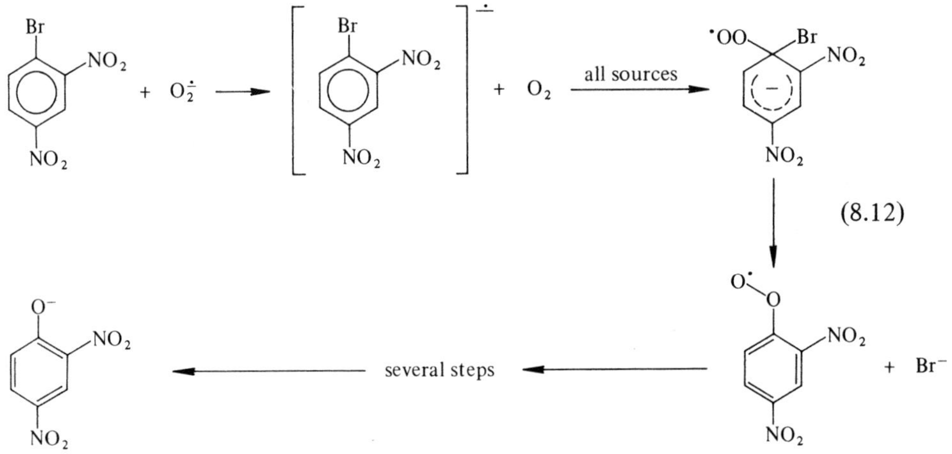

$$(8.12)$$

References

1. LeBerre, A., Berguer, Y.: Bull. Soc. Chim. France *1966*, 2363, 2368.
2. Johnson, E. L., Pool, K. H., Hamm, R. E.: Anal. Chem. *38*, 183 (1966).
3. Mayed, E. A., Bard, A. J.: J. Am. Chem. Soc. *95*, 6223 (1973).
4. Valentine, J. S., Curtis, A. B.: J. Am. Chem. Soc. *97*, 224 (1975).
5. Johnson, R. A., Nidy, E. G.: J. Org. Chem. *40*, 1680 (1975).
6. San Filippo, Jr., J., Chern, C.-I., Valentine, J. S.: J. Org. Chem. *40*, 1678 (1975).
7. Corey, E. J., Nicolaou, K. C., Shibasaki, M., Machida, Y., Shiner, C. S.: Tetrahedron Let. *1975*, 3183.
8. Corey, E. J., Nicolaou, K. C., Shibasaki, M.: J. Chem. Soc. Chem. Commun. *1975*, 658.
9. Gibian, M. J., Ungermann, T.: J. Org. Chem. *41*, 2500 (1976).
10. San Filippo, Jr., J., Romano, L. J., Chern, C.-I., Valentine, J. S.: J. Org. Chem. *41*, 586 (1976).
11. Johnson, R. A.: Tetrahedron Let. *1976*, 331.
12. San Filippo, Jr., J., Chern, C.-I., Valentine, J. S.: J. Org. Chem. *41*, 1077 (1976).
13. Rosenthal, I., Frimer, A.: Tetrahedron Let. *1976*, 2805.
14. Frimer, A., Rosenthal, I.: ibid. *1976*, 2809.

9. Reactions of Other Nucleophiles

9.1 Introduction

In the preceding chapters we have dealt with a variety of nucleophilic species which have been utilized under phase transfer catalytic conditions. In each of these chapters, we believe that enough information has been accumulated or the chemistry is individual enough to justify this approach. There are, however, a number of cases in which the reactivity is more or less that which might have been anticipated or there are relatively few examples of the process. We have, of necessity, combined them in this chapter in the hope that the information will thereby be available.

As will be discerned from a perusal of this chapter, the reactions of halide ions, fluoride, chloride, bromide and iodide and the pseudo-halide ions azide, and nitrite, among others, are presented here. In addition, some elimination reactions (hydroxide nucleophile/base) and several organometallic species are discussed. We have also mentioned briefly the phase transfer catalyzed initiation of several addition polymerization processes.

9.2 Halide Ions

All four commonly occurring halide ions (fluoride [1–5], chloride [5–11], bromide [5, 8–10], and iodide [5, 7–9, 10, 12–15] have been phase-transferred and in the process, quaternary ions [1, 6–8, 10, 12–15], crowns [2, 4, 8, 9, 13], cryptates [3, 13] and resins [5] have all been utilized. Most of the processes reported are essentially Finkelstein reactions [16]. In a typical phase transfer of fluoride utilizing crown ether as catalyst, an acetonitrile solution of benzyl bromide is stirred with a catalytic amount of 18-crown-6 and solid potassium fluoride. The product, benzyl fluoride (see Eq. 9.1), is isolated in quantitative yield [2].

$$C_6H_5-CH_2-Br + KF \xrightarrow{crown} C_6H_5-CH_2-F + KBr \qquad (9.1)$$

Reactions of Other Nucleophiles

In liquid-liquid phase transfer processes, either a quaternary ammonium ion [6, 7] or a crown ether [9] is utilized to solubilize the halide in a nonpolar solution of the substrate. The reservoir in such cases is an aqueous solution of the alkali metal halide.

The groups which have been displaced by halide ion include other halides, methanesulfonates, brosylates and tosylates. The methanesulfonate group appears to be particularly convenient in the secondary alkyl halide case, affording higher yields of halides and less elimination than the bromides. The methanesulfonates of optically active secondary alcohols react with chloride and bromide nucleophiles affording the secondary halides which are largely inverted [10]. The corresponding reaction with iodide ion leads to nearly racemic product. Although the secondary alkyl substrates are generally less reactive than are the primary ones, the nucleophilicity order I > Br > Cl > F remains the same in either case [10]. Crown complexed alkali metal halides also appear to be more dissociated and more reactive than the corresponding tetrabutylammonium halides [8]. The reactions of phase-transferred halide ions are summarized in Table 9.1.

One interesting variant of the phase-transfer synthesis of chlorides involves the use of concentrated aqueous HCl as chloride reservoir. Water insoluble primary alcohols were stirred and heated with HCl in the presence of hexadecyltributylphosphonium bromide. The reactions were slow but ultimately yielded well [11]. The data are included in Table 9.1.

One might legitimately question the advantage of the phase-transfer method in the synthesis of halides, particularly iodides, when the Finkelstein reaction of sodium iodide in acetone with an organic substrate is an efficient, high yield reaction. Where comparative data are available [17], there seems to be little practical advantage to the phase transfer method. It should be borne in mind, however, that the phase transfer technique offers the possibility of using solvents other than acetone or 2-butanone in these reactions and halides other than iodide can obviously be used. The latter point is particularly important in the case of fluoride ion, for which fewer alternatives exist.

As expected for a Finkelstein reaction, phase-transferred halide ions behave as bases as well as nucleophiles. Bromocyclohexane is reported to react with 18-crown-6 activated potassium fluoride to quantitatively yield cyclohexene [2]. The corresponding reaction of chlorocyclohexane in which fluoride ion is associated with a resin-bound quaternary ion yields only 50% cyclohexene [5]. Quantitative alkene formation is observed, however, when sec-octyl bromide is treated with potassium fluoride in the presence of hexadecyltributylphosphonium ion [1]. A more complicated situation is encountered in the reaction of 2-chloro-2-methylcyclohexanone with fluoride ion. In this case, a mixture of elimination and substitution products is obtained (69% and 31% respectively, see Eq. 9.2) [2]. The elimination reactions of phase-transferred halide ions are included in Table 9.1.

$$\text{2-chloro-2-methylcyclohexanone} \xrightarrow[\text{crown}]{\text{KF}} \text{2-methylcyclohexenone} + \text{2-fluoro-2-methylcyclohexanone} \quad (9.2)$$

Table 9.1. Reactions of halide ions

Substrate	Reagent/catalyst	Product (% yield)	Ref.
a) Fluoride			
CH_3COCl	KF/18-C-6	CH_3COF (100)	[2]
$(CH_3)_3CCH_2Br$	KF/HDTBP	0	[1]
$(CH_3)_3SiCH_2CO_2C_2H_5$	TBAF/THF	$H_5C_2OCO-CH_2^-$	[19]
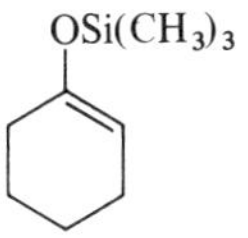 1-(trimethylsilyloxy)cyclohexene	BTMAF	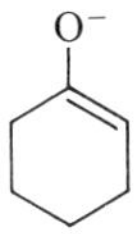 cyclohexenolate (67)	[18]
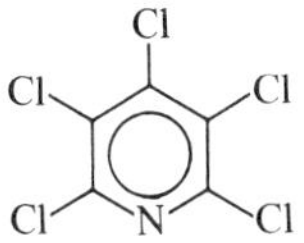 2,3,4,5,6-pentachloropyridine	KF/[2.2.2]cryptate	$2\text{-}C_5Cl_4FN$ (65); $2,6\text{-}C_5Cl_3F_2N$ (12)	[3]
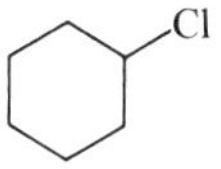 cyclohexyl chloride	Resin-$\overset{+}{N}(CH_3)_3F^-$	Cyclohexene (50)	[5]
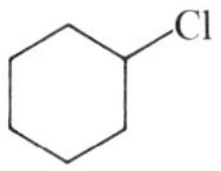 cyclohexyl chloride	KF/HDTBP	0	[1]
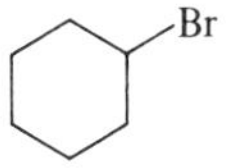 cyclohexyl bromide	KF/18-C-6	Cyclohexene (100)	[2]
$n\text{-}C_6H_{13}Cl$	KF/HDTBP	$n\text{-}C_6H_{13}F$ (80)	[1]
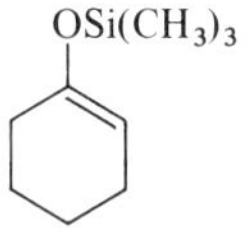 1-(trimethylsilyloxy)cyclohexene	F^-/BTMAF/THF	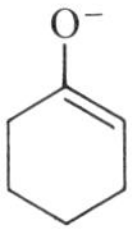 cyclohexenolate (79)	[18]
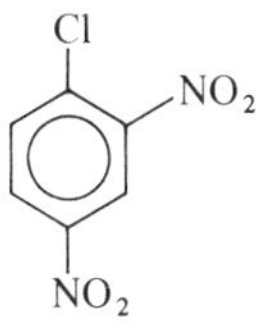 1-chloro-2,4-dinitrobenzene	KF/18-C-6, CH_3CN	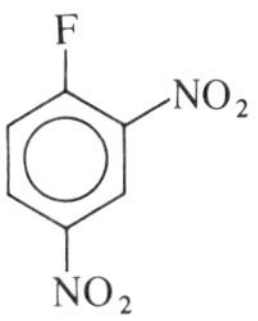 1-fluoro-2,4-dinitrobenzene (100)	[2]
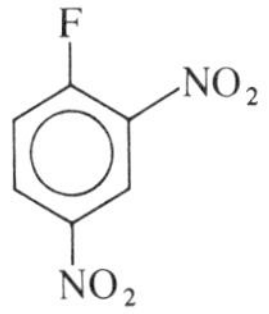 1-fluoro-2,4-dinitrobenzene	KF/18-C-6, Meisenheimer	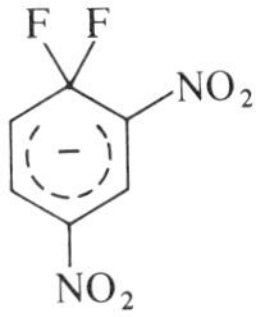 difluoro Meisenheimer complex	[2]

Reactions of Other Nucleophiles

Table 9.1 (continued)

Substrate	Reagent/catalyst	Product (% yield)	Ref.
$C_6H_5CH_2Cl$	KF/HDTBP	$C_6H_5CH_2F$ (90)	[1]
$C_6H_5CH_2Cl$	KF/18-C-6	$C_6H_5CH_2F$ (100)	[2]
$C_6H_5CH_2Cl$	Resin-$\overset{+}{N}(Me_3)F^-$	$C_6H_5CH_2F$ (95)	[5]
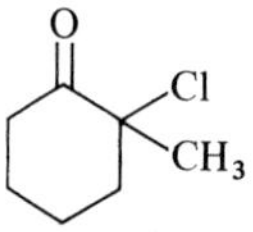 2-chloro-2-methylcyclohexanone	2:1 KF/18-C-6	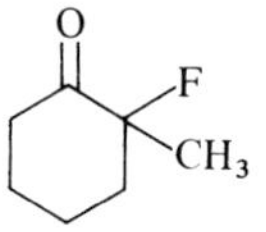 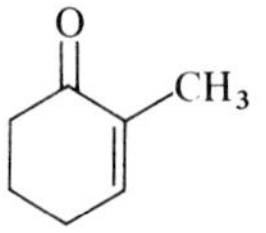 2-fluoro-2-methylcyclohexanone + 2-methylcyclohex-2-enone	[2]
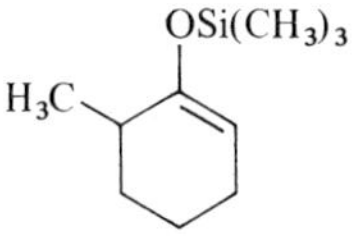 1-methyl-6-(trimethylsilyloxy)cyclohexene	BTMAF/THF	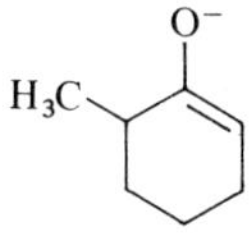 2-methylcyclohexenolate (89)	[18]
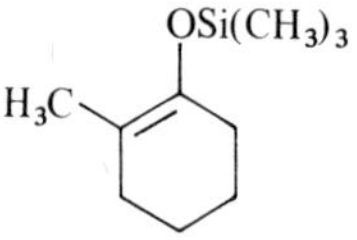 2-methyl-1-(trimethylsilyloxy)cyclohexene	BTMAF/THF	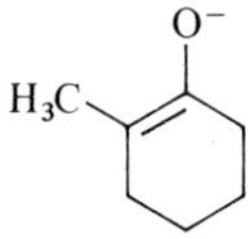 2-methylcyclohexenolate (59)	[18]
$(CH_3)_3SiCH_2CO_2C_2H_5$	TBAF/THF	$^-CH_2CO_2C_2H_5$	[19]
n-$C_8H_{17}Cl$	KF/HDTBP	n-$C_8H_{17}F$ (82)	[1]
	Resin-$\overset{+}{N}(CH_3)_3F^-$	n-$C_8H_{17}F$ (87)	[5]
n-$C_8H_{17}Br$	KF/HDTBP	n-$C_8H_{17}F$ (64)	[1]
	Resin-$\overset{+}{N}(CH_3)_3F^-$	n-$C_8H_{17}F$ (82)	[5]
	KF/18-C-6/CH_3CN	n-$C_8H_{17}F$ (92)	[2]
n-$C_8H_{17}OSO_2CH_3$	KF/HDTBP	n-$C_8H_{17}F$ (90)	[1]
	Resin-$\overset{+}{N}(CH_3)_3F^-$	n-$C_8H_{17}F$ (92)	[5]
	DC-18-C-6	n-$C_8H_{17}F$ (65)	[9]
	KF/$C_{11}H_{23}$-substituted cryptand 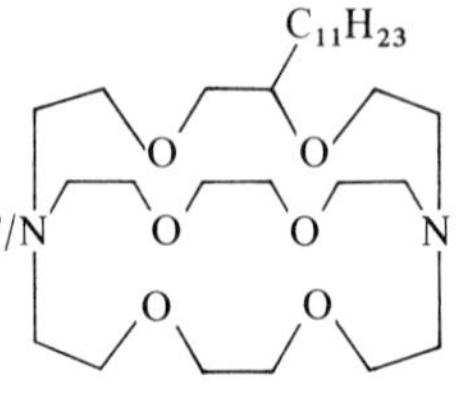	n-$C_8H_{17}F$ (85)	[13]
	KF/HDTBP	n-$C_8H_{17}F$ (99)	[13]

Table 9.1 (continued)

Substrate	Reagent/catalyst	Product (% yield)	Ref.
n-C_6H_{13}CH(CH_3)Cl	KF/HDTBP	n-C_6H_{13}CHFCH_3 (20) Alkene (66)	[1]
n-C_6H_{13}CH(CH_3)Br	KF/HDTBP	Alkene (100)	[1]
	Resin-$\overset{+}{N}$(CH_3)$_3$F$^-$	Alkene (20)	[5]
	KF/18-C-6/CH_3CN	n-C_6H_{13}–CHFCH_3 (32) Octenes (68)	[2]
n-C_6H_{13}CH(CH_3)OSO_2CH_3	KF/HDTBP	n-C_6H_{13}CH(CH_3)F (54) Alkene (26)	[1]
	Resin-$\overset{+}{N}$(CH_3)$_3$F$^-$	n-C_6H_{13}CH(CH_3)F (70)	[5]
n-C_6H_{13}CH(CH_3)OSO_2CH_3 (Optically active)	KF/HDTBP	n-C_6H_{13}CHFCH_3 (57) $[\alpha]^{20} = +6.2°$	[10]
n-C_6H_{13}CH(CH_3)OTs	Resin-$\overset{+}{N}$(CH_3)$_3$F$^-$	n-C_6H_{13}CH(CH_3)F (62)	[5]
C_6H_5CO–CH_2Br	Resin-$\overset{+}{N}$(CH_3)$_3$F$^-$	C_6H_5CO–CH_2F (99)	[5]
n-$C_{12}H_{25}$Cl	KF/HDTBP	n-$C_{12}H_{25}$F (77)	[1]
n-$C_{16}H_{33}$Br	Resin-$\overset{+}{N}$(CH_3)$_3$F$^-$	n-$C_{16}H_{33}$F (77)	[5]

i-Bu–CH–NH
O=C   C=O
  NH O–$CH_2C_6H_5$
  CH_2
  CO–OC_2H_5
	TBAF/THF	i-Bu–CH–NH (90)
		O=C C=O
		     N
		     CH_2
		     CO–OC_2H_5
			[20]

CH_2–OSi(C_6H_5)$_2$–t-C_4H_9
  O–CH
CH_2
  O–CH
     CH–O
     CH–O^{CH_2}
     CH_2OH
	TBAF/THF	CH_2O^- (90)
		  O–CH
		CH_2
		  O–CH
		     CH–O
		     CH–O^{CH_2}
		     CH_2OH
			[21]

b) Chloride

Substrate	Reagent/catalyst	Product (% yield)	Ref.
n-C_4H_9OH	HCl/HDTBP	n-C_4H_9Cl (65)	[11]
n-C_6H_{13}OH	HCl/HDTBP	n-C_6H_{13}Cl (95)	[11]
n-C_8H_{17}OH	HCl/HDTBP	n-C_8H_{17}Cl (94)	[11]
n-C_8H_{17}Cl	Na^{36}Cl	n-$C_8H_{17}{}^{36}$Cl (equil)	[7]
n-C_8H_{17}Br	Resin-$\overset{+}{N}$(CH_3)$_3$Cl$^-$	n-C_8H_{17}Cl (100)	[5]
n-$C_8H_{17}OSO_2CH_3$	NaCl/DC-18-C-6	n-C_8H_{17}Cl (75)	[9]

Reactions of Other Nucleophiles

Table 9.1 (continued)

Substrate	Reagent/catalyst	Product (% yield)	Ref.
$n\text{-}C_8H_{17}OSO_2CH_3$	KCl/DC-18-C-6	$n\text{-}C_8H_{17}Cl$ (89)	[9]
(R) $n\text{-}C_6H_{13}CH(CH_3)OSO_2CH_3$	KCl/TCMAC	89% S-chloride (83) 11% R-chloride	[10]
$n\text{-}C_{10}H_{21}Br$	NaCl/TCMAC	$n\text{-}C_{10}H_{21}Cl$ (86)	[6]
$n\text{-}C_{12}H_{25}OH$	HCl/HDTBP	$n\text{-}C_{12}H_{25}Cl$ (94)	[11]
$n\text{-}C_{16}H_{33}OH$	HCl/HDTBP	$n\text{-}C_{16}H_{33}Cl$ (97)	[11]
c) Bromide			
$n\text{-}C_8H_{17}Cl$	Resin-$\overset{+}{N}(CH_3)_3Br^-$	$n\text{-}C_8H_{17}Br$ (95)	[5]
$n\text{-}C_8H_{17}OSO_2CH_3$	NaBr/DC-18-C-6	$n\text{-}C_8H_{17}Br$ (88)	[9]
	KBr/DC-18-C-6	$n\text{-}C_8H_{17}Br$ (96)	[9]
$n\text{-}C_8H_{17}I$	KBr/DC-18-C-6	$n\text{-}C_8H_{17}Br$ (40)	[9]
$n\text{-}C_6H_{13}CH(CH_3)OSO_2CH_3$	KBr/DC-18-C-6	$n\text{-}C_6H_{13}CH(CH_3)Br$ (100)	[9]
R–$n\text{-}C_6H_{13}CH(CH_3)OSO_2CH_3$	KBr/HDTBP	72% S-bromide (78) 28% R-bromide	[10]
d) Iodide			
CH_2Cl_2	2.4 eq. NaI/HDTBP	CH_2ICl (13) CH_2I_2 (67)	[12]
CH_2Cl_2	3.0 eq. NaI/HDTBP	CH_2ICl (7) CH_2I_2 (75)	[12]
	0.2 eq. NaI/HDTBP	CH_2ICl (45) CH_2I_2 (14)	[12]
CH_3OTs	NaI/acetone	CH_3I (~100)	[17]
$C_6H_5CH_2OTs$	NaI/acetone	$C_6H_5CH_2I$ (~100)	[17]
$n\text{-}C_8H_{17}Cl$	KI/cryptate[a]	$n\text{-}C_8H_{17}I$ (77)	[13]
$n\text{-}C_8H_{17}Cl$	KI/cryptate[b]	$n\text{-}C_8H_{17}I$ (85)	[13]
$n\text{-}C_8H_{17}Cl$	KI/HDTBP	$n\text{-}C_8H_{17}I$ (80)	[13]
$n\text{-}C_8H_{17}Br$	KI/cryptate[a]	$n\text{-}C_8H_{17}I$ (100)	[13]
$n\text{-}C_8H_{17}Br$	KI/cryptate[b]	$n\text{-}C_8H_{17}I$ (92)	[13]
$n\text{-}C_8H_{17}Br$	KI/[2.2.2]-cryptate	$n\text{-}C_8H_{17}I$ (90)	[13]
$n\text{-}C_8H_{17}Br$	KI/HDTBP	$n\text{-}C_8H_{17}I$ (93)	[13]
$n\text{-}C_8H_{17}Br$	KI/DC-18-C-6	$n\text{-}C_8H_{17}I$ (100)	[9, 13]
$n\text{-}C_8H_{17}Br$	Resin-$\overset{+}{N}(CH_3)_3I^-$	$n\text{-}C_8H_{17}I$ (80)	[5]
$n\text{-}C_8H_{17}Br$	KI/polypode[a]	$n\text{-}C_8H_{17}I$ (82)	[15]
$n\text{-}C_8H_{17}Br$	KI/polypode[b]	$n\text{-}C_8H_{17}I$ (85)	[15]

Table 9.1 (continued)

Substrate	Reagent/catalyst	Product (% yield)	Ref.
$n\text{-}C_8H_{17}Br$	KI/polypode[c]	$n\text{-}C_8H_{17}I$ (45)	[15]
$n\text{-}C_8H_{17}Br$	KI/polypode[d]	$n\text{-}C_8H_{17}I$ (42)	[15]
$n\text{-}C_8H_{17}Br$	KI/polypode[e]	$n\text{-}C_8H_{17}I$ (45)	[15]
$n\text{-}C_8H_{17}Br$	KI/polypode[f]	$n\text{-}C_8H_{17}I$ (51)	[15]
$n\text{-}C_8H_{17}Br$	KI/polypode[g]	$n\text{-}C_8H_{17}I$ (<1)	[15]
$n\text{-}C_8H_{17}Br$	KI/polypode[h]	$n\text{-}C_8H_{17}I$ (<1)	[15]
$n\text{-}C_8H_{17}Br$	KI/HTBPC	$n\text{-}C_8H_{17}I$ (72)	[15]
$n\text{-}C_8H_{17}OSO_2CH_3$	NaI/DC-18-C-6	$n\text{-}C_8H_{17}I$ (100)	[9]
	KI/DC-18-C-6	$n\text{-}C_8H_{17}I$ (100)	[9]
$n\text{-}C_8H_{17}I$	$Na^{137}I$/TCMAC	Equilibration, 5 min	[7]
cyclohexyl-OH, CH$_2$-N(NO)-CO-CH$_3$	NaI/NaOH/TCMAC	cyclohexylidene=CHI (72)	[14]
$n\text{-}C_6H_{13}CH(CH_3)Br$	KI/HDTBP	$n\text{-}C_6H_{13}CH(CH_3)I$ (89)	[13]
$n\text{-}C_6H_{13}CH(CH_3)Br$	KI/cryptate[b]	$n\text{-}C_6H_{13}CH(CH_3)I$ (86)	[13]
$R\text{-}n\text{-}C_6H_{13}CH(CH_3)OSO_2CH_3$	KI/HDTBP	$n\text{-}C_6H_{13}CH(CH_3)I$ (74) (nearly racemic)	[10]
$n\text{-}C_{10}H_{21}Br$	NaI/TCMAC	$n\text{-}C_{10}H_{21}I$ (96)	[6]
$n\text{-}C_{16}H_{33}Cl$	NaI/Q$^+$	$n\text{-}C_{16}H_{33}I$ (99)	[6]

Cryptates:

a: R = $n\text{-}C_{14}H_{29}$
b: R = $n\text{-}C_{11}H_{27}$

Polypodes:
a–d Refer to 1,3,5-triazoles:
a R = N[(CH$_2$CH$_2$O)$_4$–n-C$_4$H$_9$]$_2$
b R = N[(CH$_2$CH$_2$O)$_4$–n-C$_8$H$_{17}$]$_2$
c R = (OCH$_2$CH$_2$)$_3$O–n-C$_4$H$_9$.
d R = (OCH$_2$CH$_2$)$_3$O–n-C$_8$H$_{17}$.
e–h Refer to $n\text{-}C_{10}H_{21}$–O–CH$_2$–C[CH$_2$(OCH$_2$CH$_2$)$_3$OR′]$_3$:
e R′ = H.
f R′ = CH$_2$CH$_2$OH.
g R′ = n-C$_8$H$_{17}$.
h R′ = n-C$_8$H$_{17}$–OCH$_2$CH$_2$.

Reactions of Other Nucleophiles

An important reaction of quaternary ammonium fluoride ions is the displacement of an enolate ion from a silyl enol ether. This reaction is ordinarily conducted stoichiometrically. Mention of the method is also made in Chap. 10 (see Eq. 10.3). The trimethylsilyl enol ether of cyclohexanone, for example, can be prepared easily and is stable [18]. Fluoride ion associated with tetrabutylammonium cation attacks silicon forming a very stable silicon-fluorine bond and stoichiometrically liberating cyclohexanone enolate which can then be monoalkylated. Several examples of this reaction (see Eq. 9.3) are included in Table 9.1.

$$\text{cyclohexenyl-OSi(CH}_3)_3 \xrightarrow{Bu_4NF} \text{cyclohexenyl-O}^- + (CH_3)_3SiF \xrightarrow{RX} \text{2-R-cyclohexanone} \tag{9.3}$$

Recently, Ykman and Hall have reported a study of crown activated halide ions in the presence of β-chloro-α,β-unsaturated esters [4]. In the presence of dicyclohexyl-18-crown-6, substitution of chloro or carbomethoxy, or elimination was observed. These transformations are recorded in Table 9.2.

9.3 Azide Ions

Azide ions have been added to a variety of acyl chlorides under phase transfer conditions and react as anticipated. By this method, acid chlorides have been transformed into acyl azides in good yield as formulated in equation 9.4 [22]. The acyl azides thus produced can be further transformed into the corresponding isocyanates by a Curtius rearrangement [23].

$$R-CO-Cl \xrightarrow{Q^+N_3^-} R-CO-N_3 + Q^+Cl^- \tag{9.4}$$

There is a report that a vinyl azide was isolated in the base catalyzed (under phase transfer conditions) decomposition of an N-nitrosoacetamide. The generality of the latter method has not been determined [14]. The formation of azides and the derived isocyanates are summarized in Table 9.3.

The reaction of azide ion with polyvinyl chloride has recently been shown to be catalyzed by such cationic surfactants as tetrabutylammonium chloride and laurylpyridinium chloride in nonpolar solvents like dioxane, tetrahydrofuran and cyclohexanone. The anionic surfactant sodium laurylbenzenesulfonate was not an effective catalyst for this reaction, (see Eq. 9.5) [24].

$$\sim\!\!CH-CH_2-CH-CH_2\!\!\sim \xrightarrow[Q^+Cl^-]{NaN_3} \sim\!\!CH-CH_2-CH-CH_2\!\!\sim \atop \underset{Cl\quad\quad Cl}{} \quad \underset{N_3\quad\quad Cl}{} \tag{9.5}$$

Table 9.2. Reactions of activated halides with β-chloro-α,β-unsaturated esters [4]

Substrate	Reagent/Catalyst	Product (% yield)
(Cl)(H)C=C(CO$_2$CH$_3$)(CO$_2$CH$_3$)	KF/DC-18-C-6	(F)(H)C=C(CO$_2$CH$_3$)(CO$_2$CH$_3$) (nr)
(Cl)(Cl)C=C(CO$_2$CH$_3$)(CO$_2$CH$_3$)	KF/DC-18-C-6	(Cl)(Cl)C=C(Cl)(CO$_2$CH$_3$) (70)
(CH$_3$OCO)(Cl)C=C(COOCH$_3$)(COOCH$_3$)	KCl/DC-18-C-6	CH$_3$OOC−C≡C−COOCH$_3$ (7)
(NC)(Cl)C=C(COOCH$_3$)(COOCH$_3$)	KCl/DC-18-C-6	NC−C≡C−COOCH$_3$ (16)
CH$_3$OCO−CHCl−CHClCO$_2$CH$_3$	KCl/DC-18-C-6	E + Z CH$_3$OCOCH=CClCO$_2$CH$_3$ (58)

Table 9.3. Preparation of azides in nonpolar media

			Ref.
CH$_3$COCl	TBA$^+$N$_3^-$	CH$_3$CON$_3$ (52)	[22]
CH$_3$CH$_2$COCl	TBA$^+$N$_3^-$	CH$_3$CH$_2$CON$_3$ (61)	[22]
n-C$_3$H$_7$COCl	TBA$^+$N$_3^-$	n-C$_3$H$_7$CO−N$_3$ (73)	[22]
Cl−(CH$_2$)$_3$CO−Cl	TBA$^+$N$_3^-$	Cl−(CH$_2$)$_3$CO−N$_3$ (87)	[22]
(CH$_3$)$_2$CH−CH$_2$−COCl	TBA$^+$N$_3^-$	(CH$_3$)$_2$CH−CH$_2$CO−N$_3$ (86)	[22]
(CH$_3$)$_3$C−CO−Cl	TBA$^+$N$_3^-$	(CH$_3$)$_3$C−CO−N$_3$ (60)	[22]
C$_6$H$_5$−CO−Cl	NaN$_3$/MTNAC	C$_6$H$_5$−N=C=O (90)	[23]
C$_6$H$_5$−CO−Cl	TBA$^+$N$_3^-$	C$_6$H$_5$−CO−N$_3$ (89)	[22]
cyclohexane with OH and CH$_2$−N(NO)−CO−CH$_3$	NaN$_3$/NaOH/TCMAC	cyclohexylidene=N$_3$ (56)	[14]
CH$_3$(CH$_2$)$_{14}$−CO−Cl	NaN$_3$/MTNAC	CH$_3$−(CH$_2$)$_{14}$−N=C=O (90)	[23]
CH$_3$(CH$_2$)$_{16}$−CO−Cl	NaN$_3$/MTNAC	CH$_3$(CH$_2$)$_{16}$−N=C=O (93)	[23]

9.4 Nucleophile Induced Elimination Reactions

Elimination reactions have been induced under phase transfer conditions by a variety of techniques [1, 2, 4, 5, 25–28]. Predominant among these is the dehydrohalogenation of alkyl halides or sulfonates [1, 2, 5] and the dehalogenation of vicinal dihalides [25]. In the former case, some elimination is observed as a side reaction of the intended S_N2 halogen exchange reaction. The direct displacement of halide by halide (the

Finkelstein reaction) is favored for primary alkyl halides and also by sulfonate leaving groups. For example, in the case of resin-bound fluoride ion in pentane solution, 2-bromooctane is transformed into a mixture of octenes and fluoride in a ratio of 73%:20%. In contrast, 2-octyl mesylate yields the same products in ratios which are essentially reversed (25%:70%; see Eqs. 9.6 and 9.7) [5].

$$n\text{-}C_6H_{13}CH(CH_3)\text{-}Br \xrightarrow{\text{Resin-}\overset{+}{N}R_3F^-} n\text{-}C_6H_{13}\text{-}CHF\text{-}CH_3 \quad (20\%)$$
$$+ \text{ octenes} \quad (73\%) \quad (9.6)$$

$$n\text{-}C_6H_{13}\text{-}CH(CH_3)\text{-}OSO_2CH_3 \xrightarrow{\text{Resin-}\overset{+}{N}R_3F^-} n\text{-}C_6H_{13}CH(CH_3)F \quad (70\%)$$
$$+ \text{ octenes} \quad (25\%) \quad (9.7)$$

Likewise, 1-bromooctane is converted by 18-crown-6 complexed KF in acetonitrile solution into 1-fluorooctane (92%) and the by-product olefin in only 8% yield. Bromocyclohexane, on the other hand, is converted quantitatively into cyclohexene under the same conditions [2]. Although most of the systems which can give mixtures of E-2 and S_N2 products, do so, the sulfonates seem to most favor substitution and the fluoride nucleophile/bromide nucleofuge pair tends to favor elimination.

Vicinal dehalogenation has been reported under phase transfer conditions [25]. In this case, a series of vicinal dibromides was treated with an aqueous solution of sodium iodide and sodium thiosulfate in the presence of a quaternary ammonium catalyst. 1,2-Dibromooctane was thereby converted to 1-octene in 92% yield (see Eq. 9.8) [25]. It was found that the elimination reaction could be effected by sodium thiosulfate itself, but that this method was slower than when iodide ion was present. Only a catalytic amount of iodide ion was required in the reaction, it being continuously regenerated by reaction with thiosulfate.

$$n\text{-}C_6H_{13}\text{-}CHBr\text{-}CH_2Br \xrightarrow[\text{HTBPB}]{\text{NaI-Na}_2S_2O_3} n\text{-}C_6H_{13}\text{-}CH=CH_2 \quad (9.8)$$

Another elimination technique which deserves mention is the deoxygenation of epoxides by trimethylsilylpotassium [26]. 18-Crown-6 apparently catalyzes the reaction of potassium methoxide with hexamethyldisilane in HMPT solvent. The trimethylsilylpotassium thus generated deoxygenates the epoxide with overall inversion of sterochemistry. The process is formulated in equation 9.9 and examples are recorded in Table 9.4.

$$(CH_3)_3SiK + R\text{-}\overset{\overset{\displaystyle O}{\diagup\ \diagdown}}{CH\text{-}CH}\text{-}R \rightarrow R\text{-}CH=CH\text{-}R + (CH_3)_3SiOK \quad (9.9)$$

Chloride ion has been used in several-cases to effect the conversion of α,β-unsaturated-β-chloroesters into the corresponding acetylenes under phase transfer catalytic

Table 9.4. Nucleophile induced eliminations: olefin formation

Substrate	Reagent system	Product (% yield)	Ref.
Chlorocyclohexane	KF/HTBPB	Cyclohexene (100)	[1]
Chlorocyclohexane	Resin–NMe$_3$F/pentane	Cyclohexene (50)	[5]
Bromocyclohexane	KF/18-C-6/MeCN	Cyclohexene (100)	[2]
E-3-hexene epoxide	KOCH$_3$–(TMS)$_2$/18-C-6/HMPT	Z-3-hexene (99)	[26]
Z-3-hexene epoxide	KOCH$_3$–(TMS)$_2$/18-C-6/HMPT	E-3-hexene (86)	[26]
E-3-methyl-2-pentene epoxide	KOCH$_3$–(TMS)$_2$/18-C-6/HMPT	Z-3-methyl-2-pentene (91)	[26]
Z-3-methyl-3-pentene epoxide	KOCH$_3$–(TMS)$_2$/18-C-6/HMPT	E-3-methyl-3-pentene (99)	[26]
2-chloro-2-methylcyclohexanone	KF/18-C-6/MeCN	2-methylcyclohex-2-enone (69) + S$_N$2 (31)	[2]
2-chlorooctane	KF/HTBPB	Octenes (66)	[1]
2-bromooctane	KF/HTBPB	Octenes (64)	[1]
2-bromooctane	KF/18-C-6/MeCN	Octenes (68) + S$_N$2 (32)	[2]
2-bromooctane	Resin–NMe$_3$F/pentane	Octenes (73) + S$_N$2 (20)	[5]
2-mesyloxyoctane	Resin–NMe$_3$F/pentane	Octenes (25) + S$_N$2 (70)	[5]
2-mesyloxyoctane	KF/HTBPB	Octenes (26) + S$_N$2 (71)	[1]
2-tosyloxyoctane	Resin–NMe$_3$F/pentane	Octenes (25) + S$_N$2 (62)	[5]
E-4-octene epoxide	KOCH$_3$–(TMS)$_2$/18-C-6/HMPT	Z-4-octene (96)	[26]
Z-4-octene epoxide	KOCH$_3$–(TMS)$_2$/18-C-6/HMPT	E-4-octene (93)	[26]
E-2,5-dimethyl-3-hexene epoxide	KOCH$_3$–(TMS)$_2$/18-C-6/HMPT	Z-2,5-dimethyl-3-hexene (91)	[26]
Z-2,5-dimethyl-3-hexene epoxide	KOCH$_3$–(TMS)$_2$/18-C-6/HMPT	E-2,5-dimethyl-3-hexene (99)	[26]
1,2-dibromooctane	NaI–Na$_2$SO$_3$/HTBPB	1-octene (92)	[25]
Methyl *erythro*-2,3-dibromo-3-phenylpropanoate	NaI–Na$_2$SO$_3$/HTBPB/toluene	Methyl cinnamate (85)	[25]
Methyl *threo*-2,3-dibromo-3-phenylpropanoate	NaI–Na$_2$SO$_3$/HTBPB/toluene	Methyl cinnamate (95)	[25]
Methyl *erythro*-2,3-dibromo-4-oxophenylbutanoate	NaI–Na$_2$SO$_3$/HTBPB/toluene	Methyl E-4-oxo-4-phenyl-2-butenoate (86)	[25]
Methyl *threo*-2,3-dibromo-4-oxophenylbutanoate	NaI–Na$_2$SO$_3$/HTBPB/toluene	Methyl E-4-oxo-4-phenyl-2-butenoate (84)	[25]
Meso-1,2-dibromo-1,2-diphenylethane	NaI–Na$_2$SO$_3$/HTBPB/*sym*-tetrachloroethane	E-stilbene (89)	[25]
d,l-1,2-dibromo-1,2-diphenylethane	NaI–Na$_2$SO$_3$/HTBPB/toluene	Stilbene (86)	[25]

conditions (see Eq. 9.10 and Tables 9.2 and 9.5) [4]. The yields in this reaction are rather low but the dicyclohexyl-18-crown-6-KCl catalyzed dehydrohalogenation of

$$\underset{Cl}{\overset{NC}{>}}C=C\underset{COOCH_3}{\overset{COOCH_3}{<}} \xrightarrow[DC-18-C-6]{KCl} NC-C\equiv C-COOCH_3 \quad (16\%) \qquad (9.10)$$

dimethyl 2,3-dichlorosuccinate into a mixture of chlorofumaric and chloromaleic esters occurs in higher yield.

A more direct approach to acetylenes has been reported which utilizes phase-transfer generated tetrabutylammonium hydroxide to dehydrohalogenate α-halo-olefins or *bis*-dehydrohalogenate vicinal dihalides. An example of the latter is the reaction of 1,2-dibromo-1-phenylethane in pentane with 50% aqueous sodium hydroxide in the presence of tetrabutylammonium bisulfate. The product, phenylacetylene, is produced in 87% yield in less than an hour by this method (see Eq. 9.11) [27]. The advantage of this approach is apparent when one considers that in the traditional method, β-bromostyrene (from the dehydrohalogenation and decarboxylation of cinnamic acid dibromide) is heated with KOH at over 200° and the phenylacetylene distills as formed in 67% yield [28].

$$C_6H_5-CHBr-CH_2Br \xrightarrow[TBAB]{NaOH} C_6H_5-C\equiv CH \qquad (9.11)$$

The formation of dichloroacetylene from trichloroethylene in nonpolar solution [29] in the presence of epoxides is an example which involves a general method for generating hydroxide ion in solution [30]. Typically, a halide ion such as chloride reacts with ethylene oxide to give 2-chloroethyl alkoxide. This reaction is the microscopic reverse of hydroxide catalyzed epoxide formation from a 2-chloroalkoxide (see Eq. 9.12).

$$Cl^- + \overset{O}{\overset{\triangle}{CH_2-CH_2}} + H_2O \rightleftharpoons ClCH_2CH_2OH + HO^- \qquad (9.12)$$

Table 9.5. Nucleophile induced eliminations: alkyne formation

Substrate	Reagent/Catalyst	Product (% yield)	Ref.
$Cl_2C=CHCl$	Q^+Cl^-/epichlorohydrin	$Cl-C\equiv C-Cl$ (nr)	[28]
$CH_2Br-CHBr-CH(OC_2H_5)_2$	NaOH/TBAB	$HC\equiv C-CH(OC_2H_5)_2$ (80)	[27]
$CCl_2=CH-CH(OC_2H_5)_2$	NaOH/TBAB	$Cl-C\equiv C-CH(OC_2H_5)_2$ (70)	[27]
$NC-CCl=C(CO_2CH_3)_2$	KCl/DC-18-C-6	$NC-C\equiv C-COOCH_3$ (16)	[4]
$C_6H_5-CHBrCH_2Br$	NaOH/TBAB	$C_6H_5-C\equiv C-H$ (87)	[27]
$CH_3OCO-CCl=C(CO_2CH_3)_2$	KCl/DC-18-C-6	$CH_3OCO-C\equiv CCO_2CH_3$ (7)	[4]
$4-CH_3-C_6H_4-CCl=CH_2$	NaOH/TBAB	$4-CH_3-C_6H_4-C\equiv CH$ (77)	[27]
$C_6H_5-CHBr-CHBr-C_6H_5$	NaOH/TBAB	$C_6H_5-C\equiv C-C_6H_5$ (75)	[27]

In the formation of dichloroacetylene, chloride ion (from sodium chloride or a quaternary ammonium chloride) adds to an epoxide such as epichlorohydrin and the resulting alkoxide dehydrohalogenates trichloroethylene. The overall process is formulated in equation 9.13.

$$R-\overset{O}{\overset{\diagup\diagdown}{CH-CH_2}} + Cl^- + Cl_2C=CHCl \rightarrow R-CHOH-CH_2Cl$$
$$+ Cl-C\equiv C-Cl + Cl^- \qquad (9.13)$$

Numerous reactions have been investigated which involve alkoxides generated *in situ* by the reaction of halide with an epoxide. The kinship of this chemistry to the phase transfer method is manifest, but a detailed discussion of this body of work is beyond the scope of this book. In any event, the halide/epoxide technique has recently been reviewed [30].

9.5 Nitrite Ion

The reactions of halide and azide ions (see above), cyanide ion (see Chap. 7) and thiocyanate ions (see Sect. 13.7) have all been discussed. In the general context of halides and pseudohalides, it should be noted that nitrite ion has been successfully phase-transferred [31, 32]. The yields for the formation of nitroalkanes in either crown [31] or quaternary ion catalyzed [32] processes are fair to good for primary alkanes and poor for secondary systems. Bromocyclohexane, for example, afforded only traces of nitrocyclohexane when treated with nitrite ion, the by-products resulting presumably from oxygen attack (nitrite ester formation) or elimination. The conversion of *n*-octyl bromide into the corresponding nitro-compound is formulated in equation 9.14 and several examples of the transformation are recorded in Table 9.6.

$$n\text{-}C_8H_{17}-Br \xrightarrow[\text{crown}]{KNO_2} n\text{-}C_8H_{17}-NO_2 \qquad (9.14)$$

Table 9.6. Preparation of nitroalkanes from alkali metal nitrites under phase transfer conditions

Substrate	Product	% Yield	References
$c\text{-}C_6H_{11}Br$	$c\text{-}C_6H_{11}NO_2$	0–3	[31]
$C_6H_5CH_2Cl$	$C_6H_5CH_2NO_2$	34	[31]
$n\text{-}C_8H_{17}Br$	$n\text{-}C_8H_{17}NO_2$	≤ 20	[32]
$n\text{-}C_8H_{17}Br$	$n\text{-}C_8H_{17}NO_2$	65–70	[31]
$n\text{-}C_8H_{17}I$	$n\text{-}C_8H_{17}NO_2$	50–55	[31]
$C_6H_5CH_2CH_2Br$	$C_6H_5CH_2CH_2NO_2$	32	[31]
$C_6H_5CH_2CH_2CH_2Br$	$C_6H_5CH_2CH_2CH_2NO_2$	51	[31]

9.6 Hydrolysis Reactions

Pedersen reported in 1967 that the dicyclohexyl-18-crown-6 complex of potassium hydroxide was soluble in toluene and in this medium could readily hydrolyze the very sterically hindered ester methyl mesitoate [33]. He later reported that this complex could also hydrolyze the tertiary-butyl ester [34]. Similarly, Lehn found that the [2.2.2]-cryptate complex of potassium hydroxide was even more effective in this saponification reaction under directly comparable conditions [35]. The hydrolysis reaction is formulated in equation 9.15. This method has recently been applied in the hydrolysis of ^{13}C-labelled methyl tetradecanoate [36]. Starks found that tetradecanoate anion acted as a catalyst poison and impeded further hydrolysis [7]. This is therefore one of the few examples where crown ether catalysis is clearly superior to quaternary ammonium ion catalysis.

$$\text{Mes-COOCH}_3 \xrightarrow[\text{complexing agent}]{\text{KOH}} \text{Mes-COOH} + \text{CH}_3\text{OH} \qquad (9.15)$$

Starks found that tricaprylylmethylammonium chloride catalyzed the hydrolysis of n-dodecanesulfonyl chloride by aqueous hydroxide. Quantitative yields of the sulfonic acid were obtained in the presence of the quaternary ammonium catalyst whereas little reaction was observed in its absence [6, 7]. The hydrolysis of trichloromethylbenzene to benzoic acid (see Eq. 9.16) was likewise catalyzed by a quaternary ammonium salt [37].

$$\text{C}_6\text{H}_5-\text{CCl}_3 \xrightarrow[\text{TCMAC}]{\text{NaOH}} \text{C}_6\text{H}_5-\text{COOH} \qquad (9.16)$$

9.7 Anionic Polymerization Initiation

The phase transfer method offers the possibility of conducting polymerization reactions in nonpolar, aprotic solvents using anionic initiators which ordinarily are not soluble in such media and are often too inactive to effect polymerization. We have discussed in Sect. 6.4 the condensation polymerization of bisphenol A with phosgene under phase-transfer conditions. The formation of polycarbonates by this method is one of the earliest and most important (from the industrial point of view) applications of the method.

There has undoubtedly been much work conducted on the influence of phase transfer agents in various polymerization reactions, particularly in industrial laboratories, but relatively little of this has appeared in the general literature. One of the first such reports was that of Pedersen, who found that anhydrous formaldehyde was readily polymerized in aprotic solvents by crown-complexed KOH [34].

Boileau, Kaempf, Schué and coworkers have studied the cryptate mediated anionic addition polymerization of several systems including ethylene oxide [38], propylene sulfide [39–40], isobutylene sulfide [40], isoprene [38], methyl methacrylate [38], hexamethyl trisiloxane [40], ε-caprolactone [41], styrene [38, 40, 41], α-methylstyrene [41], 1,1-diphenylethylene [41] and β-propiolactone [42]. The polymerization of the latter compound induced by dibenzo-18-crown-6 complexed sodium acetate has also been reported [43]. In general, it was found that the polymer-

Table 9.7. Polymerizations initiated by activated anions

Monomer	Initiation system	References
Formaldehyde	KOH/aprotic/DC-18-C-6	[34]
Ethylene oxide	Carbazyl potassium/THF/[2.2.2]-cryptate	[38]
β-propiolactone	NaOAc/CH_2Cl_2/DB-18-C-6	[43]
	NaCl/toluene/[2.2.2]-cryptate	[42]
	KCl/toluene/[2.2.2]-cryptate	[42]
	KOAc/toluene/[2.2.2]-cryptate	[42]
	KOAc/chloroform/[2.2.2]-cryptate	[42]
	KCNS/toluene/[2.2.2]-cryptate	[42]
	KCNS/chloroform/[2.2.2]-cryptate	[42]
	KOH/toluene/[2.2.2]-cryptate	[42]
Propylene sulfide	Na^+seed^-/THF/[2.1.1]-cryptate	[39]
	Na^+seed^-/THF/[2.2.2]-cryptate	[39]
	Na^+seed^-/THF/$[2_0.2_0.2_s]$-cryptate	[39]
	Cs^+seed^-/THF/[3.2.2]-cryptate	[39]
	$Na^+C_{10}H_8^-$/THF/[2.2.2]-cryptate	[38]
	KCNS/THF/[2.2.2]-cryptate	[40]
	NaBϕ$_4$/THF/[2.2.1]-cryptate	[40]
Isobutylene sulfide	KOAc/benzene/[2.2.B]-cryptate	[40]
Isoprene	n-C_4H_9Li/THF/[2.2.2]-cryptate	[38]
	Na°/THF/[2.2.1]-cryptate	[41]
	K°/benzene/[2.2.2]-cryptate	[41]
	Cs°/toluene/[2.2.2]-cryptate	[41]
Methyl methacrylate	$Li^+C_{10}H_8^-$/THF/[2.1.1]-cryptate	[38]
	$Na^+C_{10}H_8^-$/THF/[2.1.1]-cryptate	[38]
	$K^+C_{10}H_8^-$/THF/[2.2.2]-cryptate	[38]
Hexamethyl cyclo-trisiloxane	KOH/benzene/[2.2.2]-cryptate	[40]
ε-caprolactone	KO-t-Bu/toluene/[2.2.2]-cryptate	[42]
Styrene	Li°/benzene/[2.1.1]-cryptate	[38]
	Li°/THF/[2.1.1]-cryptate	[38]
	Na°/THF/[2.1.1]-cryptate	[41]
	Ca°/THF/[2.2.2]-cryptate	[41]
	K°/dioxane/[2.2.2]-cryptate	[41]
	NaO-t-Amyl/benzene/[2.2.2]-cryptate	[40]
α-methylstyrene	K°/benzene/[2.2.2]-cryptate	[41]
1,1-diphenylethylene	K°/toluene/[2.2.2]-cryptate	[41]

Reactions of Other Nucleophiles

izations in the presence of crown or cryptate activated nucleophiles were faster and often led to higher molecular weight polymers. In Table 9.7 are recorded the initiators, complexing agents and solvents which have been utilized in these polymerization processes.

Benzyltriethylammonium chloride has been found to catalyze the polymerization of 2,2-bis-(4-hydroxyphenyl)-1,1-dichloroethylene when conducted under two-phase conditions [44].

9.8 Organometallic Systems

We have previously discussed the reactions of phase-transferred metallic species which are used as oxidizing agents. Permanganate ion (Sect. 11.2), chromate ion (Sect. 11.3), and a broad range of species containing Os, Mo, W, Se, V, Cr, Ti, Ce, Ni, Mn, Co, Pt, Fe, and Pb (Sect. 11.5) have all been successfully phase-transferred. In addition to oxidizing agents, the phase transfer of metallic ions has been useful in other ways. For example, pentacarbonyl hydroxides and fluorides of tungsten and chromium have been prepared with the aid of crown ethers [45].

Phase transfer catalysis has been utilized in the *ortho*-metallation of thiobenzophenone derivatives. Triiron dodecacarbonyl reacts with thiones in benzene solution in the presence of benzyltriethylammonium chloride and aqueous sodium hydroxide to give the complex shown in equation 9.17. Thiobenzophenone, 4,4'-dimethylthiobenzophenone, 4,4'-dimethoxythiobenzophenone and thio-Micheler ketone were all successfully metallated in 70%, 36%, 80% and 76% yields respectively.

$$\text{Ar-CS-Ar} + \text{Fe}_3(\text{CO})_{12} \xrightarrow[\text{BTEAC}]{2\text{N NaOH, C}_6\text{H}_6} \text{[complex]} \quad (9.17)$$

It is assumed by the authors that $\text{HFe}_3(\text{CO})_{11}^-$ "is an important species in the reaction" [46].

Diiron dodecacarbonyl has also been used as a reagent for the reduction of nitrobenzenes to anilines [47]. The two-phase reduction reaction which is believed to involve phase transfer of the $\text{HFe}_3(\text{CO})_{11}^-$ anion occurs rapidly at room temperature and requires one half equivalent of $\text{Fe}_3(\text{CO})_{12}$. The reaction takes place as formulated in equation 9.18 and affords yields of 85%, 92%, 88%, and 60% with 4-methyl-, 4-methoxy-, 4-chloro- and 4-acetylnitrobenzenes. It is interesting that these reactions are so successful in benzene solution, a medium in which BTEAC is an ineffectual catalyst under other circumstances [48].

$$\text{R-C}_6\text{H}_4\text{-NO}_2 + 1/2\,\text{Fe}_3(\text{CO})_{12} \xrightarrow[\text{BTEAC, C}_6\text{H}_6]{\text{Aq. NaOH}} \text{R-C}_6\text{H}_4\text{-NH}_2 \quad (9.18)$$

Organometallic Systems

π-Allyl complexes of cobalt have been prepared under phase transfer catalytic conditions according to equation 9.19 [49]. In a two-phase system consisting of benzene

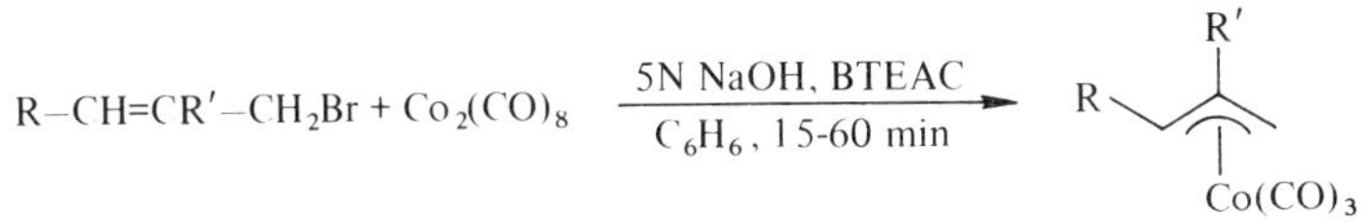

$$R-CH=CR'-CH_2Br + Co_2(CO)_8 \xrightarrow[C_6H_6, 15-60 \text{ min}]{5N \text{ NaOH, BTEAC}} R\underset{Co(CO)_3}{\overset{R'}{\diagup\!\!\!\diagdown}} \quad (9.19)$$

and aqueous sodium hydroxide, allyl bromides react with dicobalt octacarbonyl to afford good yields of the complexes. The synthesis of these complexes according to equation 9.19 is recorded in Table 9.8.

Table 9.8. Phase transfer synthesis of π-allyl complexes [49]

R	R'	% Yield
H	H	80
H	CH_3	73
CH_3	H	80
C_6H_5	H	72

Cobalt cluster compounds have also been prepared from tri- and tetrahalides under phase transfer conditions according to equation 9.20 [49].

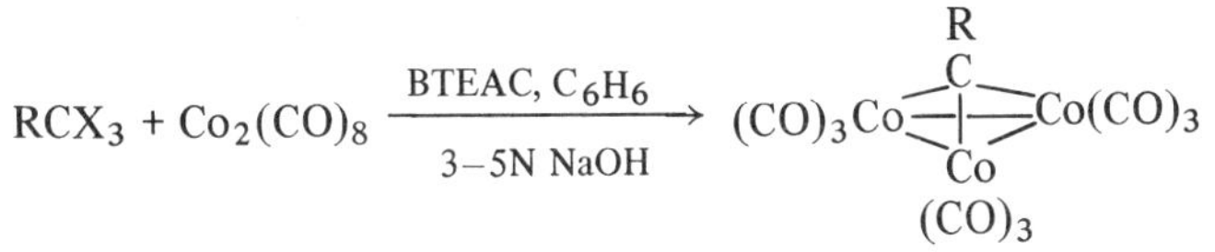

$$RCX_3 + Co_2(CO)_8 \xrightarrow[3-5N \text{ NaOH}]{BTEAC, C_6H_6} (CO)_3Co\underset{Co(CO)_3}{\overset{R}{\diagup\!\!\!\diagdown}}Co(CO)_3 \quad (9.20)$$

The cobalt cluster compounds which have been prepared by this method are recorded in Table 9.9; in which R refers to the substituent in equation 9.20.

Table 9.9. Phase transfer synthesis of cluster compounds [49]

Reactant	R	% Yield
CCl_4	Cl	42
CBr_4	Br	11
$C_6H_5-CCl_3$	C_6H_5	53
$CCl_3COOC(CH_3)_3$	$COOC(CH_3)_3$	30
CCl_3CH_2OH	CH_2OH	trace

Reactions of Other Nucleophiles

One of the most interesting and useful applications of phase transfer catalysis in organometallic chemistry is the catalytic carbonylation reaction of alkyl halides [50]. Palladium tetrakis triphenylphosphine catalyzes the carbonylation of benzyl chloride in a two phase system containing xylene and aqueous sodium hydroxide according to equation 9.21. Phenylacetic acid (4000 moles/mole Pd) is produced in

$$RX + CO + 2\ NaOH \xrightarrow[Pd[P(C_6H_5)_3]_4]{BTEAC/xylene} RCOONa + NaX + H_2O \qquad (9.21)$$

83% yield by this method. Moreover, p-dibromobenzene gave 4-bromobenzoic acid with a selectivity of 95% at 90% conversion.

18-Crown-6 polyether has been utilized to solubilize K_2PtCl_4 in chloroform and effect reaction of this nucleophile with a phosphazene polymer [51].

9.9 Isotopic Exchange

Starks showed in his first paper on phase transfer catalysis that 2-octanone undergoes facile H-D exchange with 5% sodium deuteroxide in heavy water [7]. The tetrabutylammonium bisulfate catalyzed hydrogen-deuterium exchange reaction in a variety of thiazoles has also been reported [52]. In this case, a systematic study revealed that the reaction rate was dependent on temperature and catalyst concentration and the position of exchange depended on the substitution pattern in the heterocycle. In general, the amount of exchange was high and the rates were large.

In work related to these phase-transfer isotopic exchange reactions, it has been found that reaction rates for several substituted sulfonium salts are influenced by micellar and solvent effects as well as the intrinsic stability of the derived carbanion [53].

References

1. Landini, D., Montanari, F., Rolla, F.: Synthesis *1974*, 428.
2. Liotta, C. L., Harris, H. P.: J. Am. Chem. Soc. *96*, 2250 (1974).
3. Gross, M., Peter, F.: Bull. Soc. Chim. Fr. *1975*, 871.
4. Ykman, P., Hall, H. K.: Tetrahedron Let. *1975*, 2429.
5. Cainelli, G., Manescalchi, F., Panunzio, M.: Synthesis *1976*, 472.
6. Starks, C. M., Napier, D. R.: Brit. Pat. 1,227,144.
7. Starks, C. M.: J. Am. Chem. Soc. *93*, 195 (1971).
8. Sam, D. J., Simmons, H. E.: J. Am. Chem. Soc. *96*, 2252 (1974).
9. Landini, D., Montanari, F., Pirisi, F.: J. C. S. Chem. Commun. *1974*, 879.
10. Landini, D., Quici, S., Rolla, F.: Synthesis *1975*, 430.
11. Landini, D., Montanari, F., Rolla, F.: Synthesis *1974*, 37.
12. Landini, D., Rolla, F.: Chem. Ind. (London) *1974*, 533.
13. Cinquini, M., Montanari, F., Tundo, P.: J. C. S. Chem. Commun. *1975*, 393.

14. Newman, M. S., Liang, C.: J. Org. Chem. *38,* 2438 (1973).
15. Fornasier, R., Montanari, F., Podda, G., Tundo, P.: Tetrahedron Let. *1976,* 1381.
16. Finkelstein, H.: Chem. Ber. *43,* 1528 (1910).
17. Tipson, R. S., Clapp, M. A., Cretcher, L. H.: J. Org. Chem. *12,* 133 (1947).
18. Kuwajima, I., Nakamura, E.: J. Am. Chem. Soc. *97,* 3257 (1975).
19. Nakamura, E., Shimizu, M., Kuwajima, I.: Tetrahedron Let. *1976,* 1699.
20. Pless, J.: J. Org. Chem. *39,* 2644 (1974).
21. Hanessian, S., Lavallee, P.: Can. J. Chem. *53,* 2975 (1975).
22. Brandström, A., Lamm, B., Palmertz, I.: Acta Chem. Scand. *B28,* 699 (1974).
23. Japan Kokai, 74,66,634, June 27, 1974 to General Mills.
24. Takeishi, M., Kawashima, R., Okawara, M.: Makromol. Chem. *167,* 267 (1973).
25. Landini, D., Quici, S., Rolla, F.: Synthesis *1975,* 396.
26. Dervan, P. B., Shippey, M. A.: J. Am. Chem. Soc. *98,* 1265 (1976).
27. Gorgues, A., Le Coq, A.: Tetrahedron Let. *1976,* 4723.
28. Hessler, J. C.: Org. Synth. Coll. Vol. *1,* 438 (1941).
29. Dobinson, B., Green, G. E.: Chem. Ind. London *1972,* 214.
30. Buddrus, J.: Angew. Chem. Int. Ed. *11,* 1041 (1972).
31. Zubrick, J. W., Dunbar, B. I., Durst, H. D.: Tetrahedron Let. *1975,* 71.
32. Kimura, C., Kashiwaya, K., Murai, K.: Asahi Garasu Kogyo Gijutsu Shoreikai Kenkyu Hokuku *24,* 59 (1974) [Chem. Abstr. *84,* 121027g (1976)].
33. Pedersen, C. J.: J. Am. Chem. Soc. *89,* 7017 (1967).
34. Pedersen, C. J., Bromels, M. J.: U.S. Patent 3,847,949 to E. I. du Pont de Nemours, Co. Inc. [Chem. Abstr. *82,* 73049 (1975)].
35. Dietrich, B., Lehn, J. M.: Tetrahedron Let. *1973,* 1225.
36. Patel, K. M., Pownall, H. J., Morrisett, J. D., Sparrow, J. T.: Tetrahedron Let. *1976,* 4015.
37. Menger, F. M., Rhee, J. U., Rhee, H. K.: J. Org. Chem. *40,* 3803 (1975).
38. Boileau, S., Kaempf, B., Lehn, J. M., Schué, F.: Polymer Let. *12,* 203 (1974).
39. Hemery, P., Boileau, S., Sigwalt, P., Kaempf, B.: Polymer Let. *13,* 49 (1975).
40. Boileau, S., Hemery, P., Kaempf, B., Schué, F., and Viguier, M.: Polymer Let. *12,* 217 (1974).
41. Boileau, S., Kaempf, B., Raynal, S., Lacoste, J., Schué, F.: Polymer Let. *12,* 211 (1974).
42. Deffieux, A., Boileau, S.: Macromolecules *9,* 369 (1976).
43. Slomkowski, S., Penczek, S.: Macromolecules *9,* 367 (1976).
44. Brzozowski, Z. K., Kielkiewicz, J., Goclawski, Z.: Angew. Makromol. Chem. *44,* 1 (1975).
45. Cihonski, J. C., Levinson, R. A.: Inorganic Chem. *14,* 1717 (1975).
46. Alper, H., des Roches, D.: J. Organomet. Chem. *1976,* C44;
47. des Abbayes, H., Alper, H.: J. Am. Chem. Soc. *99,* 98 (1977).
48. Herriott, A. W., Picker, D.: J. Am. Chem. Soc. *97,* 2345 (1975).
49. Alper, H., des Abbayes, H., des Roches, D.: J. Organomet. Chem. *121,* C31 (1976).
50. Cassar, L., Foa, M., Gardano, A.: J. Organomet. Chem. *121,* C55 (1976).
51. Allcock, H. R., Allen, R. W., O'Brien, J. P.: J. C. S. Chem. Comm. *1976,* 717.
52. Spillane, W. J., Dou, H. J.-M., Metzger, J.: Tetrahedron Let. *1976,* 2269.
53. Okonogi, T., Umezawa, T., Tagaki, W.: J. C. S. Chem. Commun. *1974,* 363.

10. Alkylation Reactions

10.1 Introduction

Because the formation of carbon-carbon bonds occupies a central position in synthetic organic chemistry, new alkylation methodology is arduously sought after. A continuing problem in alkylation chemistry is that those groups which activate carbon towards alkylation may activate more than one carbon. Moreover, the newly substituted carbon may be activated by the same group and undergo a second alkylation. In addition, the activating group itself may undergo alkylation, cleavage, or both. The report of a new general technique (PTC) for the alkylation of carbon acids and heteroatoms which offered the possibility of realizing new selectivity, justifiably attracted attention [1, 2].

The two-phase alkylation method is essentially similar to other phase transfer reactions; for a detailed discussion of mechanistic principles, refer to Sect. 1.4–1.7. In essence, a carbon acid in an organic solution which is in contact with an aqueous reservoir of caustic or solid potassium hydroxide can be deprotonated by the base. This may occur at the liquid-liquid interface or at the liquid-solid interface, or it may occur in the bulk organic phase. The available evidence suggests that all of these possibilities are operable and that essentially no reaction involving an organic substrate occurs in the bulk aqueous phase [3–5]. The principal difference between the two liquid-liquid mechanisms is that if deprotonation occurs at the interface [6, 7], the role of the phase transfer catalyst will be to transport the carbanion into the organic phase where it can be alkylated, whereas if reaction occurs in the bulk organic phase [8], the quaternary ammonium hydroxide ion pair must be freely soluble therein. To date, there has been no definitive study which distinguishes between these two possibilities for a wide variety of cases. It therefore seems likely that either both, or components of both, mechanisms operate in different situations.

At this time, there is relatively little information on alkylation by the solid-liquid phase transfer method, although this approach has been valuable in a variety of other nucleophilic reactions. Most of the mass of alkylation data which has been accumulated involves an organic phase in contact with 50% aqueous sodium hydroxide solution and a catalytic amount of a quaternary ammonium salt (Makosza's method) [2] or a stoichiometric amount of a quaternary ammonium ion pair (Brandström's method) [8]. In addition to these, quaternary ammonium fluorides have been used stoichiometrically to cleave silyl enol ethers [9], and crown ethers have been used to separate ion pairs [10]. In each of the examples reported in the tables, the catalyst is reported and this indicates which method was used.

Makosza's Method (catalytic):

$$R-CH_2-CN + R'X \xrightarrow[\text{2-phase } Q^+X^-]{\text{aq. NaOH}} R-\underset{R'}{\underset{|}{C}H}-CN \qquad (10.1)$$

Brandström's Method (stoichiometric):

$$R-CH_2-CN + Q^+OH^- \rightarrow R-\underset{CN}{\underset{|}{C}H}^-Q^+ \xrightarrow{R'X} R-\underset{CN}{\underset{|}{C}H}-R' \qquad (10.2)$$

Fluoride Method (stoichiometric):

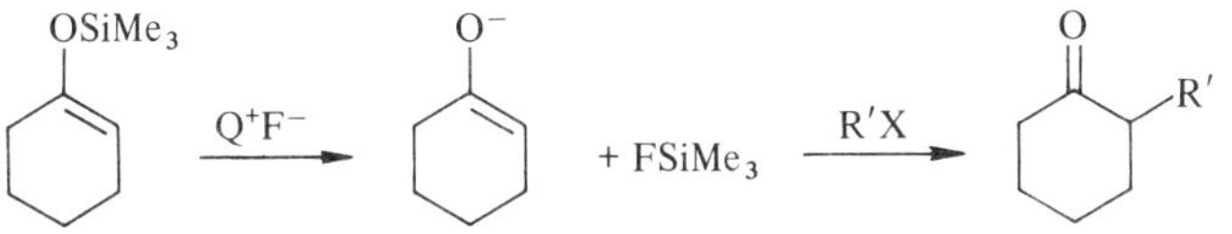

(10.3)

10.2 The Substances Alkylated

Much of the early alkylation work was conducted on nitriles, particularly phenylacetonitrile (Eq. 10.4). Numerous nitriles related to phenylacetonitrile have also

$$C_6H_5-CH_2-CN + CH_3CH_2-Cl \xrightarrow{\text{BTEAC}} C_6H_5CH(C_2H_5)CN \qquad (10.4)$$

been alkylated. These include α-methyl-[11], α-methoxy- [12], α-ethyl- [13], α-n-propyl- [14], α-i-propylphenylacetonitrile [15], and so on. The analogs such as α-naphthylacetonitrile [16] have also been studied. A brief survey of Table 10.1 will suggest to the reader the range of nitriles examined. In addition, nitriles substituted with such ancillary functions as ketone or ester have been alkylated and the alkylation of these systems is summarized in Table 10.3 (Eq. 10.5).

$$NC-CH_2-CO-OCH_3 + (CH_3)_2CHI \rightarrow NC-\underset{\underset{CH_3}{\overset{CH}{\diagup}}\diagdown CH_3}{\underset{|}{C}H}-CO-OCH_3 \qquad (10.5)\ [17]$$

One nitrile which deserves special attention is chloroacetonitrile. Chloro-substitution generally decreases the pKa of a carbon acid by one to two log units. Chloroaceto-

Alkylation Reactions

nitrile is therefore slightly more acidic than is acetonitrile, but the more interesting property of this system is that the halide can be lost after alkylation occurs. For example, chloroacetonitrile condenses with acetone under phase transfer conditions to give an α-chloroalkoxide. Alkoxide displaces chloride yielding an α,β-epoxynitrile [18]. The reaction is formulated in equation 10.6 and is successful for a broad range of aldehydes and ketones (see Table 10.1). The α-halosulfones behave similarly; see Sect. 13.8.

$$Cl-CH_2-CN \xrightarrow[PTC]{base} Cl-\bar{C}H-CN \xrightarrow{(CH_3)_2CO} \quad (10.6)$$

$$\underset{\underset{CN\ CH_3}{|\ \ |}}{Cl-CH-\overset{O^{\ominus}}{\underset{|}{C}}-CH_3} \longrightarrow \underset{\underset{CH_3}{|}}{NC-CH-\overset{O}{\overset{/\backslash}{C}}-CH_3}$$

(60%)

The carbon acids which are probably most often alkylated under classical conditions are aldehydes, ketones, and esters. The phase transfer alkylation of these systems has also been reported (results presented in Table 10.2) but these compounds deserve special comment. Aldehydes which might readily be alkylated are usually subject to aldol condensation as well (Eq. 10.7). The only aldehydes which have been alkylated

$$\underset{\underset{C_2H_5-CH-OH}{|}}{CH_3-CH-CHO} \xleftarrow[base]{CH_3CH_2CHO} CH_3CH_2CHO \xrightarrow[base]{RX} CH_3CH(R)CHO$$

aldol product   alkylation   (10.7)

under phase transfer conditions are those which produce tertiary carbanions [19]. These anions are reactive and alkylate rapidly and the high level of substitution tends to hinder nucleophilic addition to the carbonyl function. Likewise, methyl ketones afford poor yields of monoalkylation product, often contaminated by dialkylation products (Eq. 10.8). In cases where a carbonyl group is flanked by a methyl and a carbon bearing a more acidic proton (as in the benzyl group of phenylacetone) the more stabilized carbanion is generated and mono or dialkylated depending predomi-

$$C_6H_5-CO-CH_3 + (CH_3)_2C=CH-CH_2Cl \xrightarrow{40\%}$$

$$C_6H_5-CO-CH_2-CH_2-CH=C(CH_3)_2 \quad (10.8)\ [20]$$

nantly on the relative amounts of alkylating agent and base present (see Eq. 10.9) but obviously on the number of acidic hydrogens as well.

$$C_6H_5-CH_2-CO-CH_3 + (CH_3)_2C=CH-CH_2Cl \xrightarrow{74\%}$$

$$CH_3-CO-CH(C_6H_5)-CH_2CH=C(CH_3)_2 \quad (10.9)\ [21]$$

Sulfones are weakly acidic substances, but owing primarily to their inability to undergo side reactions under mild conditions, they alkylate cleanly. Likewise, hydrocarbons (Table 10.4) and amines (Table 10.5) alkylate under phase transfer conditions to give the anticipated products. In fact, the principal advantage of the phase transfer method is in the convenience and economy of the method and not in the unique nature of the chemistry.

Diactivated substrates such as diesters, dinitriles, ketosulfones, etc. (see Table 10.3) also alkylate as anticipated, but these systems often show a dependence on the position of alkylation (C vs. O) due to differences in solvent polarity. In addition to the expected chemistry of these diactivated substrates, several observations have been reported on differences in alkylation depending on solvent and the presence or absence of crown ether (see Eqs. 10.10 and 10.11) [22].

$$CH_3-CO-CH_2-CO-OC_2H_5 + CH_3CH_2OTs \xrightarrow[C_6H_6]{KOH}$$

$$CH_3-CO-\underset{C_2H_5}{CH}-CO-OC_2H_5$$

(10.10)

$$CH_3-CO-CH_2-CO-OC_2H_5 + CH_3CH_2OTs \xrightarrow[crown]{KOH/C_6H_6}$$

$$CH_3-CO-\underset{C_2H_5}{CH}-CO-OC_2H_5 + CH_3-\underset{OC_2H_5}{C}=CH-CO-OC_2H_5 \ (1:1)$$

(10.11)

10.3 Phase Transfer Alkylating Agents

Phenylacetonitrile (benzyl cyanide) was the first compound reported to undergo phase transfer alkylation and is probably the most abundantly studied example of this procedure [1]. It is a good substrate for study because it is reasonably acidic, but not readily alkylated by aqueous sodium hydroxide in the absence of catalyst at any synthetically useful rate. It has the further advantage that it can be mono- or dialkylated depending on conditions and it, like most nitriles, is not prone to rapid hydrolysis. It is a versatile synthetic intermediate, however, in the sense that the nitrile function can be hydrolyzed, reduced, or added to by organometallics after an alkylation has been carried out.

Phenylacetonitrile reacts with alkyl, allyl, and benzyl halides as one would expect (see Eq. 10.1). In the presence of excess alkylating agent and base, dialkylation is usually observed. The yield of alkylation product is usually high with normal alkyl halides (Cl, Br, I) and somewhat lower for secondary halides. There is no report of the successful alkylation of phenylacetonitrile by a tertiary alkyl halide under phase transfer conditions. This is undoubtedly due to the facile dehydrohalogenation reaction of such haloalkanes. It is probably the elimination problem which keeps 2-phenylethyl chloride from alkylating phenylacetonitrile in high yield [23], despite the fact that it is a primary halide. It is not clear why 1-phenylethyl chloride should alkylate the same substrate almost quantitatively under the same conditions [23], although the fact that the latter is a benzyl halide no doubt is part of the reason.

α-Haloesters are normally quite reactive alkylating agents and this is also the case under phase transfer conditions. The difficulty with these substrates is that the commonly available ones, i. e., methyl and ethyl bromoacetate, are hydrolyzed under the catalytic alkylation conditions. As a result, only cyclohexyl, i-propyl, t-butyl or other sterically hindered esters can be used as alkylating agents. This principle extends to the substrates as well: alkylation of esters can best be achieved if the alcohol component of the ester is bulky.

As noted above, the mono- or dialkylation of phenylacetonitrile can be controlled by the amounts of alkylating agent and base present and the reaction time. In some situations, dialkylation is desired and there are several useful applications of this technique reported. 1,4-Dibromobutane [24] and 1,5-dibromopentane [24] alkylate phenylacetonitrile according to equation 10.12 to give 1-cyano-1-phenyl-cyclopentane or -hexane in 88% and 40% yields respectively [24]. Alkylation with 2,2′-dichlorodiethyl ether affords the oxygen containing six-membered ring in almost

$$C_6H_5-CH_2-CN + Br-(CH_2)_n-Br \longrightarrow \underset{NC}{\overset{C_6H_5}{>}}\!\!\!\!\!\!\!\!\!\!\bowtie\!\!\!\!\!\!\!\!\!(CH_2)_n \qquad (10.12)$$

n = 4 or 5

70% yield [24]. 1,7-Dibromoheptane fails to close the eight-membered ring, yielding instead the monoalkylated system bearing a bromoheptyl side chain [24]. 1,3-Dibromopropane gives a mixture of monoalkylated product and cyclobutane [24], although under somewhat different conditions, 1,3-dichloropropane resulted in a 60% yield of cyclobutane [25]. A number of carbon acids also react with 1,2-dibromoethane to yield cyclopropanes. Ethyl cyanoacetate, for example, reacts with this dihalide to yield (86%) 1-carboxy-1-cyanocyclopropane according to equation 10.13 [26]. Note that in this case, alkylation is successful even though the ethyl ester suffers hydrolysis.

$$NC-CH_2CO_2Et + BrCH_2CH_2Br \longrightarrow \underset{COOH}{\overset{CN}{\triangleright\!\!\!<}} \qquad (10.13)$$

Another interesting prospect arises when dihalides are used to alkylate ketones instead of nitriles. Apparently, four-membered ring formation is disfavored to such an extent in the acenaphthone case that 1,3-dibromopropane C-alkylates and then O-alkylates after a second enolization. The product is the tetracyclic enol ether shown in equation 10.14.

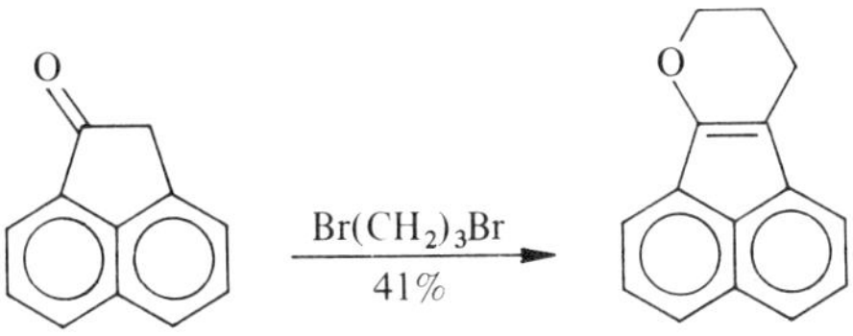

$$\text{acenaphthone} \xrightarrow[41\%]{\text{Br(CH}_2)_3\text{Br}} \text{tetracyclic enol ether} \qquad (10.14)$$

Dichloromethane [24] and benzal chloride [24] yield dimers as shown in equations 10.15 and 10.16. In the case of benzal chloride, the second alkylation could be normal

$$CH_2Cl_2 + C_6H_5-CH_2-CN \xrightarrow{69\%} \underset{\underset{C_6H_5}{|}}{NC-CH}-CH_2-\underset{\underset{C_6H_5}{|}}{CH}-CN \qquad (10.15)$$

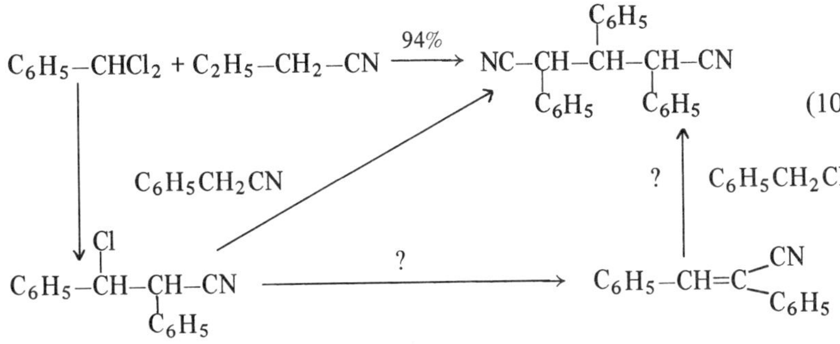

$$C_6H_5-CHCl_2 + C_2H_5-CH_2-CN \xrightarrow{94\%} NC-\underset{\underset{C_6H_5}{|}}{CH}-\overset{\overset{C_6H_5}{|}}{CH}-\underset{\underset{C_6H_5}{|}}{CH}-CN \qquad (10.16)$$

alkylation or could result from dehydrohalogenation followed by Michael addition of the anion. Makosza suggests this mechanism for the alkylation reaction with chloromethyl methyl ether [28], in which the same dimer produced in equation 10.15 is isolated. A similar mechanism can be envisaged for any alkylating agents which contain a leaving group beta to a carbanion stabilizer. This mechanism is explicitly invoked for the α-chloroacetonitrile and chloromethyl methyl ether cases but could easily apply in the other cases as well.

When phenylacetonitrile is alkylated with carbon tetrachloride, E-1,2-dicyano-1,2-diphenyl ethylene is isolated in 70% yield. Makosza suggests that this olefin is formed according to equation 10.17, in which carbon tetrachloride serves as a chlorinating agent [29].

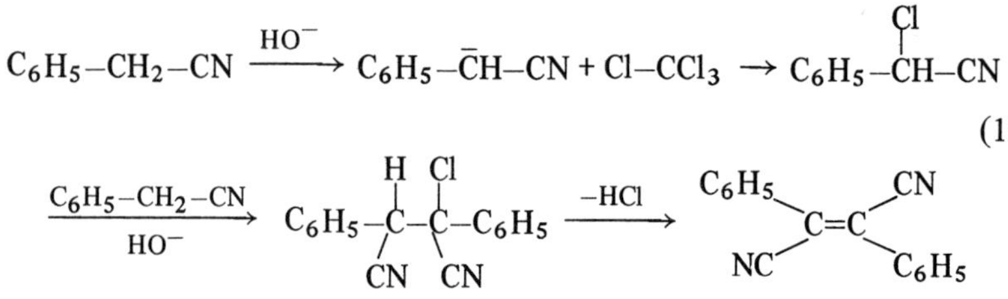

$$C_6H_5-CH_2-CN \xrightarrow{HO^-} C_6H_5-\bar{C}H-CN + Cl-CCl_3 \rightarrow C_6H_5-\overset{Cl}{\underset{|}{C}H}-CN \quad (10.17)$$

$$C_6H_5-CH_2-CN \xrightarrow{HO^-} C_6H_5-\overset{H}{\underset{CN}{C}}-\overset{Cl}{\underset{CN}{C}}-C_6H_5 \xrightarrow{-HCl} \underset{NC}{\overset{C_6H_5}{>}}C=C\underset{C_6H_5}{\overset{CN}{<}}$$

Alkyl thiocyanates react with a variety of nucleophiles to yield thioethers (see Sect. 13.4). A particularly clever and useful application of this reaction is the dialkylation of phenylacetonitrile with 1,2-dithiocyanatoethane. The product of this formal oxidation is the ethylenedithioketal of benzoyl cyanide (see Eq. 10.18) [30]. The phase transfer synthesis of the starting material from thiocyanate ion and ethylene bromide

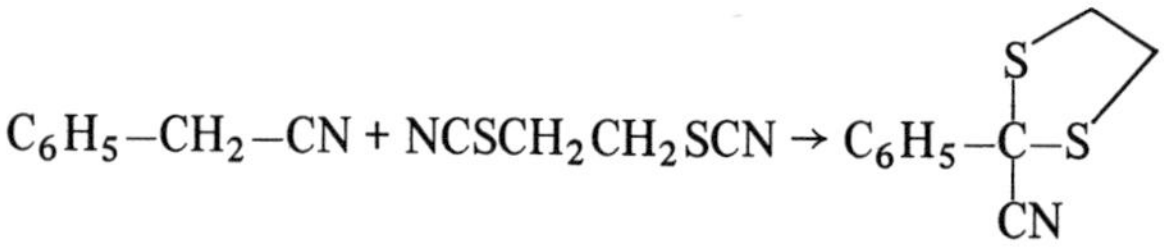

$$C_6H_5-CH_2-CN + NCSCH_2CH_2SCN \rightarrow C_6H_5-\underset{CN}{\overset{S\diagup}{\underset{|}{C}-S}} \quad (10.18)$$

makes this approach particularly attractive (see Sect. 13.5).

2-Phenylpropionitrile can be alkylated by acetylene itself [31] or by substituted acetylenes [31, 32] under phase transfer conditions (see Eqs. 10.19–10.22). The products are those resulting from addition, followed by protonation of the vinyl

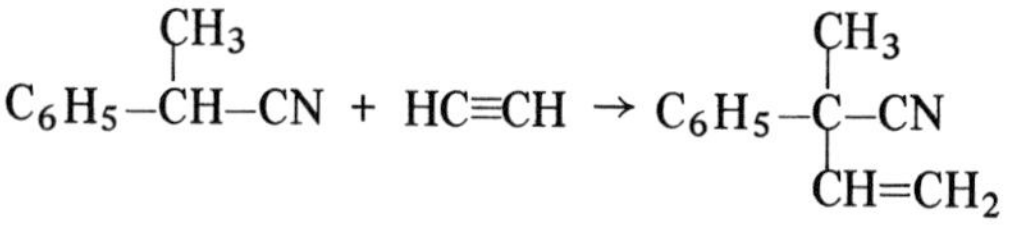

$$C_6H_5-\overset{CH_3}{\underset{|}{C}H}-CN + HC\equiv CH \rightarrow C_6H_5-\overset{CH_3}{\underset{CH=CH_2}{\overset{|}{C}}}-CN \quad (10.19)$$

$$C_6H_5-\overset{CH_2-CH_3}{\underset{|}{C}H}-CN + HC\equiv C-C_6H_5 \rightarrow C_6H_5-\overset{CH_2-CH_3}{\underset{CH=CH-C_6H_5}{\overset{|}{C}}}-CN \quad (10.20)$$

$$C_6H_5-\overset{CH_3}{\underset{|}{C}H}-CN + HC\equiv C-S-n\text{-}C_4H_9 \rightarrow C_6H_5-\overset{CH_3}{\underset{CH=CH-S-n\text{-}C_4H_9}{\overset{|}{C}}}-CN \quad (10.21)$$

$$C_6H_5-\overset{CH_3}{\underset{|}{C}H}-CN + HC\equiv C-O-C_2H_5 \rightarrow C_6H_5-\overset{CH_3}{\underset{C_2H_5O\,\,\,CH_2}{\overset{|}{\underset{\diagdown \|}{C}}}}-CN \quad (10.22)$$

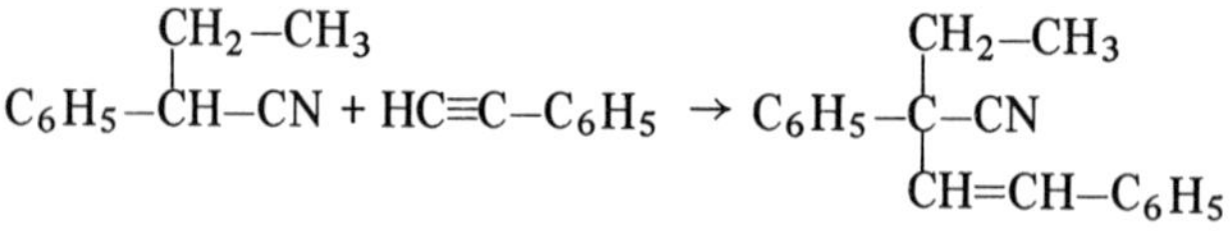

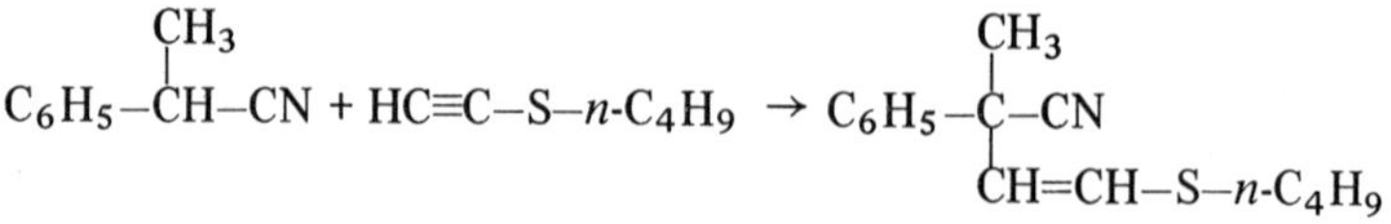

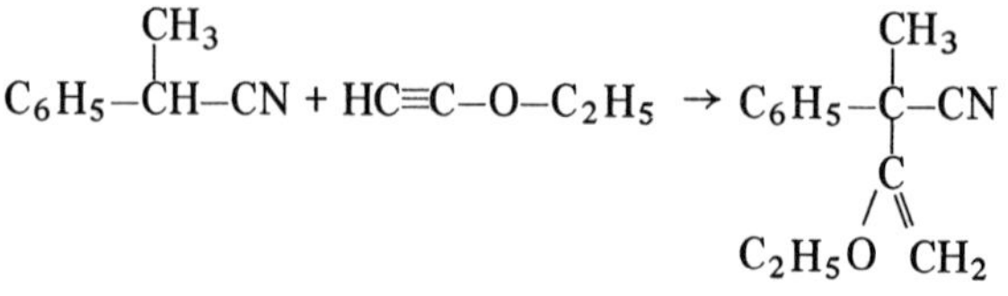

Phase Transfer Alkylating Agents

anion. In the case of acetylene itself [31], there is no orientation problem, but for substituted acetylenes, addition occurs in such a way that the most stabilized anion results. In phenylacetylene [31] (Eq. 10.20) and n-butylthioacetylene [32] (Eq. 10.21) addition occurs at the unsubstituted side of the triple bond, whereas the opposite orientation is observed for ethoxyacetylene [32] (Eq. 10.22).

It has also been shown that phase transfer conditions are effective for promoting carbanion arylation by nucleophilic aromatic substitution. 2-Phenylpropionitrile reacts with 4-nitrochlorobenzene to give 2-phenyl-2-(4-nitrophenyl)-propionitrile in 82% yield (see Eq. 10.23) [33]. Similar results were observed for several other chlorinated nitroaromatics, but two cases deserve special attention. First, the reaction of 2-phenylpropionitrile with 4-bromo-4'-nitrobenzophenone affords the aromatic substitution product, but the leaving group in this reaction is nitrite, not bromide (see Eq. 10.24) [34]. Nitrite is known to be a good leaving group in nucleophilic aromatic

$$C_6H_5-CH(CH_3)CN + 4-NO_2-C_6H_4-Cl \rightarrow C_6H_5-\underset{CN}{\overset{CH_3}{C}}-C_6H_4-NO_2 \quad (10.23)$$

$$Br-C_6H_4-\overset{O}{\underset{\|}{C}}-C_6H_4-NO_2 + C_6H_5-CH(CH_3)CN \longrightarrow$$

$$C_6H_5-\underset{CN}{\overset{CH_3}{C}}-C_6H_4-\overset{O}{\underset{\|}{C}}-C_6H_4-Br \quad (10.24)$$

substitution so this reactivity pattern is not unexpected. Secondly, the alkylation of phenylacetonitrile by 4-nitrochlorobenzene fails [33], apparently because after alkylation equivalent to the amount of catalyst present has occurred, the benzhydryl cyanide anion evidently ion pairs with the quaternary salt preventing further reaction of any other nucleophile (i.e., poisoning the catalyst).

There are relatively few reports of carbanion alkylation by aldehydes and ketones. Although the method is fraught with the difficulty of multiple condensations occurring as a side reaction, the condensation is useful in certain situations. Acetonitrile, for example, condenses cleanly and in good yield with benzaldehyde according to equation 10.25 to yield cinnamonitrile [35]. The reaction is a solid liquid phase transfer process in which the phase transfer agent is critical only in difficult cases.

$$NC-CH_3 + C_6H_5-CHO \xrightarrow{KOH} NC-CH=CH-C_6H_5 \quad (10.25)$$

10.4 Alkylation of Reissert's Compound

Reissert's compound (1-cyano-1,2-dihydroquinolinyl benzamide) is apparently deprotonated under phase transfer conditions at the one-carbon as expected. The condensation of the carbanion with an aldehyde or ketone leads to an N-benzoyl alkoxide in which oxygen acylation (N → O acyl transfer) results in five-membered ring formation. The intermediate oxazolidine decomposes with loss of cyanide to give the benzoate ester of an isoquinolinoylcarbinol as shown in equation 10.26 [36].

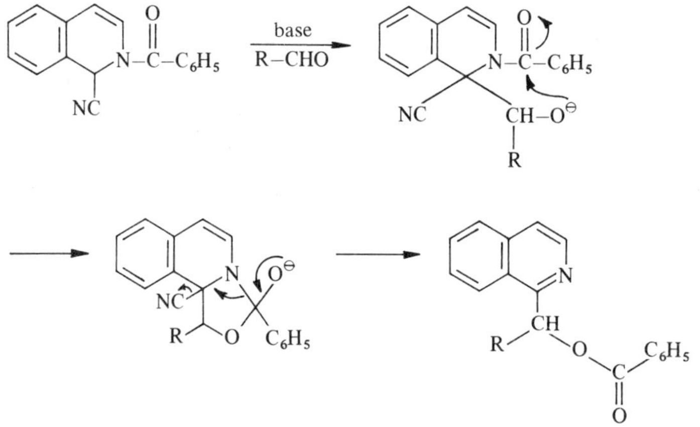

(10.26)

Note on Table Usage

In order to facilitate the use of the tables in this chapter, they have been organized as follows: The substrates are listed in order of increasing (total) number of carbons. The list of substances used as alkylating agents is likewise in order of increasing carbon number and in order of increasing complexity within a set of isomers or closely related compounds.

Table 10.1. Alkylation of nitriles

Number of carbons	Substrate	Electrophile	Product	% Yield	Catalyst	References
2	CH$_3$–CN	C$_6$H$_5$CHO	C$_6$H$_5$–CH=CH–CN	82	–	[35]
		4-CH$_3$–C$_6$H$_4$CHO	4-CH$_3$–C$_6$H$_4$–CH=CH–CN	61	–	[35]
		4-CH$_3$O–C$_6$H$_4$CHO	4-CH$_3$O–C$_6$H$_4$–CH=CH–CN	81	–	[35]
		4-Cl–C$_6$H$_4$CHO	4-Cl–C$_6$H$_4$–CH=CH–CN	57	–	[35]
		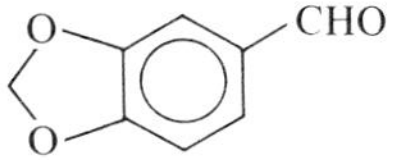 ![piperonal]	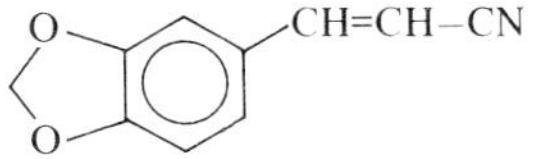 ![piperonal-acrylonitrile]	86	–	[35]
		(C$_6$H$_5$)$_2$CO	(C$_6$H$_5$)$_2$C=CH–CN	84	–	[35]
		cyclohexanone	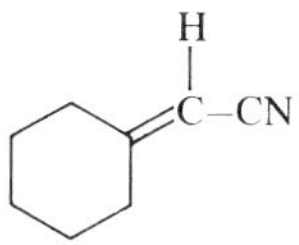 ![cyclohexylidene-acetonitrile]	50	–	[35]
2	Cl–CH$_2$–CN	C$_6$H$_5$CHO	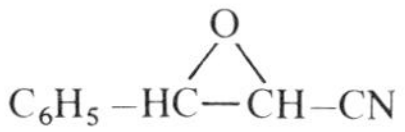 C$_6$H$_5$–HC–CH–CN (epoxide)	75	BTEAC	[18]
		CH$_3$COCH$_3$	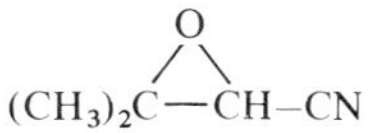 (CH$_3$)$_2$C–CH–CN (epoxide)	60	BTEAC	[18]
		C$_6$H$_5$COCH$_3$	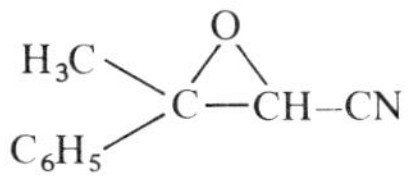 H$_3$C, C$_6$H$_5$ C–CH–CN (epoxide)	80	BTEAC	[18]
		cyclopentanone	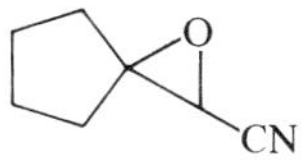 ![spiroepoxy-CN]	65	BTEAC	[18]

Alkylation Reactions

Table 10.1 (continued)

Number of carbons	Substrate	Electrophile	Product	% Yield	Catalyst	References
2	Cl–C$\underline{H_2}$–CN	cyclohexanone	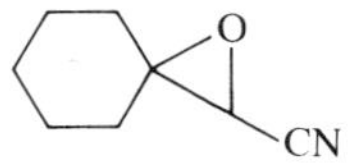 [epoxide fused to cyclohexane with CN]	79	BTEAC	[18]
		2-methylcyclohexanone	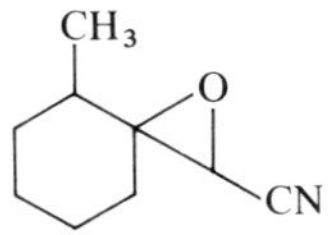 [epoxide fused to 2-methylcyclohexane with CN]	78	BTEAC	[18]
3		$(C_6H_5)_2CO$	$(C_6H_5)_2C\overset{O}{\triangle}CH-CN$	55	BTEAC	[18]
	CH$_3$CHClCN	CH$_2$=CH–CN	cyclopropane	75	BTEAC	[37]
		CH$_2$=C(CN)CH$_3$	cyclopropane	75	BTEAC	[37]
		CH$_2$=CH–CO$_2$–t-C$_4$H$_9$	cyclopropane	67	BTEAC	[37]
		CH$_2$=CH–SO$_2$C$_6$H$_5$	$\begin{cases}\text{cyclopropane} \\ CH_3-\underset{CN}{CCl}(CH_2)_2SO_2C_6H_5\end{cases}$	$\begin{matrix}7\\85\end{matrix}$	BTEAC	[37]
8	C$_6$H$_5$–C$\underline{H_2}$–CN	CH$_3$Cl	di	91	BTEAC	[13]
		CH$_3$Cl	mono	84	BTEAC	[23]
		CH$_3$Br	$\begin{cases}\text{mono} \\ \text{di}\end{cases}$	$\begin{matrix}80\\4\end{matrix}$	BTEAC	[23]
		CH$_3$I	$\begin{cases}\text{mono} \\ \text{di}\end{cases}$	$\begin{matrix}72\\14\end{matrix}$	TBA$^+$OH$^-$ (St)	[38]
		CH$_3$I	mono	95	α-phosphoryl sulfoxide	[39]

Table 10.1 (continued)

Number of carbons	Substrate	Electrophile	Product	% Yield	Catalyst	References
8	$C_6H_5-\underline{CH_2}-CN$	CH_3I	mono di	8 92	TBA^+OH^- (3:1 excess)	[38]
		C_2H_5Cl	mono	88	BTEAC	[2, 23]
		C_2H_5Br	mono	88	BTEAC	[23, 98]
		C_2H_5I	mono	93	α-phosphoryl sulfoxide	[39]
		$i\text{-}C_3H_7Br$	mono	90	TBA^+OH^- (St)	[38]
		$i\text{-}C_3H_7I$	mono	60	BTEAC	[23]
		$n\text{-}C_4H_9Cl$	mono	75	TBA^+OH^- (St)	[38]
			mono	0	α-phosphoryl sulfoxide	[39]
		$n\text{-}C_4H_9Br$	mono	87	TBA	[40]
			mono	97	TBAI	[40]
			mono	74	BTEAC	[23]
			mono	91	α-phosphoryl sulfoxide	[39]
		$n\text{-}C_4H_9I$	mono	91	α-phosphoryl sulfoxide	[39]
		$i\text{-}C_4H_9Br$	mono	47	BTEAC	[23]
		$s\text{-}C_4H_9Br$	mono	63	BTEAC	[23]
		$n\text{-}C_5H_{11}Br$	mono	72	BTEAC	[23]
		$i\text{-}C_5H_{11}Br$	mono	56	BTEAC	[23]
		$CH_3-CH=CH-CH-CHCl-CH_3$	mono	96	BTEAC	[23]
		$(CH_3)_2C(NO_2)CH_2CH_2Br$	mono (-Br)	20	BTEAC	[41]
		$n\text{-}C_6H_{13}Br$	mono	85	BTEAC	[23]
		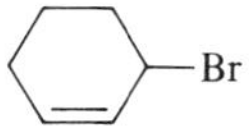 cyclohexenyl-Br	mono	90	BTEAC	[23]

147

Table 10.1 (continued)

Number of carbons	Substrate	Electrophile	Product	% Yield	Catalyst	References
8	$C_6H_5-C\underline{H}_2-CN$	$(C_2H_5)_2NCH_2CH_2Cl$	mono	76	BTEAC	[23, 42]
		$n\text{-}C_7H_{15}Br$	mono	63	BTEAC	[23]
		$C_6H_5CH_2Cl$	{ mono di	50 20 }	BTEAC	[1, 23]
		$C_6H_5CH_2Cl$	di	92	BTEAC	[13]
		$C_6H_5CHClCH_3$	mono	97	BTEAC	[23]
		$C_6H_5CH_2CH_2Cl$	mono	35	BTEAC	[23]
		$(C_6H_5)_2CHCl$	mono	94	BTEAC	[23]
		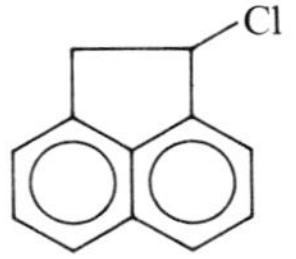 ![acenaphthyl chloride structure]	{ mono di	67 11 }	BTEAC/DMSO	[43]
		CH_2Cl_2	$C_6H_5{-}\underset{CN}{\overset{}{C}H}{-}CH_2{-}\underset{CN}{\overset{}{C}H}{-}C_6H_5$	69	BTEAC	[24]
		CCl_4	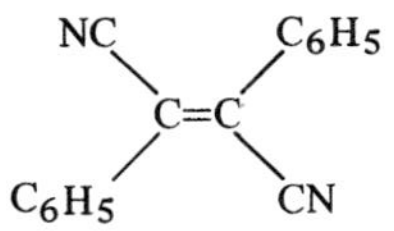 $\underset{C_6H_5}{\overset{NC}{>}}C=C\underset{CN}{\overset{C_6H_5}{<}}$	70	BTEAC	[29]
		$(CH_2-S-CN)_2$	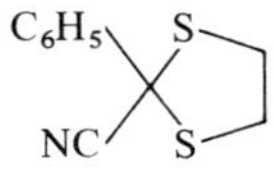 [1,3-dithiolane with C_6H_5 and CN substituents]	45	BTEAC	[30]
		$Cl(CH_2)_3Cl$	mono-chloropropyl	47	BTEAC	[24]
		$Br(CH_2)_3Br$	{ mono-bromopropyl cyclobutane	20 26 }	BTEAC	[24]

Table 10.1 (continued)

Number of carbons	Substrate	Electrophile	Product	% Yield	Catalyst	References
8	$C_6H_5-C\underline{H}_2-CN$	$Cl(CH_2)_4Cl$	cyclopentane	60	BTEAC	[25]
		$Br(CH_2)_4Br$	cyclopentane	88	BTEAC	[24]
				97	TBA^+OH^-	[38]
		$(ClCH_2CH_2)_2O$	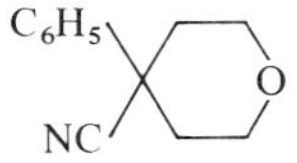 C_6H_5 — (tetrahydropyran with NC)	69	BTEAC	[24]
		$Cl(CH_2)_5Cl$	mono-chloropentyl	40	BTEAC	[24]
		$Br(CH_2)_5Br$	cyclohexane	40	BTEAC	[24]
		$Br(CH_2)_5Br$	mono-bromopentyl	12	BTEAC	[24]
		$Br(CH_2)_7Br$	mono-bromoheptyl	28	BTEAC	[24]
		$C_6H_5CHCl_2$	$C_6H_5-\underset{CN}{CH}-\underset{C_6H_5}{CH}-\underset{CN}{CH}-C_6H_5$	94	BTEAC	[24]
		$Ts-N(CH_2CH_2Cl)_2$	N-tosylpiperidine	68	BTEAC	[44]
		$ClCH_2OCH_3$	$C_6H_5-\underset{CN}{CH}-CH_2-\underset{CN}{CH}-C_6H_5$	88	BTEAC	[28]
		$Cl-\underset{CH_3}{CHOCH_3}$	$C_6H_5-\underset{CN}{CH}-\underset{CH_3}{CH}-\underset{CN}{CH}-C_6H_5$	70	BTEAC	[28]
		$BrCH_2CO_2\text{-}i\text{-}C_3H_7$	mono / di	18 / 12	BTEAC	[45]
		$BrCH_2CO_2\text{-}t\text{-}C_4H_9$	mono / di	27 / 10	BTEAC	[45]
		$ClCH_2CO_2\text{-}c\text{-}C_6H_{11}$	mono / di	17 / 10	BTEAC	[45]

Table 10.1 (continued)

Number of carbons	Substrate	Electrophile	Product	% Yield	Catalyst	References
8	$C_6H_5-CH_2-CN$	$Cl-CH_2-CN$	$C_6H_5C(CH_2CN)_2CN$	41	BTEAC	[46]
		$Cl-(CH_2)_2CN$	$C_6H_5-C(CN)(CH_2CH_2CN)_2$	45	BTEAC	[11]
		$Cl-(CH_2)_3CN$	mono	76	BTEAC	[11]
		$Cl-CH_2-\underset{CH_3}{CH}-CN$	$C_6H_5C(CN)(CH_2-\underset{CH_3}{CH}-CN)_2$	16	BTEAC	[11]
		$Cl(CH_2)_4CN$	mono	75	BTEAC	[11]
		$4-NO_2-C_6H_4Cl$	—	0	BTEAC	[33]
8	$4-F-C_6H_4-CH_2CN$	$ClCH_2CN$	di	36	BTEAC	[46]
	$4-Cl-C_6H_4-CH_2CN$	$ClCH_2CN$	di	83	BTEAC	[46]
	$4-Br-C_6H_4-CH_2CN$	$ClCH_2CN$	di	81	BTEAC	[46]
	$4-I-C_6H_4-CH_2-CN$	$ClCH_2CN$	di	79	BTEAC	[46]
	$C_6H_5-S-CH_2-CN$	CH_3Br (excess)	di	75	BTEAC	[47]
		C_2H_5Br	mono	80	BTEAC	[47]
		$CH_2=CH-CH_2Cl$ (excess)	di	80	BTEAC	[47]
		$n-C_4H_9Br$	mono	82	BTEAC	[47]
		$C_6H_5CH_2Cl$ (excess)	di	82	BTEAC	[47]
		$Br(CH_2)_2Br$	cyclopropane	47	BTEAC	[47]
		$Br(CH_2)_3Cl$	mono-chloropropyl	39	BTEAC	[47]
		$Br(CH_2)_4Br$	cyclopentane	69	BTEAC	[47]
		$Br(CH_2)_5Br$	cyclohexane	50	BTEAC	[47]
9	$C_6H_5-CH(CH_3)CN$	$ClCH_2-O-CH_2Cl$	$(C_6H_5-\underset{CH_3}{\overset{CN}{C}}-CH_2)_2O$	84	BTEAC	[28]
		C_2H_5SCN	$C_6H_5-\underset{CN}{\overset{CH_3}{C}}-SC_2H_5$	77	BTEAC	[30]

Table 10.1 (continued)

Number of carbons	Substrate	Electrophile	Product	% Yield	Catalyst	References
9	C$_6$H$_5$–CH(CH$_3$)CN	HC≡CH	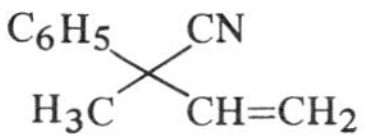 C$_6$H$_5$ CN / H$_3$C CH=CH$_2$	83	BTEAC/DMSO	[31]
		ClCH$_2$CN	mono	73	BTEAC	[11]
		Cl(CH$_2$)$_2$CN	mono	73	BTEAC	[11]
		CH$_3$CHClOCH$_3$	mono	68	BTEAC	[28]
		HC≡C–OC$_2$H$_5$	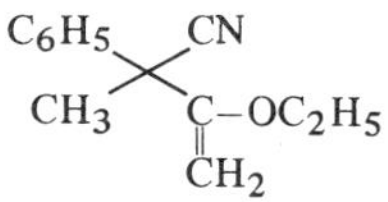 C$_6$H$_5$ CN / CH$_3$ C–OC$_2$H$_5$ / CH$_2$	64	BTEAC	[32]
		Cl(CH$_2$)$_3$CN	mono	83	BTEAC	[11]
		Cl(CH$_2$)$_4$CN	mono	60	BTEAC	[11]
		CH$_3$–CHO–i-C$_3$H$_7$ / Cl	mono	75	BTEAC	[28]
		BrCH$_2$CO$_2$–i-C$_3$H$_7$	mono	74	BTEAC	[14]
		n-C$_5$H$_{11}$–SCN	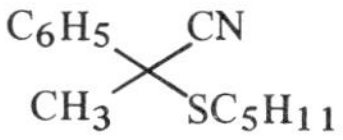 C$_6$H$_5$ CN / CH$_3$ SC$_5$H$_{11}$	68	BTEAC	[30]
		Cl–CH$_2$–CO$_2$–t-C$_4$H$_9$	mono	76	BTEAC	[14]
		HC≡C–S–n-C$_4$H$_9$	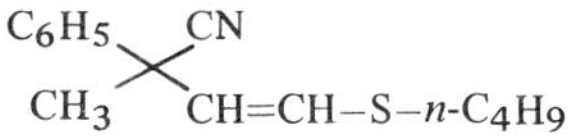 C$_6$H$_5$ CN / CH$_3$ CH=CH–S–n-C$_4$H$_9$	74	BTEAC	[32]
		C$_6$H$_5$CH$_2$Cl	mono	82	BTEAC	[15]
		BrCH$_2$CO$_2$–c-C$_6$H$_{11}$	mono	78	BTEAC	[14]
		4-NO$_2$–C$_6$H$_4$Cl	mono (–Cl)	82	BTEAC	[33]
		2-NO$_2$–C$_6$H$_4$Cl	mono (–Cl)	53	BTEAC	[33]
		3,4-dichloronitrobenzene	mono (–Cl)	92	BTEAC	[33]

Table 10.1 (continued)

Number of carbons	Substrate	Electrophile	Product	% Yield	Catalyst	References
9	$C_6H_5-\underline{CH}(CH_3)CN$	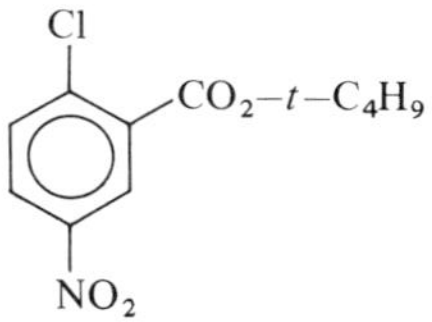 2-chloro-3-(CO_2-t-C_4H_9)-5-nitro-benzene (Cl, CO_2-t-C_4H_9, NO_2)	mono (–Cl)	88	BTEAC	[33]
		2-chloro-5-nitrotoluene	mono (–Cl)	88	BTEAC	[33]
		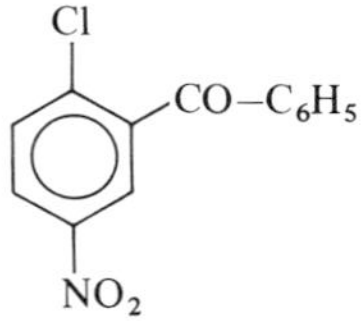 2-chloro-3-(CO–C_6H_5)-5-nitrobenzene (Cl, CO–C_6H_5, NO_2)	mono (–Cl)	67	BTEAC	[34]
		4-Cl–C_6H_4–CO–C_6H_4–NO_2-4	mono (–NO_2)	75	BTEAC	[34]
		4-Br–C_6H_4–CO–C_6H_4–NO_2-4	mono (–NO_2)	69	BTEAC	[34]
		4-CH_3–C_6H_4–CO–C_6H_4–NO_2-4	mono (–NO_2)	93	BTEAC	[34]
		C_6H_5–CO–C_6H_4–NO_2-4	mono (–NO_2)	82	BTEAC	[34]
		4-CH_3O–C_6H_4–CO–C_6H_4–NO_2-4	mono (–NO_2)	64	BTEAC	[34]
		2,4-dinitrochlorobenzene	mono (–Cl)	75	BTEAC	[48]
		3,4-dichloronitrobenzene	mono (–Cl)	92	BTEAC	[48]
		2,5-dichloronitrobenzene	mono (–Cl)	85	BTEAC	[48]
9	$C_6H_5-\underset{\underline{OCH_3}}{\underline{CH}}-CN$	C_2H_5Br	mono	75	BTEAC	[12]
		n-C_3H_7Br	mono	73	BTEAC	[12]
		i-C_3H_7Br	mono	45	BTEAC	[12]

Table 10.1 (continued)

Number of carbons	Substrate	Electrophile	Product	% Yield	Catalyst	References
9	C₆H₅–CH(OCH₃)–CN	CH₂=CH–CH₂Br	mono	73	BTEAC	[12]
		(CH₃)₂C=CH–CH₂Cl	mono	78	BTEAC	[12]
		C₆H₅–CH₂Cl	mono	70	BTEAC	[12]
		BrCH₂Br	mono-bromomethyl	70	BTEAC	[12]
		Br(CH₂)₂Br	mono-bromoethyl	12	BTEAC	[12]
		4-NO₂–C₆H₄–Cl	mono (–Cl)	42	BTEAC	[12]
10	C₆H₅–CH(C₂H₅)–CN	CH₃Cl	mono	90	BTEAC	[13]
		C₂H₅Br	mono	70	BTEAC	[13]
		HC≡CH	C₆H₅\\C₂H₅/C(CN)(CH=CH₂)	80	BTEAC/DMSO	[31, 99]
		CH₂=CHCH₂Cl	mono	93	BTEAC	[13]
		i-C₃H₇Br	mono	43	BTEAC	[13]
		n-C₃H₇Br	mono	59	BTEAC	[13]
		n-C₄H₉Br	mono	29	BTEAC	[13]
		i-C₄H₉Br	mono	40	BTEAC	[13]
		i-C₅H₁₁Br	mono	62	BTEAC	[13]
		(C₂H₅)₂NCH₂CH₂Cl	mono	41	BTEAC	[13]
		n-C₆H₁₃Br	mono	47	BTEAC	[13]
		ClCH₂CO₂-t-C₄H₉	mono	77	BTEAC	[14]
		n-C₇H₁₅Br	mono	66	BTEAC	[13]
		C₆H₅CH₂Cl	mono	94	BTEAC	[13, 15]
		HC≡CC₆H₅	C₆H₅\\C₂H₅/C(CN)(CH=CHC₆H₅)	94	BTEAC	[31]

Table 10.1 (continued)

Number of carbons	Substrate	Electrophile	Product	% Yield	Catalyst	References
10	$C_6H_5-\underset{\underset{C_2H_5}{\|}}{CH}-CN$	[chloro-indane structure]	mono	71	BTEAC/DMSO	[43]
		$(C_6H_5)_2CHCl$	mono	97	BTEAC	[13]
		CH_2Cl_2	mono-chloromethyl	67	BTEAC	[49]
		CH_2Br_2	mono-bromomethyl	71	BTEAC	[49]
		$Br(CH_2)_3Br$	$C_6H_5-\underset{\underset{CN}{\|}}{\overset{\overset{C_2H_5}{\|}}{C}}-CH_2-CH=CH_2$	38	BTEAC	[49]
			mono-bromopropyl	5		
		$Cl(CH_2)_4Cl$	mono-chlorobutyl	65	BTEAC	[49]
		$Br(CH_2)_4Br$	mono-bromobutyl	47	BTEAC	[49]
		$Br(CH_2)_5Br$	mono-bromopentyl	49	BTEAC	[49]
			dinitrile	23		
		$C_6H_5CHCl_2$	mono	69	BTEAC	[49]
		2,4-dinitrochlorobenzene	mono (—Cl)	61	BTEAC	[48]
		3,4-dichloronitrobenzene	mono (—Cl)	61	BTEAC	[48]
		2,5-dichloronitrobenzene	mono (—Cl)	75	BTEAC	[48]
		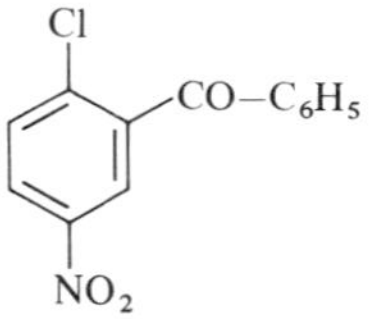 [2-chloro-5-nitro-benzophenone structure with Cl, $CO-C_6H_5$, NO_2]	mono (Cl)	90	BTEAC	[34]

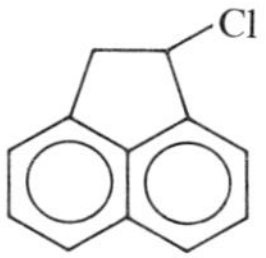

Table 10.1 (continued)

Number of carbons	Substrate	Electrophile	Product	% Yield	Catalyst	References
10	$C_6H_5-\underset{\underset{C_2H_5}{\vert}}{CH}-CN$	$4\text{-}Cl\text{-}C_6H_4-CO-C_6H_4-NO_2\text{-}4$ $4\text{-}Br-C_6H_4-CO-C_6H_4-NO_2\text{-}4$ $C_6H_5-CO-C_6H_4-NO_2\text{-}4$	mono ($-NO_2$) mono ($-NO_2$) mono ($-NO_2$)	70 61 89	BTEAC BTEAC BTEAC	[34] [34] [34]
10	$C_6H_5-\underset{\underset{CN}{\vert}}{CH}-CH_2CN$	$Cl-CH_2-CN$	$C_6H_5C(CN)(CH_2CN)_2$	74	BTEAC	[46]
	$C_6H_5-\underset{\underset{OCH_2CH_2Cl}{\vert}}{CH}-CN$	—	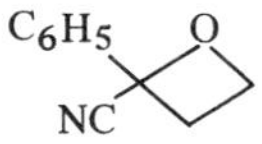 $C_6H_5\underset{NC}{\overset{O}{\diamond}}$	50	BTEAC	[12]
	$C_6H_5-\underset{\underset{N(CH_3)_2}{\vert}}{CH}-CN$	C_2H_5Br $i\text{-}C_3H_7Br$ $CH_2=CH-CH_2Cl$ $n\text{-}C_4H_9Br$ $C_6H_5-CH_2Cl$	mono mono mono mono mono	56 62 75 75 82	BTEAC BTEAC BTEAC BTEAC BTEAC	[50] [50] [50] [50] [50]
	indole with CH_2CN at position 3 (α), N-H	$C_6H_5-CH_2Cl$	1-benzyl (at 60 °C) 1,α-tribenzyl (at 100 °C)	no yield reported	BTMAC BTMAC	[51] [51]
11	$C_6H_5-\underset{\underset{n\text{-}C_3H_7}{\vert}}{CH}-CN$	$HC\equiv C-S-n\text{-}C_4H_9$	$\underset{n\text{-}C_3H_7}{\overset{C_6H_5}{>}}\underset{CH=CH-S-n\text{-}C_4H_9}{\overset{CN}{<}}$	68	BTEAC	[32]

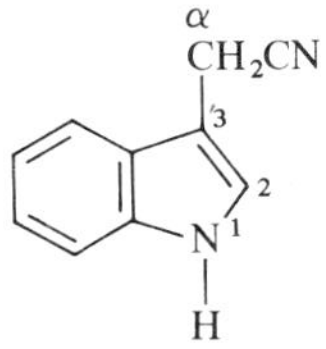

Table 10.1 (continued)

Number of carbons	Substrate	Electrophile	Product	% Yield	Catalyst	References
11	$C_6H_5-\underline{CH}-CN$ $\|$ $n-C_3H_7$	$ClCH_2CO_2-t-C_4H_9$ $C_6H_5CH_2Cl$ $4-NO_2-C_6H_4Cl$	mono mono mono (—Cl)	74 94 63	BTEAC BTEAC BTEAC	[14] [15] [33]
		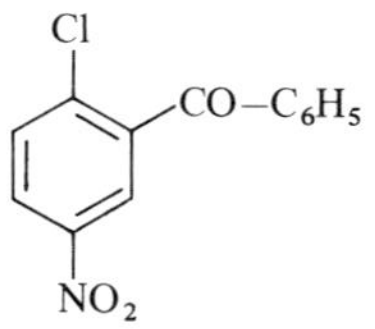 2-Cl, 5-NO_2-benzophenone (Cl, CO-C_6H_5, NO_2 substituted benzene)	mono (—Cl)	60	BTEAC	[34]
		$4-Cl-C_6H_4-CO-C_6H_4-NO_2-4$	mono (—NO_2)	78	BTEAC	[34]
	$C_6H_5-\underline{CH}-CN$ $\|$ $i-C_3H_7$	$HC{\equiv}CH$	$C_6H_5\underset{i-C_3H_7}{\overset{CN}{\diagup\!\!\!\diagdown}}CH{=}CH_2$	82	BTEAC/DMSO	[31]
		$Cl-CH_2OCH_3$	mono	7	BTEAC	[28]
		$HC{\equiv}C-S-n-C_4H_9$	$C_6H_5\underset{i-C_3H_7}{\overset{CN}{\diagup\!\!\!\diagdown}}CH{=}CH-S-n-C_4H_9$	74	BTEAC	[32]
		$BrCH_2CO_2-t-C_4H_9$ $C_6H_5CH_2Cl$ $HC{\equiv}C-C_6H_5$	mono mono $C_6H_5\underset{i-C_3H_7}{\overset{CN}{\diagup\!\!\!\diagdown}}CH{=}CH-C_6H_5$	56 85 83	BTEAC BTEAC BTEAC	[14] [15] [31]

Table 10.1 (continued)

Number of carbons	Substrate	Electrophile	Product	% Yield	Catalyst	References
11	$i\text{-}C_3H_7$ $C_6H_5-\underline{CH}-CN$	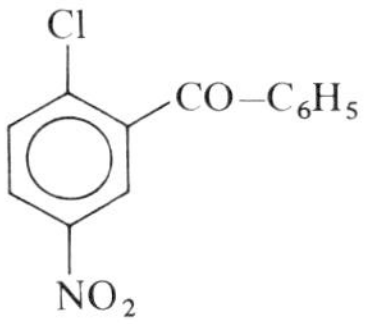 2-Cl, 5-NO$_2$-C$_6$H$_3$-CO-C$_6$H$_5$	mono (–Cl)	60	BTEAC	[34]
		2,4-dinitrochlorobenzene	mono (–Cl)	20	BTEAC	[48]
		3,4-dichloronitrobenzene	mono (–Cl)	30	BTEAC	[48]
	$OCH_2-CH=CH_2$ $C_6H_5-\underline{CH}-CN$	CH_3CH_2Br	mono	44	BTEAC	[12]
	$n\text{-}C_4H_9$ $C_6H_5-\underline{CH}-CN$	$BrCH_2CO_2\text{-}t\text{-}C_4H_9$	mono	69	BTEAC	[14]
		$C_6H_5CH_2Cl$	mono	57	BTEAC	[15]
12	$C_6H_5-\overset{H}{\underset{|}{C}}-CN$ $\quad\;\; CH_2$ $\quad\;\; CH(CH_3)_2$	$C_6H_5CH_2Cl$	mono	80	BTEAC	[15]
12	$C_6H_5-\overset{H}{\underset{|}{C}}-CN$ $\quad\;\; CH$ $\quad CH_3\;\;\;C_2H_5$	$C_6H_5CH_2Cl$	mono	56	BTEAC	[15]

Alkylation Reactions

Table 10.1 (continued)

Number of carbons	Substrate	Electrophile	Product	% Yield	Catalyst	References
12	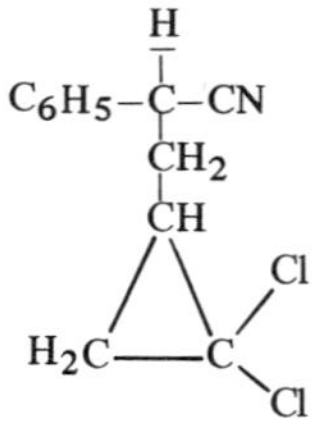 C$_6$H$_5$–CH(H)–CN with CH$_2$–CH–CCl$_2$ (cyclopropane ring, H$_2$C)	(C$_6$H$_5$)$_2$CHCl	mono	50	BTEAC	[29]
	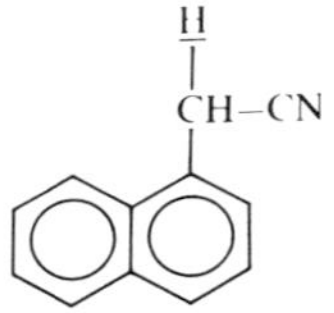 CH–CN attached to naphthyl	CH$_3$Br	mono	92	BTEAC	[16]
		C$_2$H$_5$Br	mono	82	BTEAC	[16]
		n-C$_3$H$_7$Br	mono	90	BTEAC	[16]
		i-C$_3$H$_7$Br	mono	83	BTEAC	[16]
		CH$_2$=CHCH$_2$Cl	mono	88	BTEAC	[16]
		n-C$_4$H$_9$Br	mono	91	BTEAC	[16]
		(C$_2$H$_5$)$_2$NCH$_2$CH$_2$Cl	mono	85	BTEAC	[16]
		C$_6$H$_5$CH$_2$Cl	mono	95	BTEAC	[16]
		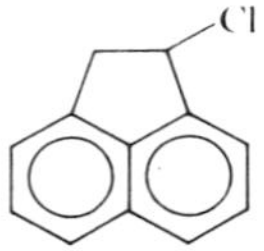 Cl-substituted acenaphthyl	mono / di	58 / 7	BTEAC/DMSO	[43]
		CH$_2$Br$_2$	dinitrile	80	BTEAC	[16]
		Br(CH$_2$)$_2$Br	cyclopropane	67	BTEAC	[16]
		Br(CH$_2$)$_3$Br	cyclobutane	21	BTEAC	[16]
		Br(CH$_2$)$_4$Br	cyclopentane	80	BTEAC	[16]
		Br(CH$_2$)$_5$Br	cyclohexane	17	BTEAC	[16]

Table 10.1 (continued)

Number of carbons	Substrate	Electrophile	Product	% Yield	Catalyst	References
12	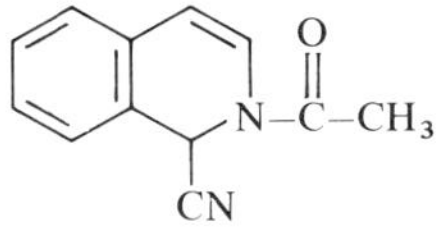 1-(1-cyano)-2-acetyl-1,2-dihydroisoquinoline (N-acetyl, 1-CN on isoquinoline)	cyclohexanone	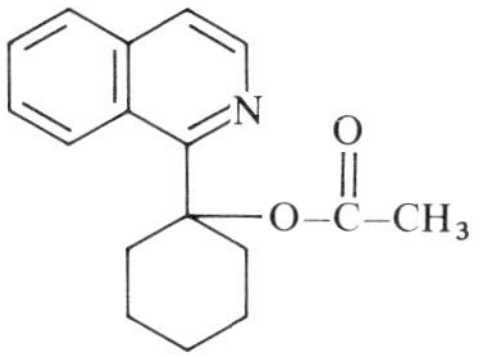 1-isoquinolyl-1-(1-acetoxycyclohexyl) derivative	30	BTEAC	[36]
13	$C_6H_5-CH(CN)-n\text{-}C_5H_{11}$	$HC{\equiv}CH$	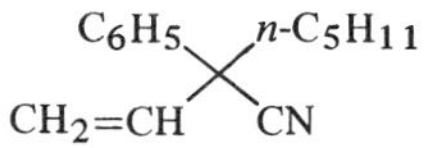 $CH_2{=}CH-C(C_6H_5)(n\text{-}C_5H_{11})CN$	88	BTEAC/DMSO	[31]
		$ClCH_2CO_2\text{-}t\text{-}C_4H_9$	mono	64	BTEAC	[14]
		$C_6H_5CH_2Cl$	mono	94	BTEAC	[15]
		$ClCH_2CO_2\text{-}c\text{-}C_6H_{11}$	mono	26	BTEAC	[14]
	$C_6H_5-CH(CN)-CH_2-CH_2-CH(CH_3)_2$	$C_6H_5CH_2Cl$	mono	78	BTEAC	[15]
	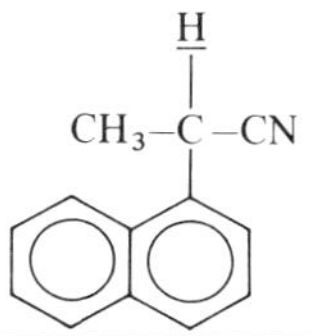 $CH_3-CH(CN)-$(1-naphthyl)	CH_3Br	mono	64	BTEAC	[16]
		C_2H_5Br	mono	61	BTEAC	[16]

Table 10.1 (continued)

Number of carbons	Substrate	Electrophile	Product	% Yield	Catalyst	References
14	$(C_6H_5)_2C\underline{H}CN$	CH_3Br	mono	94	BTEAC	[52]
		C_2H_5Br	mono	92	BTEAC	[52]
		$ClCH_2OCH_3$	mono	80	BTEAC	[28]
		$ClCH_2CN$	mono	88	BTEAC	[11]
		$Cl(CH_2)_2CN$	mono	90	BTEAC	[11]
		$CH_2=CH-CH_2Br$	mono	95	BTEAC	[11, 52]
		$CH_3-\underset{Cl}{C}H-O-CH_3$	mono	50	BTEAC	[28]
		$HC{\equiv}C-O-C_2H_5$	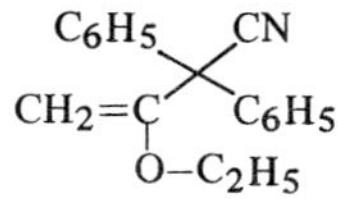 $\underset{CH_2=C}{\overset{C_6H_5}{}}\!\!\!\!\!\!\!{\underset{O-C_2H_5}{\overset{CN}{\diagdown}}}$ C_6H_5	77	BTEAC	[32]
		$Cl-(CH_2)_3CN$	mono	74	BTEAC	[11]
		$n-C_4H_9Br$	mono	94	BTEAC	[52]
		$Cl-CH_2-O-i-C_3H_7$	mono	88	BTEAC	[28]
		$n-C_5H_{11}Br$	mono	94	BTEAC	[52]
		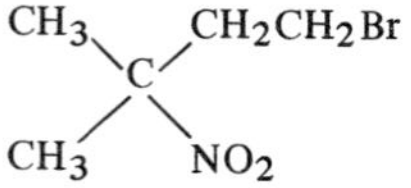 $CH_3\diagdown\diagup CH_2CH_2Br$ C $CH_3\diagup\diagdown NO_2$	mono (−Br)	82	BTEAC	[41]
		$Cl(CH_2)_4CN$	mono	78	BTEAC	[11]
		$n-C_6H_{13}Br$	mono	92	BTEAC	[52]
		$Cl-CH_2-CO_2-t-C_4H_9$	mono	96	BTEAC	[14]
		$HC{\equiv}C-S-n-C_4H_9$	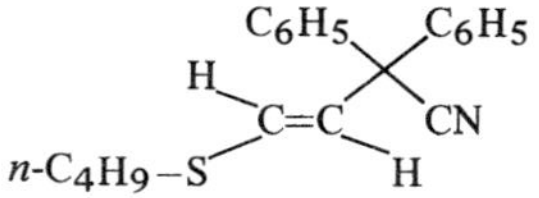 $\underset{n-C_4H_9-S}{\overset{C_6H_5}{\diagdown}}\!\!\!\!\!C{=}C\!\!\!\!\!\underset{H}{\overset{C_6H_5,\ CN}{\diagup}}$	55	BTEAC	[32]

Table 10.1 (continued)

Number of carbons	Substrate	Electrophile	Product	% Yield	Catalyst	References
14	$(C_6H_5)_2CHCN$	$(C_2H_5)_2NCH_2CH_2Cl$	mono	85	BTEAC	[52]
		$C_6H_5CH_2Cl$	mono	98	BTEAC	[52]
		$C_6H_5CHClCH_3$	mono	96	BTEAC	[52]
		$(C_6H_5)_2CHCl$	mono	96	BTEAC	[52]
		CCl_4	$(C_6H_5)_2\underset{CN}{\overset{CN}{C}}-C(C_6H_5)_2$	91	BTEAC	[29]
		CH_2Cl_2	mono-chloromethyl	82	BTEAC	[49]
		CH_2Br_2	mono-bromomethyl	93	BTEAC	[49]
		$ClCH_2OCH_2Cl$	dimer	90	BTEAC	[28]
		$Br(CH_2)_2Br$	mono-bromoethyl	91	BTEAC	[49]
		$Br(CH_2)_3Br$	mono-bromopropyl	87	BTEAC	[49]
			dinitrile	64	BTEAC	[49]
		$Br(CH_2)_4Br$	mono-bromobutyl	79	BTEAC	[49]
			dinitrile	13	BTEAC	[49]
		$Br(CH_2)_5Br$	mono-bromopentyl	72	BTEAC	[49]
			dinitrile	16		
		4-chloronitrobenzene	mono (—Cl)	71	BTEAC	[33]
		2,4-dinitrochlorobenzene	mono (—Cl)	91	BTEAC	[48]
		3,4-dichloronitrobenzene	mono (—Cl)	60	BTEAC	[48]
	$C_6H_5-\underset{n\text{-}C_6H_{13}}{\overset{CN}{\underset{|}{C}}-H}$	$C_6H_5CH_2Cl$	mono	75	BTEAC	[15]
14	C_6H_5CHCN $(CH_2)_2$ $N(C_2H_5)_2$	C_2H_5Br	mono	84	BTEAC	[50]
		$n\text{-}C_4H_9Br$	mono	80	BTEAC	[50]
		$i\text{-}C_5H_{11}Br$	mono	66	BTEAC	[50]

Table 10.1 (continued)

Number of carbons	Substrate	Electrophile	Product	% Yield	Catalyst	References
14	$\text{C}_6\text{H}_5-\underset{\underset{\text{N(C}_2\text{H}_5)_2}{\overset{\mid}{(\text{CH}_2)_2}}}{\overset{\text{CN}}{\overset{\mid}{\text{C}}}}-\text{H}$	$\text{BrCH}_2\text{CO}_2-t\text{-C}_4\text{H}_9$	mono	80	BTEAC	[14]
		$(\text{C}_2\text{H}_5)_2\text{NCH}_2\text{CH}_2\text{Cl}$	mono	34	BTEAC	[50]
		$\text{C}_6\text{H}_5\text{CH}_2\text{Cl}$	mono	88	BTEAC	[50]
		$\text{HC}\equiv\text{C}-\text{C}_6\text{H}_5$	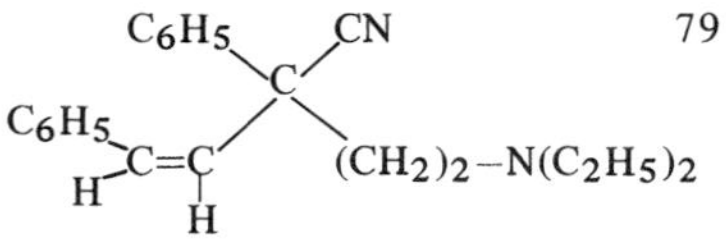 $\underset{\text{H}}{\overset{\text{C}_6\text{H}_5}{\text{C}}}=\underset{\text{H}}{\overset{\overset{\text{CN}}{\mid}}{\underset{\mid}{\text{C}}}}-(\text{CH}_2)_2-\text{N(C}_2\text{H}_5)_2$	79	BTEAC/DMSO	[31]
	$\text{C}_2\text{H}_5-\overset{\text{CN}}{\underset{\mid}{\text{C}}}-\text{H}$ (naphthyl)	$\text{C}_2\text{H}_5\text{Br}$	mono	31	BTEAC	[16]
		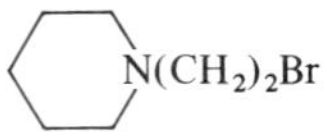 $\text{N(CH}_2)_2\text{Br}$ (piperidine)	mono	42	BTEAC	[16]
	$\text{C}_6\text{H}_5-\underset{\underset{\text{CN}}{\mid}}{\overset{\overset{\text{O}-\text{C}_6\text{H}_5}{\mid}}{\text{C}}}-\text{H}$	$\text{C}_2\text{H}_5\text{Br}$	mono	63	BTEAC	[12]
		$n\text{-C}_3\text{H}_7\text{Br}$	mono	66	BTEAC	[12]
		$i\text{-C}_3\text{H}_7\text{Br}$	mono	30	BTEAC	[12]
		$\text{CH}_2=\text{CH}-\text{CH}_2\text{Br}$	mono	71	BTEAC	[12]
		$(\text{CH}_3)_2\text{C}=\text{CH}-\text{CH}_2\text{Cl}$	mono	69	BTEAC	[12]
		CH_2Br_2	mono-bromomethyl	70	BTEAC	[12]
		$\text{Br(CH}_2)_2\text{Br}$	mono-bromoethyl	64	BTEAC	[12]
		$\text{Br(CH}_2)_3\text{Br}$	mono-bromopropyl	15	BTEAC	[12]

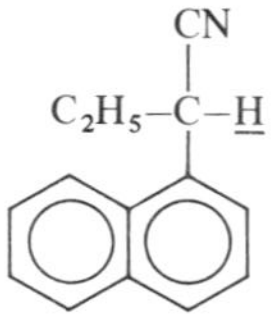

Table 10.1 (continued)

Number of carbons	Substrate	Electrophile	Product	% Yield	Catalyst	References
14	$C_6H_5-\underset{\underset{H}{|}}{\overset{\overset{CN}{|}}{C}}$	C_2H_5Br	mono	81	BTEAC	[53]
		$n-C_3H_7Br$	mono	71	BTEAC	[53]
		$CH_2=CH-CH_2Cl$	mono	75	BTEAC	[53]
	$CH_3-CH-O-n-C_4H_9$	CH_2Br_2	mono-bromomethyl	73	BTEAC	[53]
15	$C_6H_5-\underset{\underset{C_6H_5}{\overset{|}{CH_2}}}{\overset{\overset{CN}{|}}{C}}-CN$	$HC{\equiv}C-OC_2H_5$	$\underset{CH_2=C}{C_6H_5}\underset{OC_2H_5}{\overset{CN}{\diagdown}}\overset{CH_2-C_6H_5}{\diagup}$	74	BTEAC	[32]
		$HC{\equiv}C-C_6H_5$	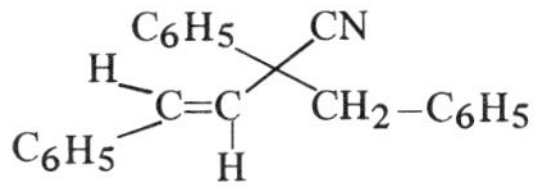 $\underset{C_6H_5}{\overset{C_6H_5}{\diagdown}}C=C\underset{H}{\overset{CH_2-C_6H_5}{\diagup}}$ H	98	BTEAC	[31]
		$HC{\equiv}C-S-n-C_4H_9$	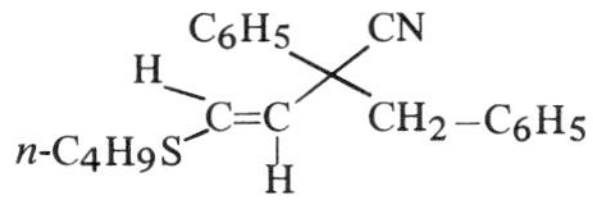 $\underset{n-C_4H_9S}{\overset{C_6H_5}{\diagdown}}C=C\underset{H}{\overset{CH_2-C_6H_5}{\diagup}}$	78	BTEAC	[32]
		$ClCH_2CO_2-t-C_4H_9$	mono	95	BTEAC	[14]
		4-chloronitrobenzene	mono (−Cl)	87	BTEAC	[33]
		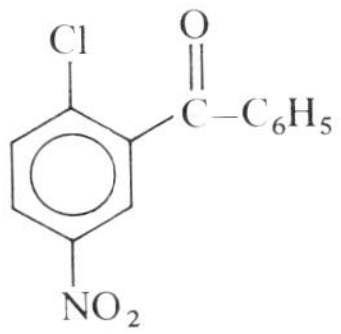 ![structure: 2-chloro-5-nitrobenzoyl phenyl ketone]	mono (−Cl)	62	BTEAC	[34]

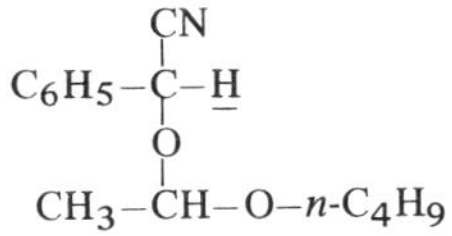

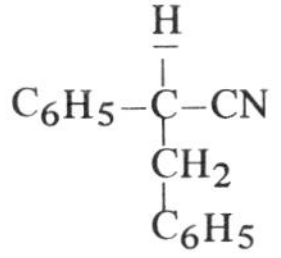

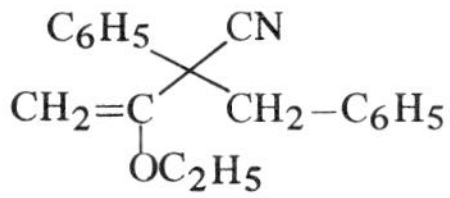

Alkylation Reactions

Table 10.1 (continued)

Number of carbons	Substrate	Electrophile	Product	% Yield	Catalyst	References
15	C₆H₅–C(H)(CH₂C₆H₅)–CN	4-Cl–C₆H₄–C(=O)–C₆H₄–NO₂-4	mono (–NO₂)	67	BTEAC	[34]
		4-Br–C₆H₄–C(=O)–C₆H₄–NO₂-4	mono (–NO₂)	90	BTEAC	[34]
		4-CH₃–C₆H₄–C(=O)–C₆H₄–NO₂-4	mono (–NO₂)	98	BTEAC	[34]
		C₆H₅–C(=O)–C₆H₄–NO₂-4	mono (–NO₂)	92	BTEAC	[34]
		4-CH₃O–C₆H₄–C(=O)–C₆H₄–NO₂-4	mono (–NO₂)	90	BTEAC	[34]
		2,4-dinitrochlorobenzene	mono (–Cl)	86	BTEAC	[48]
		3,4-dichloronitrobenzene	mono (–Cl)	82	BTEAC	[48]
		2,5-dichloronitrobenzene	mono (–Cl)	98	BTEAC	[48]
	C₆H₅–C(H)(O–CH₂–C₆H₅)–CN	C₂H₅Br	mono	71	BTEAC	[12]
	NC–C(H)(naphthyl)–CH₂–CH=CH₂	CH₂=CH–CH₂Cl	mono	80	BTEAC	[16]

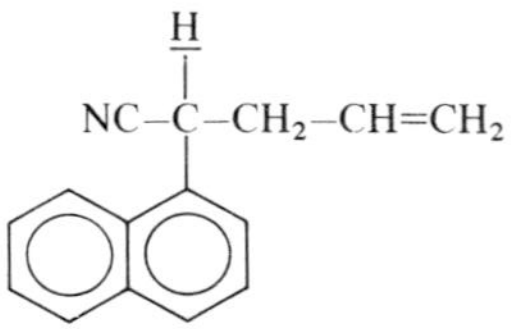

Table 10.1 (continued)

Number of carbons	Substrate	Electrophile	Product	% Yield	Catalyst	References
16	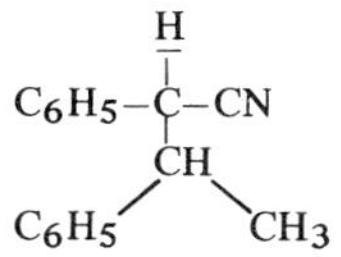 C₆H₅–C(H)(CN)–CH(CH₃)(C₆H₅)	HC≡C–S–n-C₄H₉	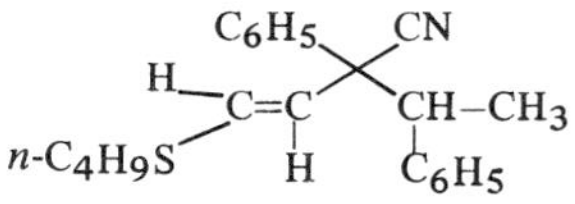 n-C₄H₉S–C(H)=C(C₆H₅)(CN)–CH(CH₃)(C₆H₅) (with H)	64	BTEAC	[32]
17	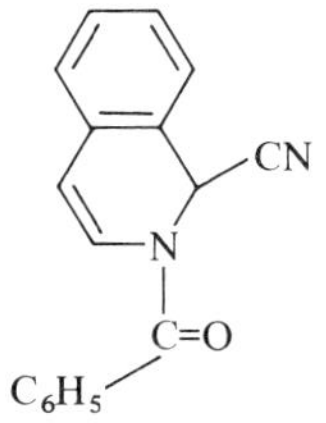 1-cyano-2-benzoyl-1,2-dihydroisoquinoline	C₂H₅Br	mono	76	BTEAC	[54]
		n-C₃H₇Br	mono	82	BTEAC	[54]
		n-C₄H₉Br	mono	78	BTEAC	[54]
		C₆H₅CH₂Cl	mono	78	BTEAC	[54]
		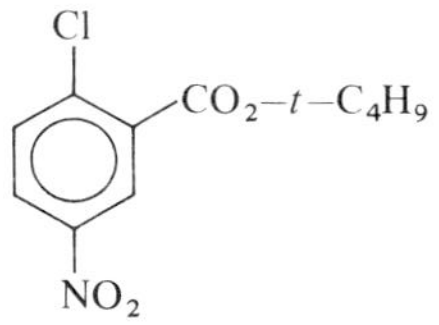 2-Cl-5-NO₂-C₆H₃-CO₂-t-C₄H₉	mono	46	BTEAC	[54]
		C₆H₅CHO	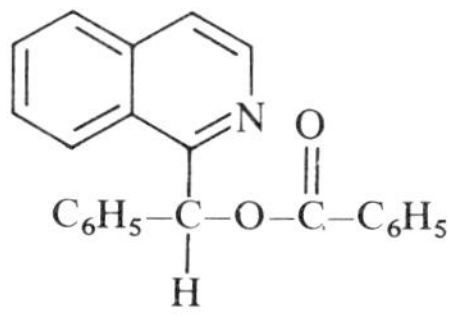 isoquinoline-N-oxide C₆H₅–CH(–O–C(=O)–C₆H₅)–	89	BTEAC	[36]

165

Table 10.1 (continued)

Number of carbons	Substrate	Electrophile	Product	% Yield	Catalyst	References
17	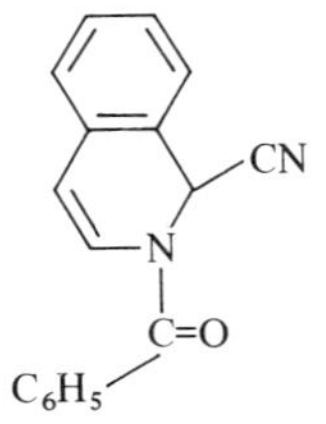 [isoquinoline with CN and N-C(=O)C$_6$H$_5$ structure]	2-CH$_3$-C$_6$H$_4$-CHO	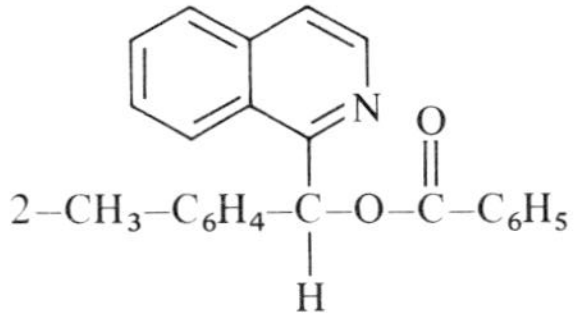 2-CH$_3$-C$_6$H$_4$-C(H)(O-C(=O)-C$_6$H$_5$)-[isoquinolinyl]	85	BTEAC	[36]
		2-Cl-C$_6$H$_4$-CHO	as above	89	BTEAC	[36]
		4-Cl-C$_6$H$_4$-CHO	as above	84	BTEAC	[36]
		2-furyl-CHO	as above	79	BTEAC	[36]
		i-C$_3$H$_7$-CHO	as above	83	BTEAC	[36]
		C$_6$H$_5$COCH$_3$	as above	33	BTEAC/HMPT	[36]
		CH$_3$COCH$_3$	as above	78	BTEAC	[36]
		cyclohexanone	as above	61	BTEAC	[36]
18	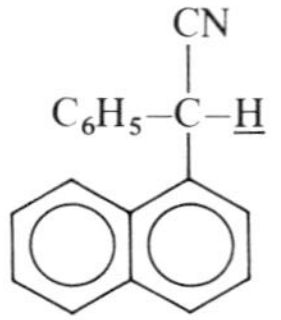 C$_6$H$_5$–C(CN)(H)–[naphthyl]	CH$_3$Br	mono	95	BTEAC	[16]
		C$_2$H$_5$Br	mono	91	BTEAC	[16]
		i-C$_3$H$_7$Br	mono	50	BTEAC	[16]
		CH$_2$=CH–CHCl	mono	99	BTEAC	[16]
		n-C$_4$H$_9$Br	mono	88	BTEAC	[16]
		(C$_2$H$_5$)$_2$NCH$_2$CH$_2$Br	mono	66	BTEAC	[16]
		Cl(CH$_2$)$_2$CO$_2$-i-C$_3$H$_7$	mono	48	BTEAC	[16]

The Substances Alkylated

Table 10.1 (continued)

Number of carbons	Substrate	Electrophile	Product	% Yield	Catalyst	References
18	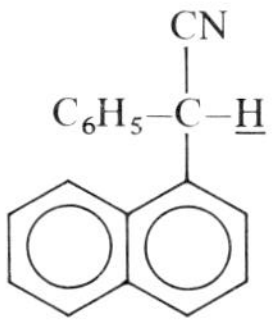 C₆H₅—C(CN)(H)—(1-naphthyl)	C₆H₅CH₂Cl	mono	99	BTEAC	[16]
		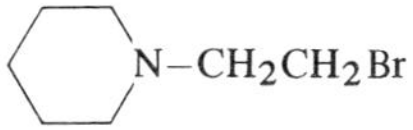 N-piperidyl-CH₂CH₂Br	mono	42	BTEAC	[16]
		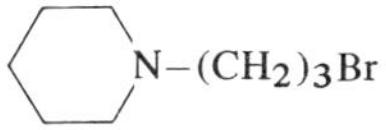 N-piperidyl-(CH₂)₃Br	mono	99	BTEAC	[16]
		(C₆H₅)₂CHCl	mono	50	BTEAC	[16]
		CH₂Br₂	mono-bromomethyl	85	BTEAC	[16]
		Br(CH₂)₂Br	mono-bromoethyl	88	BTEAC	[16]
		Br(CH₂)₄Br	mono-bromobutyl	92	BTEAC	[16]
19	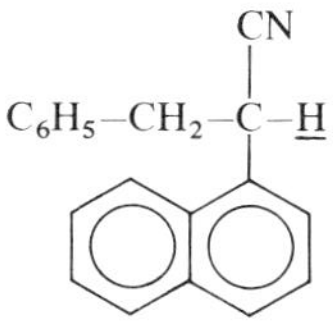 C₆H₅—CH₂—C(CN)(H)—(1-naphthyl)	CH₃Br	mono	69	BTEAC	[16]
		C₆H₅CH₂Cl	mono	78	BTEAC	[16]
20	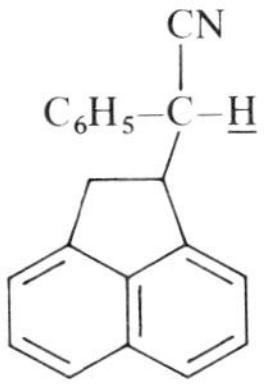 C₆H₅—C(CN)(H)—(acenaphthyl)	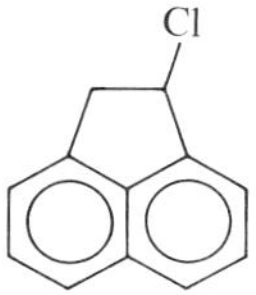 chloro-acenaphthyl	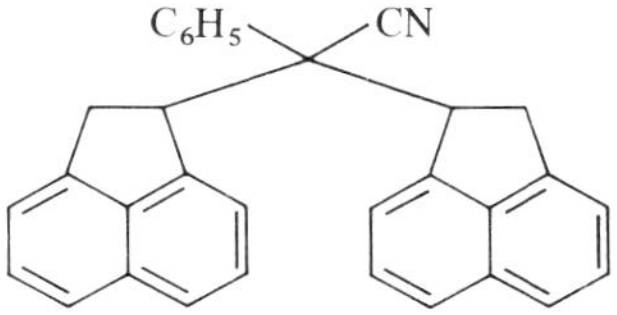 bis(acenaphthyl) product with C₆H₅ and CN	31	BTEAC/DMSO	[43]

Table 10.1 (continued)

Number of carbons	Substrate	Electrophile	Product	% Yield	Catalyst	References
21	C₆H₅–C(H)(CN)–CH(C₆H₅)₂	HC≡C–O–C₂H₅	C₆H₅–C(CN)(–CH(C₆H₅)₂)–CH=CH–OC₂H₅ (CH₂=C–OC₂H₅ shown)	100	BTEAC	[32]
		HC≡C–S–n-C₄H₉	C₆H₅–C(CN)(–CH(C₆H₅)₂)–C(S-n-C₄H₉)=CH₂	96	BTEAC	[32]

Table 10.2. Alkylation of simple aldehydes, esters, ketones and sulfones

Number of carbons	Substrate	Electrophile	Product	% Yield	Catalyst	References
2	$CH_3SO_2CH_3$	C_6H_5CHO	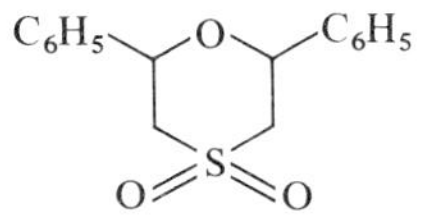 (1,3-dioxathiane dioxide with C_6H_5 groups)	48	BTEAC	[55]
3	$(CH_3)_3\overset{+}{S}=O\ I^-$	C_6H_5CHO	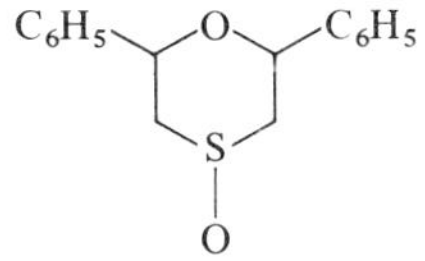 (1,3-oxathiane oxide with C_6H_5 groups)	12	TBAI	[56]
4	$(CH_3)_2CH-\underline{CH}-CHO$	CH_3I	mono	15	TBAI	[19]
		$CH_2=CH-CH_2Cl$	mono	56	TBAI	[19]
		$CH_2=CH-CH_2Br$	mono	35	TBAI	[19]
		$CH_3-CH=CH-CH_2Cl$	mono	54	TBAI	[19]
		$C_6H_5CH_2Cl$	mono	75	TBAI	[19]
5	cyclopentanone	$(CH_3)_2C=CH-CH_2Cl$	mono	19	BTEAC	[20]
5	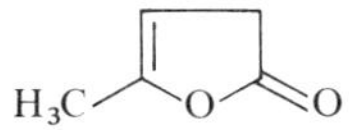 (β-methyl butenolide)	CH_3I	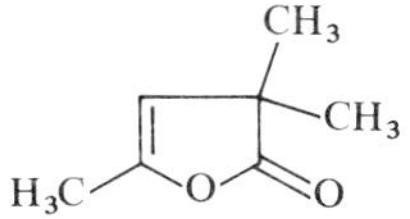 (dimethyl butenolide)	12	TBA^+OH^- (St)	[58]
			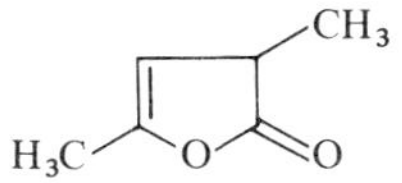 (methyl butenolide isomer)	8		
			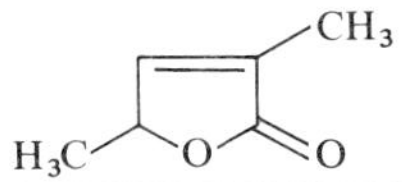 (methyl butenolide isomer)	6		

Table 10.2 (continued)

Number of carbons	Substrate	Electrophile	Product	% Yield	Catalyst	References
5	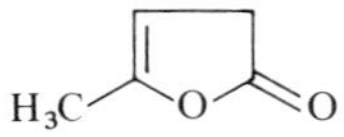 H₃C-furanone	[CH₃O]₂SO₂	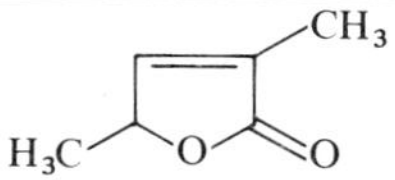 methylated furanone	7	TBA⁺OH⁻ (St)	[58]
5	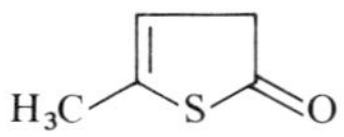 H₃C-thiolanone	CH₃I	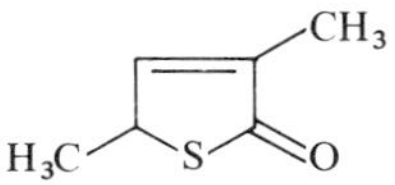 CH₃-thiolanone ketone product; H₃C-thiophene-OCH₃	45; 6	TBA⁺OH⁻ (St)	[58, 59]
5	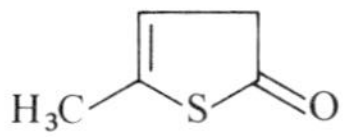 H₃C-thiolanone	[CH₃O]₂SO₂	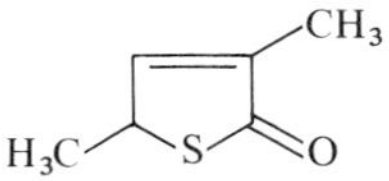 CH₃-thiolanone ketone product; H₃C-thiophene-OCH₃	3.5; 80	TBA⁺OH⁻ (St)	[58, 59]
5	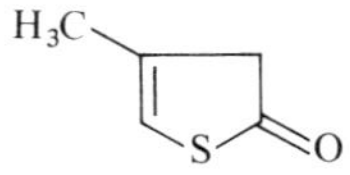 H₃C-thiolanone	CH₃I	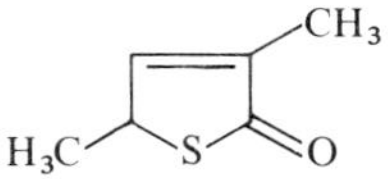 CH₃-thiolanone ketone product; H₃C-thiophene-OCH₃	41; 4.5	TBA⁺OH⁻ (St)	[58]
6	cyclohexanone	(CH₃)₂C=CH–CH₂Cl	mono	46	BTEAC	[20]

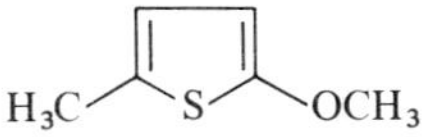

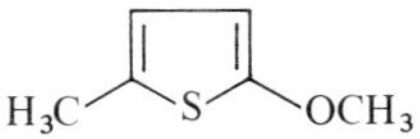

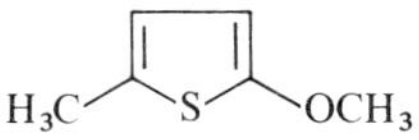

Table 10.2 (continued)

Number of carbons	Substrate	Electrophile	Product	% Yield	Catalyst	References
6	ClCH$_2$CO$_2$-t-C$_4$H$_9$	CH$_2$=CH–CN	 NC–△–CO$_2$-t-C$_4$H$_9$	45 E:Z 71:29	BTEAC	[37]
7	2-methylcyclohexanone	(CH$_3$)$_2$C=CH–CH$_2$Cl	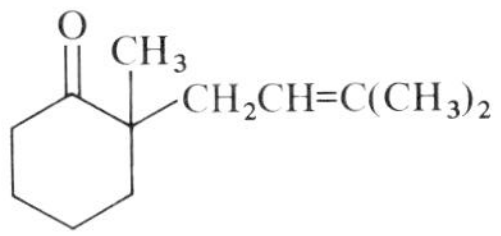 (2-methyl-2-(3-methylbut-2-enyl)cyclohexanone)	30	BTEAC	[20]
7	3-methylcyclohexanone	(CH$_3$)$_2$C=CH–CH$_2$Cl	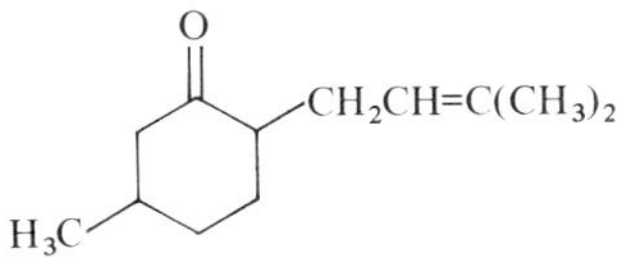 (2-(3-methylbut-2-enyl)-5-methylcyclohexanone)	50	BTEAC	[20]
7	4-methylcyclohexanone	(CH$_3$)$_2$C=CH–CH$_2$Cl	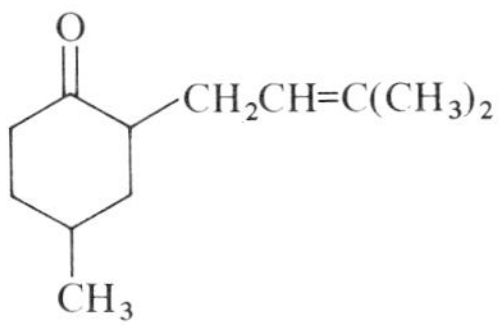 (2-(3-methylbut-2-enyl)-4-methylcyclohexanone)	51	BTEAC	[20]
7	C$_6$H$_5$SO$_2$CH$_3$	C$_6$H$_5$CHO	C$_6$H$_5$SO$_2$–CH=CH–C$_6$H$_5$	86	BTEAC	[60]
		4-Cl–C$_6$H$_4$CHO	as above	98	BTEAC	[60]
		C$_6$H$_5$–CH=CH–CHO	as above	30	BTEAC	[60]
		β-naphthyl–CHO	as above	85	BTEAC	[60]
7	C$_6$H$_5$SO$_2$CHCl$_2$	C$_2$H$_5$Br	mono	72	BTEAC	[61]
		C$_6$H$_5$CH$_2$Cl	mono	84	BTEAC	[61]

Table 10.2 (continued)

Number of carbons	Substrate	Electrophile	Product	% Yield	Catalyst	References
7	$C_6H_5SO_2CHBr_2$	$C_6H_5CH_2Cl$	mono	75	BTEAC	[61]
7	$(CH_3)_3Si-CH_2-CO_2C_2H_5$	crotonaldehyde	$(CH_3)_3SiO$ $CH_3CH=CH-CHCH_2-CO_2C_2H_5$	69	KF/DC-18-C-6 or TBAF	[9]
		E-2-hexenal	as above	82	KF/DC-18-C-6 or TBAF	[9]
		C_6H_5-CHO	as above	76	KF/DC-18-C-6 or TBAF	[9]
		citral	as above	79	KF/DC-18-C-6 or TBAF	[9]
		$C_6H_5CH=CH-CHO$	as above	81	KF/DC-18-C-6 or TBAF	[9]
		benzophenone	as above	88	KF/DC-18-C-6 or TBAF	[9]
8	$C_6H_5-\overset{O}{\underset{\|}{C}}-CH_3$	CH_3I	mono di	13 69	α-phosphoryl-sulfoxide	[39]
		$(CH_3)_2C=CH-CH_2Cl$	mono	40	BTEAC	[20]
8	$n-C_4H_9-\overset{H}{\underset{C_2H_5}{C}}-CHO$	$CH_2=CH-CH_2Cl$	C-alkylation	85	TBAI	[19]
		$(CH_3)_2CHBr$	O-alkylation	21	TBAI	[19]
		$CH_3CH=CH-CH_2Cl$	C-alkylation O-alkylation	65 25	TBAI	[19]
		$(CH_3)_2C=CH-CH_2Cl$	C-alkylation O-alkylation	60 30	TBAI	[19]

Table 10.2 (continued)

Number of carbons	Substrate	Electrophile	Product	% Yield	Catalyst	References
8	$n\text{-}C_4H_9-\overset{H}{\underset{|}{C}}-CHO$ $\phantom{n\text{-}C_4H_9-}C_2H_5$	$C_6H_5CH_2Cl$	C-alkylation O-alkylation	55 35	TBAI	[19]
8	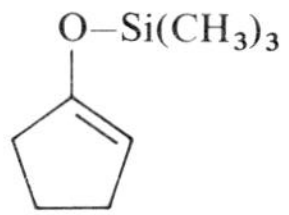 $O-Si(CH_3)_3$ (cyclopentenyl)	$CH_2=CH-CH_2Br$	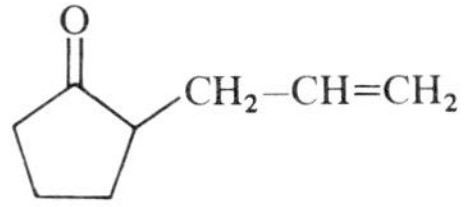 cyclopentanone with $CH_2-CH=CH_2$	56	BTMAF	[57]
8	(same cyclopentenyl silyl ether)	$C_6H_5CH_2Br$	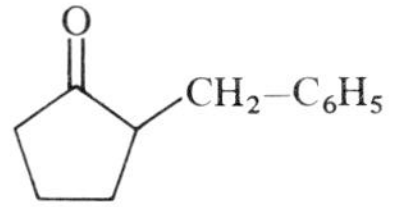 cyclopentanone with $CH_2-C_6H_5$	67	BTMAF	[57]
8	$C_6H_5-\overset{H}{\underset{|}{N}}-\overset{O}{\underset{\|}{C}}-CH_3$	$(CH_3O)_2SO_2$ $(C_2H_5O)_2SO_2$ $n\text{-}C_3H_7Br$ $n\text{-}C_4H_9I$ $C_6H_5CH_2Cl$	mono mono mono mono mono	84 90 82 82 95	BTEAC BTEAC BTEAC BTEAC BTEAC	[62] [62] [62] [62] [62]
8	$3-CH_3-C_6H_4-\overset{H}{\underset{\|}{N}}-\overset{O}{\underset{\|}{CH}}$	$(CH_3O)_2SO_2$	mono	90	BTEAC	[62]
8	$C_6H_5-SO_2CH_2CH_3$	C_6H_5CHO	$C_6H_5-SO_2-\underset{CH_3}{\overset{\|}{C}}=\underset{C_6H_5}{\overset{H}{C}}$	44	BTEAC	[60]

173

Alkylation Reactions

Table 10.2 (continued)

Number of carbons	Substrate	Electrophile	Product	% Yield	Catalyst	References
8	4-CH$_3$C$_6$H$_4$SO$_2$CH$_2$Cl	C$_6$H$_5$CH$_2$Cl	E-4-CH$_3$-C$_6$H$_4$-SO$_2$-CH=CH-C$_6$H$_5$	60	BTAEC	[61]
		CH$_3$COCH$_3$	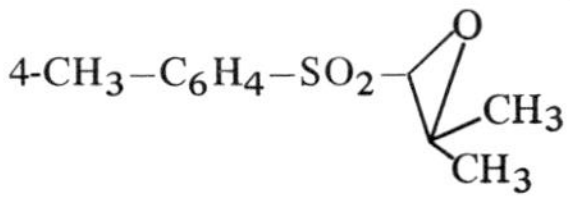 4-CH$_3$-C$_6$H$_4$-SO$_2$—⟨CH$_3$,CH$_3$⟩(O)	91	BTEAC	[61]
		(CH$_3$)$_2$CH–CHO	as above	65	BTEAC	[61]
		cyclohexanone	as above	90	BTEAC	[61]
		benzaldehyde	as above	60	BTEAC	[61]
		benzophenone	as above	90	BTEAC	[61]
8	4-CH$_3$-C$_6$H$_4$SO$_2$C$\underline{\text{H}}$Br	C$_2$H$_5$Br	mono	67	BTEAC	[61]
		n-C$_4$H$_9$Br	mono	68	BTEAC	[61]
		Br(CH$_2$)$_2$Br	cyclopropane	73	BTEAC	[61]
8	4-CH$_3$-C$_6$H$_4$SO$_2$C$\underline{\text{H}}$Cl$_2$	Br(CH$_2$)$_4$Br	dimer	87	BTEAC	[61]
8	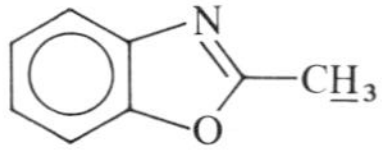 (2-methylbenzoxazole)	C$_6$H$_5$CHO	mono	50	BTEAC	[63]
		4-Cl–C$_6$H$_4$CHO	mono	26	BTEAC	[63]
		2-Cl–C$_6$H$_4$CHO	mono	no yield reported	BTEAC	[63]
		2-CH$_3$–C$_6$H$_4$CHO	mono		BTEAC	[63]
		4-CH$_3$–C$_6$H$_4$CHO	mono		BTEAC	[63]
		4-CH$_3$O–C$_6$H$_4$CHO	mono		BTEAC	[63]

Table 10.2 (continued)

Number of carbons	Substrate	Electrophile	Product	% Yield	Catalyst	References
8	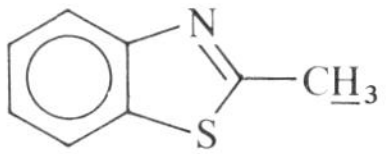 2-methylbenzothiazole	C_6H_5CHO	mono	60–80	BTEAC	[63]
		4-Cl-C_6H_4CHO	mono	60–80	BTEAC	[63]
		2-CH_3-C_6H_4CHO	mono	60–80	BTEAC	[63]
		4-CH_3-C_6H_4CHO	mono	60–80	BTEAC	[63]
		4-CH_3O-C_6H_4CHO	mono	60–80	BTEAC	[63]
9	C_6H_5-CH_2-C(O)-CH_3	CH_3I	mono	92	TBA^+OH^-	[38]
		CH_3I	mono	95	α-phosphoryl sulfoxide	[39]
		CH_3Br	mono	87	BTEAC	[21]
		C_2H_5Br	mono	90	BTEAC	[21]
			mono	100	n-Bu_3N	[40]
		n-C_3H_7Br	mono	84	BTEAC	[21]
		i-C_3H_7Br	mono	43	BTEAC	[21]
		CH_2=CH-CH_2Br	mono	85	BTEAC	[21]
		CH_2=CH-CH_2Br (excess)	di	78	BTEAC	[21]
		n-C_4H_9Br	mono	87	BTEAC	[21, 64]
			mono	86	DC-18-C-6	[65]
			mono	94	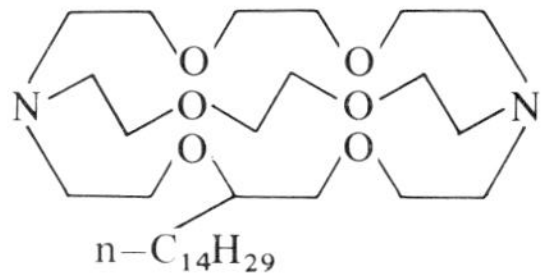 cryptand with n–$C_{14}H_{29}$	[66]
		n-$C_5H_{11}Br$	mono	90	HDTBP	[66]
		i-$C_5H_{11}Br$	mono	93	DC-18-C-6	[66]
		n-$C_5H_{11}Br$	mono	73	BTEAC	[21]
		i-$C_5H_{11}Br$	mono	79	BTEAC	[21]

Table 10.2 (continued)

Number of carbons	Substrate	Electrophile	Product	% Yield	Catalyst	References
9	$C_6H_5-CH_2-\overset{O}{\underset{\|}{C}}-CH_3$	$(CH_3)_2C=CH-CH_2Cl$	mono	74	BTEAC	[21]
		$n-C_6H_{13}Br$	mono	77	BTEAC	[21]
		$(C_2H_5)_2NCH_2CH_2Cl$	mono	45	BTEAC	[21]
		$Cl(CH_2)_2CO_2-i-C_3H_7$	mono	65	BTEAC	[21]
		$C_6H_5CH_2Cl$	mono	77	BTEAC	[21]
		$C_6H_5CH_2Cl$ (excess)	di	78	BTEAC	[21]
		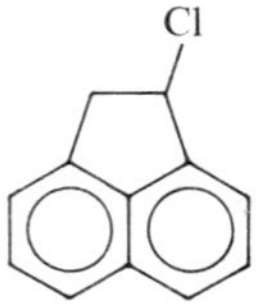 (chloroindane structure)	mono	72	BTEAC/DMSO	[43]
		$Br(CH_2)_2Br$	cyclopropane	54	BTEAC	[64, 67]
		$Br(CH_2)_3Cl$	$C_6H_5-CH-\overset{O}{\underset{\|}{C}}-CH_3$ $(CH_2)_3Cl$	67	BTEAC	[67]
		$Br-(CH_2)_3Br$	$C_6H_5-CH-\overset{O}{\underset{\|}{C}}-CH_3$ $(CH_2)_3Br$	8	BTEAC	[67]
			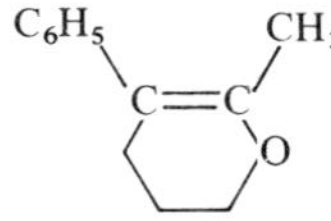 (dihydropyran structure with C_6H_5, CH_3)	37		

(cont. next page)

Table 10.2 (continued)

Number of carbons	Substrate	Electrophile	Product	% Yield	Catalyst	References
9	$C_6H_5-\overset{H}{\underset{|}{C}}H-\overset{O}{\overset{\|}{C}}-CH_3$	$Br(CH_2)_3Br$	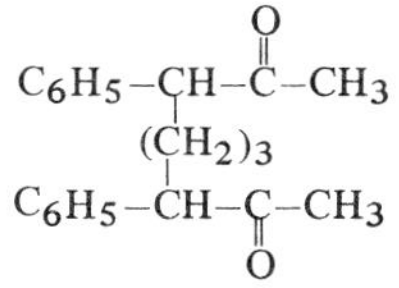 $C_6H_5-CH-\overset{O}{\overset{\|}{C}}-CH_3$ $\quad\quad\quad (CH_2)_3$ $C_6H_5-CH-\overset{O}{\overset{\|}{C}}-CH_3$	20	BTEAC (cont.)	[67]
		$Br(CH_2)_4Br$	$C_6H_5-\overset{O}{\overset{\|}{C}}-\overset{}{\underset{}{C}}-CH_3$ $\quad\quad CH_2\;\;CH_2$ $\quad\quad CH_2-CH_2$	38	BTEAC	[67]
			$C_6H_5-CH-\overset{O}{\overset{\|}{C}}-CH_3\;\;C_6H_5$ $\quad\quad (CH_2)_4-CH-COCH_3$	21		
		$Br(CH_2)_5Br$	$C_6H_5-CH-\overset{O}{\overset{\|}{C}}-CH_3$ $\quad\quad (CH_2)_5Br$	43	BTEAC	[67]
			$C_6H_5-CH-\overset{O}{\overset{\|}{C}}-CH_3$ $\quad\quad (CH_2)_5$ $C_6H_5-CH-\overset{O}{\overset{\|}{C}}-CH_3$	26	BTEAC	[67]
		$4\text{-}CH_3C_6H_4SO_2N_3$	$C_6H_5-\underset{N_2}{\overset{}{C}}-\overset{O}{\overset{\|}{C}}-CH_3$	100	TBA^+OH^- (St)	[80]

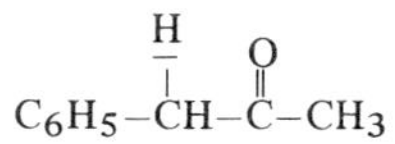

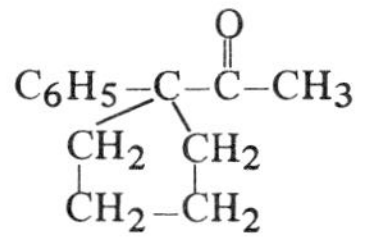

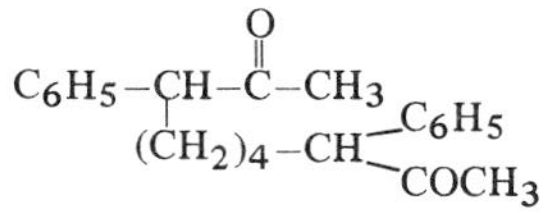

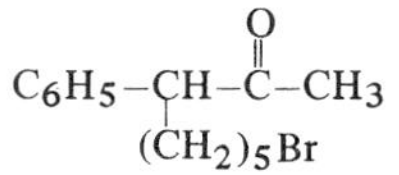

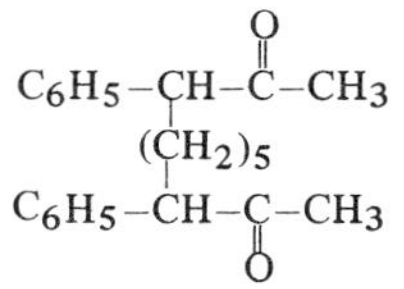

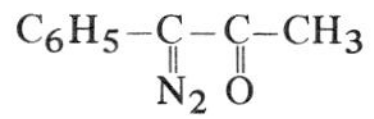

Table 10.2 (continued)

Number of carbons	Substrate	Electrophile	Product	% Yield	Catalyst	References
9	4-CH$_3$–C$_6$H$_4$–C(O)–C$\underline{H}_3$	(CH$_2$)$_2$C=CH–CH$_2$Cl	mono	20	BTEAC	[20]
9	4-CH$_3$O–C$_6$H$_4$–C(O)–C$\underline{H}_3$	(CH$_2$)$_2$C=CH–CH$_2$Cl	mono	10	BTEAC	[20]
9	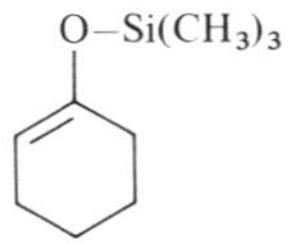 O–Si(CH$_3$)$_3$ cyclohexenyl	CH$_3$I	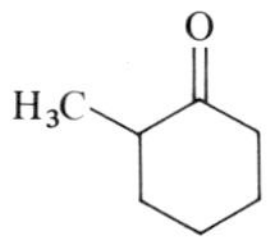 2-methylcyclohexanone	79	BTMAF	[57]
9	O–Si(CH$_3$)$_3$ cyclohexenyl	C$_6$H$_5$CH$_2$Br	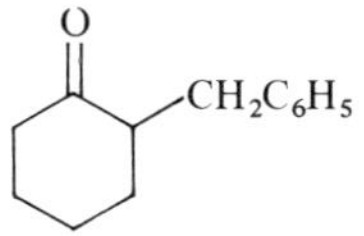 2-benzylcyclohexanone	63	BTMAF	[57]
9	C$_6$H$_5$–SO$_2$C$\underline{H}_2$N(CH$_3$)$_2$	C$_6$H$_5$CHO	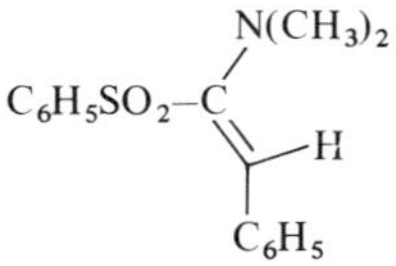 C$_6$H$_5$SO$_2$–C(N(CH$_3$)$_2$)=CH–C$_6$H$_5$	58	BTEAC	[60]
9	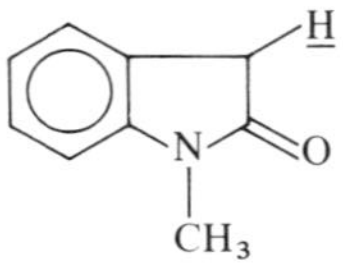 N-methylindolin-2-one	C$_6$H$_5$CH$_2$Cl Br(CH$_2$)$_2$Br Br(CH$_2$)$_4$Br	di cyclopropane cyclopentane	73 72 54	BTEAC BTEAC BTEAC	[68] [68] [68]
10	C$_6$H$_5$–C$\underline{H}$(CH$_3$)–C(O)–CH$_3$	C$_6$H$_5$CH$_2$Cl	mono	60	BTEAC	[21]

Table 10.2 (continued)

Number of carbons	Substrate	Electrophile	Product	% Yield	Catalyst	References
10	3-methyl-1-(trimethylsilyloxy)cyclohexene	BrCH$_2$CO$_2$CH$_3$	as above	56	BTMAF	[57]
10	(same)	n-C$_4$H$_9$I	2-methyl-6-n-butylcyclohexanone	40	BTMAF	[57]
10	(same)	C$_6$H$_5$CH$_2$Br C$_6$H$_5$CH=CH–CH$_2$Br	as above as above	89 78	BTMAF BTMAF	[57] [57]
10	3-methyl-1-(trimethylsilyloxy)cyclohexene	BrCH$_2$CO$_2$CH$_3$	as above	42	BTMAF	[57]
10	(same)	n-C$_4$H$_9$I	2-methyl-2-n-butylcyclohexanone	37	BTMAF	[57]
10	(same)	C$_6$H$_5$CH$_2$Br	as above	59	BTMAF	[57]
10	1-tetralone	CH$_2$=CH–CH$_2$Br i-C$_3$H$_7$Br	di { mono O-alkylation	89 43 43	BTEAC BTEAC	[69] [69]
		C$_6$H$_5$CH$_2$Cl Br(CH$_2$)$_2$Br	di cyclopropane	80 34	BTEAC BTEAC	[69] [69]
11	C$_6$H$_5$–CH(C$_2$H$_5$)–CO–CH$_3$	CH$_2$=CH–CH$_2$Br Cl(CH$_2$)$_2$–CO$_2$–i-C$_3$H$_7$ C$_6$H$_5$CH$_2$Cl	mono mono mono	50 68 73	BTEAC/DMSO BTEAC/DMSO BTEAC/DMSO	[21] [21] [21]

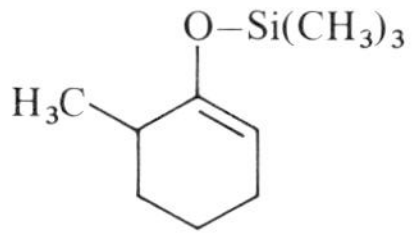

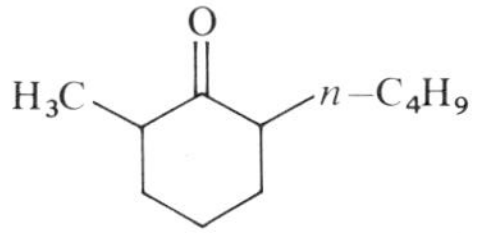

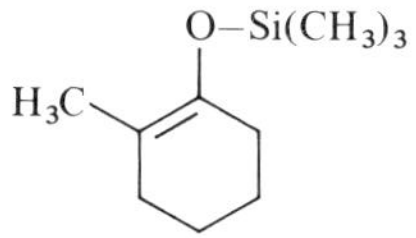

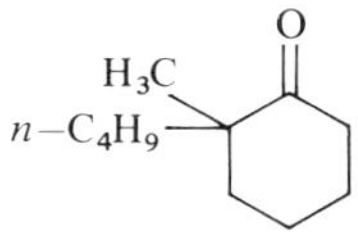

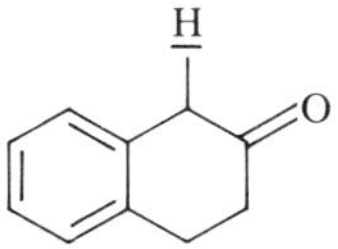

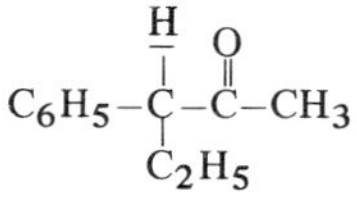

Table 10.2 (continued)

Number of carbons	Substrate	Electrophile	Product	% Yield	Catalyst	References
11	$C_6H_5-SO_2-CH_2-CH=C(CH_3)_2$	C_6H_5CHO	$C_6H_5-SO_2-\underset{CH=C(CH_3)_2}{C}=CH-C_6H_5$	25	BTEAC	[60]
12	$C_6H_5-\underset{n-C_3H_7}{\overset{H}{\underset{\mid}{C}}}-\overset{O}{\overset{\|}{C}}-CH_3$	$Cl(CH_2)_2CO_2-i-C_3H_7$ $C_6H_5CH_2Cl$	mono mono	72 68	BTEAC/DMSO BTEAC/DMSO	[21] [21]
12	$C_6H_5-\underset{CH_2-CH=CH_2}{\overset{H}{\underset{\mid}{C}}}-\overset{O}{\overset{\|}{C}}-CH_3$	$C_6H_5CH_2Cl$	mono	73	BTEAC/DMSO	[21]
12	$C_6H_5-\overset{H}{\underset{\mid}{C}}HCO_2-t-C_4H_9$	C_2H_5Br $CH_2=CH-CH_2Br$ $n-C_4H_9Br$ $Cl(CH_2)_2CO_2-i-C_3H_7$ $C_6H_5CH_2Cl$	mono mono mono mono mono	53 46 45 70 70	BTEAC/DMSO BTEAC/DMSO BTEAC/DMSO BTEAC/DMSO BTEAC/DMSO	[71] [71] [71] [71] [71]
		$Br(CH_2)_3Cl$	$C_6H_5\underset{(CH_2)_3Cl}{C}H-CO_2-t-C_4H_9$	53	BTEAC	[71]
		$Br(CH_2)_3Br$	$C_6H_5\underset{(CH_2)_3Br}{C}H-CO_2-t-C_4H_9$	16	BTEAC	[71]
		$Br(CH_2)_4Br$	$C_6H_5\underset{(CH_2)_4Br}{C}H-CO_2-t-C_4H_9$	43	BTEAC	[71]

Table 10.2 (continued)

Number of carbons	Substrate	Electrophile	Product	% Yield	Catalyst	References
12	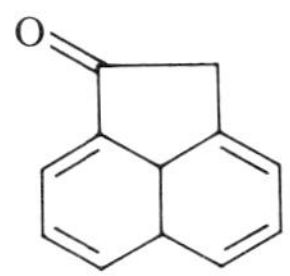	CH$_3$Br	di	40	BTEAC/DMSO	[27]
		C$_2$H$_5$Br	di	50	BTEAC/DMSO	[27]
		n-C$_3$H$_7$Br	di	57	BTEAC/DMSO	[27]
		CH$_2$=CH–CH$_2$Cl	di	54	BTEAC/DMSO	[27]
		n-C$_4$H$_9$Br	di	58	BTEAC/DMSO	[27]
		n-C$_5$H$_{11}$Br	di	59	BTEAC/DMSO	[27]
		Cl(CH$_2$)$_2$CO$_2$-i-C$_3$H$_7$	di	80	BTEAC/DMSO	[27]
		C$_6$H$_5$CH$_2$Cl	di	63	BTEAC/DMSO	[27]
			mono	53	BTEAC/DMSO	[43]
			di	16		
		(C$_6$H$_5$)$_2$CHCl	mono	54	BTEAC/DMSO	[27]
		Br(CH$_2$)$_2$Br	cyclopropane	55	BTEAC/DMSO	[27]
		Br(CH$_2$)$_3$Br	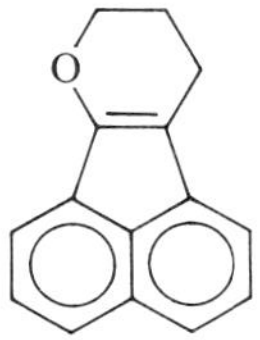	41	BTEAC/DMSO	[27]
		Br(CH$_2$)$_4$Br	cyclopentane	44	BTEAC/DMSO	[27]
		Br(CH$_2$)$_5$Br	cyclohexane	36	BTEAC/DMSO	[27]
13	C$_6$H$_5$–SO$_2$–CH$_2$C$_6$H$_5$	C$_6$H$_5$CHO	C$_6$H$_5$–SO$_2$–C(C$_6$H$_5$)=CHC$_6$H$_5$	49	BTEAC	[60]

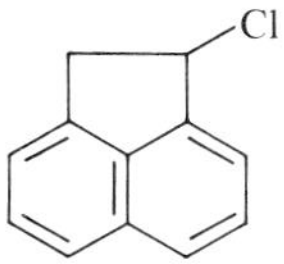

Table 10.2 (continued)

Number of carbons	Substrate	Electrophile	Product	% Yield	Catalyst	References
13	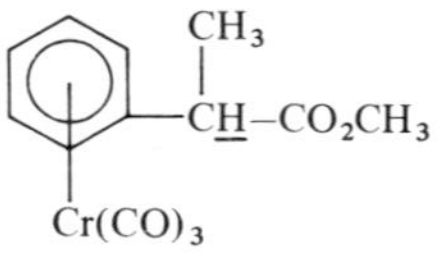 CH₃-CH(CH₃)-CO₂CH₃ on C₆H₅-Cr(CO)₃	CH₃I	mono	70	CTMAC	[70]
		CH₂=CH-CH₂Br	mono	100	CTMAC	[70]
		CH≡C-CH₂Br	mono	100	CTMAC	[70]
		C₆H₅CH₂Br	mono	100	CTMAC	[70]
13	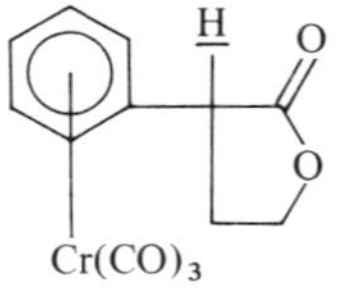 α-H γ-butyrolactone on C₆H₅-Cr(CO)₃	CH₃I	mono	40	CTMAC	[70]
		CH₂=CH-CH₂Br	mono	90	CTMAC	[70]
		CH≡C-CH₂Br	mono	100	CTMAC	[70]
		C₆H₅CH₂Br	mono	60	CTMAC	[70]
14	C₆H₅-CH₂-C(=O)-C₆H₆	C₂H₅Br	mono	77	BTEAC	[72]
		i-C₃H₇Br	{ mono / O-alkylation	40 / 23	BTEAC	[72]
		CH₂=CH-CH₂Cl	mono	94	BTEAC	[72]
		n-C₄H₉Br	mono	75	BTEAC	[72]
		n-C₆H₁₃Br	mono	77	BTEAC	[72]
		Cl(CH₂)₂CO₂-i-C₃H₇	mono	84	BTEAC	[72]
		C₆H₅-CH₂-Cl	mono	70	BTEAC	[72]
		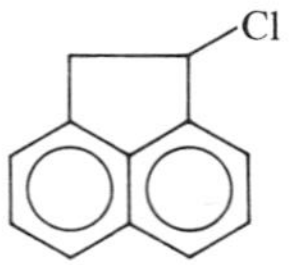 acenaphthyl chloride	mono	55	BTEAC	[43]

Table 10.2 (continued)

Number of carbons	Substrate	Electrophile	Product	% Yield	Catalyst	References
14	$C_6H_5-CH_2-\overset{O}{\underset{\|}{C}}-C_6H_5$	$(C_6H_5)_2CHCl$	mono	88	BTEAC	[72]
		CH_2Br_2	mono-bromomethyl / dimer	23 / 44	BTEAC	[72]
		$Br(CH_2)_2Br$	mono-bromoethyl / cyclopropane	9 / 51	BTEAC	[72]
		$Br(CH_2)_3Br$	mono-bromopropyl / dimer	10 / 19	BTEAC	[72]
		$Br(CH_2)_4Br$	mono-bromobutyl / dimer / cyclopentane	20 / 22 / 19	BTEAC	[72]
		$Br(CH_2)_5Br$	mono-bromopentyl	36	BTEAC	[72]
14	$C_6H_5-\underset{(CH_2)_5Br}{CH}-\overset{O}{\underset{\|}{C}}-CH_3$	—	$C_6H_5\underset{\|}{\overset{O}{\|}}\text{-cyclohexyl-}CH_3$ (1-benzoyl-1-methylcyclohexane)	20	BTEAC	[67]
14	$C_6H_5-\underset{i\text{-}C_5H_{11}}{\overset{H}{\underset{\|}{C}}}-\overset{O}{\underset{\|}{C}}-CH_3$	$C_6H_5CH_2Cl$	mono	60	BTEAC/DMSO	[21]

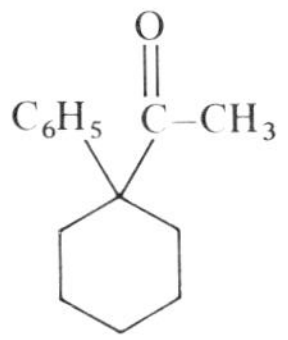

Alkylation Reactions

Table 10.2 (continued)

Number of carbons	Substrate	Electrophile	Product	% Yield	Catalyst	References
14	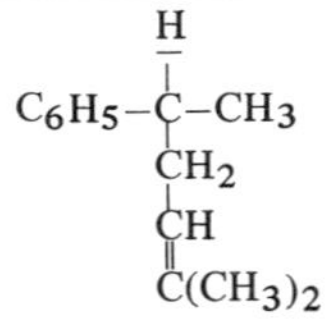 C$_6$H$_5$–C(H)(CH$_3$)–CH$_2$–CH–C(CH$_3$)$_2$	C$_6$H$_5$CH$_2$Cl	mono	76	BTEAC/DMSO	[21]
14	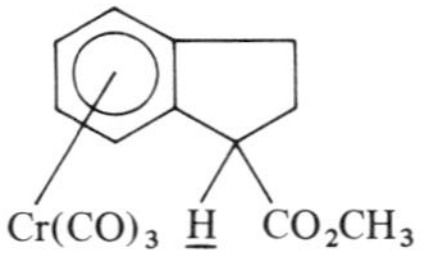 (indane-Cr(CO)$_3$ complex with CO$_2$CH$_3$)	CH$_3$I	mono	100	CTMAC	[70]
		CH$_2$=CHCH$_2$Br	mono	100	CTMAC	[70]
		HC≡C–CH$_2$Br	mono	100	CTMAC	[70]
		C$_6$H$_5$CH$_2$Br	mono	100	CTMAC	[70]
15	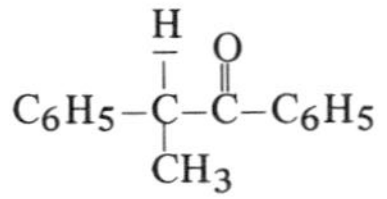 C$_6$H$_5$–C(H)(C(O)CH$_3$)–C$_6$H$_5$	Cl(CH$_2$)$_2$CO$_2$–i–C$_3$H$_7$	mono	68	BTEAC	[72]
15	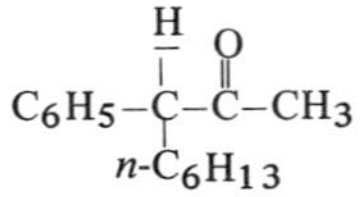 C$_6$H$_5$–C(H)(C(O)CH$_3$)–CH$_3$, n-C$_6$H$_{13}$	C$_6$H$_5$CH$_2$Cl	mono	65	BTEAC/DMSO	[21]
15	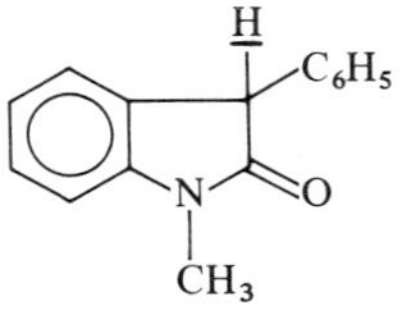 (N-methyl-3-phenyl-oxindole, H, C$_6$H$_5$)	C$_2$H$_5$Br	mono	76	BTEAC	[68]
		n-C$_4$H$_9$Br	mono	68	BTEAC	[68]
		(C$_2$H$_5$)$_2$NCH$_2$CH$_2$Cl	mono	31	BTEAC	[68]
		C$_6$H$_5$CH$_2$Cl	mono	89	BTEAC	[68]
		BrCH$_2$Br	mono	55	BTEAC	[68]
		Br(CH$_2$)$_2$Br	dimer	50	BTEAC	[68]

Table 10.2 (continued)

Number of carbons	Substrate	Electrophile	Product	% Yield	Catalyst	References
15	$C_6H_5-CH_2-\overset{O}{\overset{\|}{C}}-CH_2-C_6H_5$	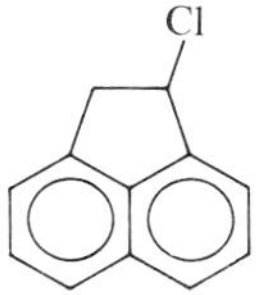 (1-chloro-acenaphthylene)	mono	45	BTEAC/DMSO	[43]
16	$C_6H_5-\overset{H}{\overset{\|}{C}}-\overset{O}{\overset{\|}{C}}-C_6H_5$ $\quad\ \ C_2H_5$	$Cl(CH_2)_2CO_2-i\text{-}C_3H_7$ $C_6H_5CH_2Cl$	mono mono	84 78	BTEAC BTEAC	[72] [72]
16	$C_6H_5-\overset{H}{\overset{\|}{C}}-\overset{O}{\overset{\|}{C}}-CH_3$ $\quad\ \ CH_2$ $\quad\ \ C_6H_5$	$C_6H_5CH_2Cl$	mono	80	BTEAC/DMSO	[21]
17	$C_6H_5-\overset{H}{\overset{\|}{C}}-\overset{O}{\overset{\|}{C}}-C_6H_5$ $\quad\ \ CH_2-CH=CH_2$	$CH_2=CH-CH_2Cl$ $C_6H_5CH_2Cl$	mono mono	85 56	BTEAC BTEAC	[72] [72]
18	$(C_6H_5)_2\overset{H}{\overset{\|}{C}}-CO_2-t\text{-}C_4H_9$	$CH_2=CH-CH_2Br$ $C_6H_5CH_2Cl$	mono mono	56 63	BTEAC/DMSO BTEAC/DMSO	[71] [71]
19	$C_6H_5\overset{O}{\overset{\|}{C}}HCC_6H_5$ $(CH_2)_5Br$		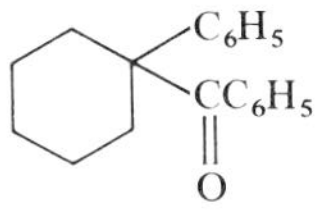 (1,1-diphenyl-cyclohexyl ketone)	23	BTEAC	[72]

Alkylation Reactions

Table 10.2 (continued)

Number of carbons	Substrate	Electrophile	Product	% Yield	Catalyst	References
20	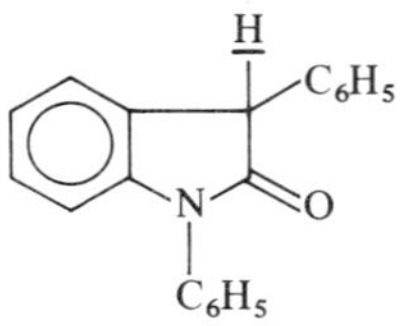 3-H,3-C₆H₅-1-C₆H₅-indolin-2-one	$C_6H_5CH_2Cl$	mono	65	BTEAC	[68]
21	3-H,3-C₆H₅-1-(CH₂-C₆H₅)-indolin-2-one	$n\text{-}C_4H_9Br$ $C_6H_5CH_2Cl$ $BrCH_2Br$	mono mono mono	50 63 48	BTEAC BTEAC BTEAC	[68] [68] [68]
21	$C_6H_5\text{-}\underset{C_6H_5\text{-}CH_2}{\overset{H}{C}}\text{-}\overset{O}{C}\text{-}C_6H_5$	$Cl(CH_2)_2CO_2\text{-}i\text{-}C_3H_7$ $C_6H_5CH_2Cl$	mono mono	93 49	BTEAC BTEAC	[72] [72]

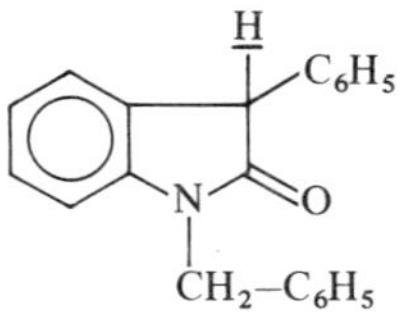

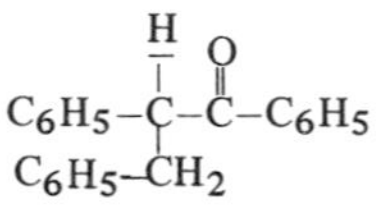

The Substances Alkylated

Table 10.3. Alkylation of diactivated substrates

Number of carbons	Substrate	Electrophile	Product	% Yield	Catalyst	References
3	H₂C-(CN)₂	CS₂	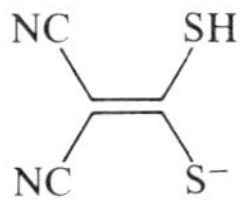 NC-C(SH)=C(CN)-S⁻	86	TBA⁺OH⁻ (St)	[73]
		Br(CH₂)₂Br	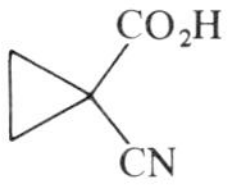 1-CO₂H, 1-CN cyclopropane	49	BTEAC	[26]
		n-C₄H₉Br	di	87	TCMAC	[3, 76]
		n-C₆H₁₃Br	di	90	TCMAC	[3, 76]
		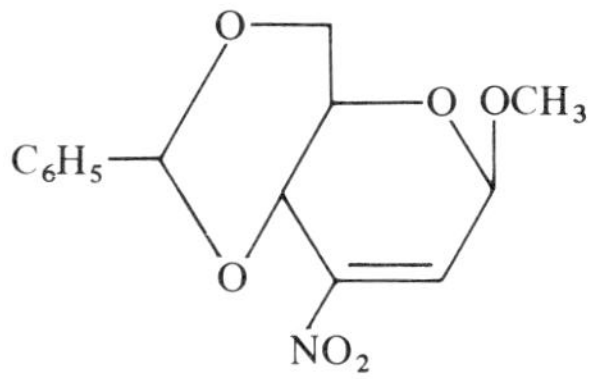 (sugar-NO₂ + OCH₃ pyranose)	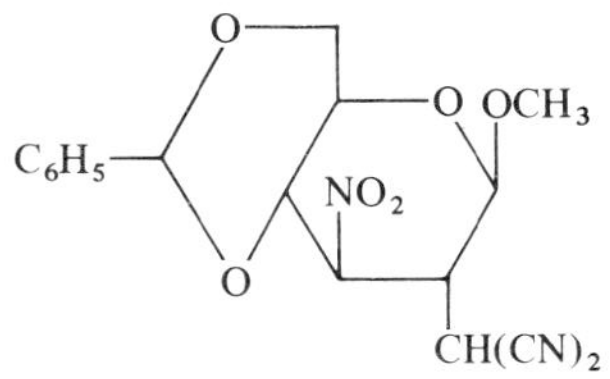 CH(CN)₂ substituted product	81	HDTBP	[74]
		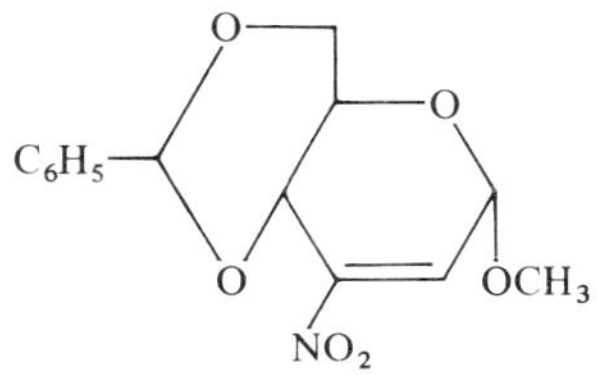 (sugar-NO₂ + OCH₃ pyranose)	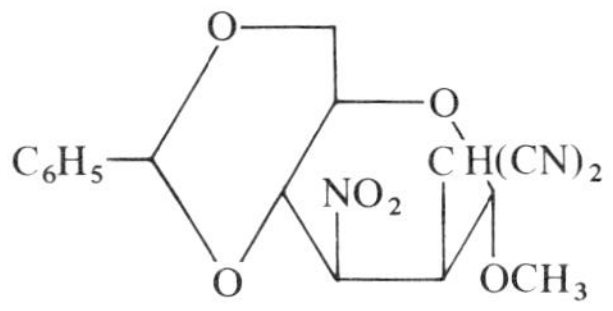 CH(CN)₂ substituted product	40	HDTBP	[75]

Alkylation Reactions

Table 10.3 (continued)

Number of carbons	Substrate	Electrophile	Product	% Yield	Catalyst	References
3	$H_2C(CN)_2$		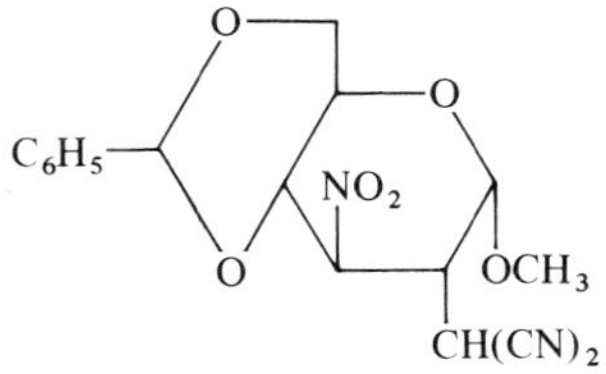 (sugar derivative with C_6H_5, NO_2, OCH_3, $CH(CN)_2$)	25	HDTBP	[75]
4	$NC-CH_2-\overset{O}{\underset{\|}{C}}-OCH_3$	CH_3I	mono / di	49 / 26	TBA^+OH^- (St)	[117]
		CS_2	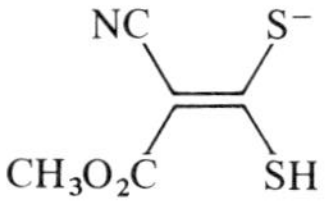 $\underset{CH_3O_2C}{\overset{NC}{>}}C=C\underset{SH}{\overset{S^-}{<}}$	65	TBA^+OH^- (St)	[73]
		C_2H_5I	mono / di	72 / 14	TBA^+OH^- (St)	[117]
		$i\text{-}C_3H_7I$	mono	94	TBA^+OH^- (St)	[117]
		$n\text{-}C_4H_9I$	mono / di	86 / 7	TBA^+OH^- (St)	[117]
5	$CH_3-\overset{O}{\underset{\|}{C}}-CH_2-\overset{O}{\underset{\|}{C}}-CH_3$	CH_3I	mono / di mixture	98	TBA^+OH^- (St)	[77]
		C_2H_5I	mono / di	72 / 16	TBA^+OH^- (St)	[77]

Table 10.3 (continued)

Number of carbons	Substrate	Electrophile	Product	% Yield	Catalyst	References
5	CH₃–C(=O)–CH₂–C(=O)–CH₃	i-C₃H₇I	mono O-alkylation	50 49	TBA⁺OH⁻ (St)	[77]
		n-C₄H₉I	mono O-alkylation	87 13	TBA⁺OH⁻ (St)	[77]
		CH₃–C(Cl)=CH–CH₂–Cl	mono	54	DBDMA	[78]
	[substrate structure with C₆H₅, OCH₃, NO₂, O, O]	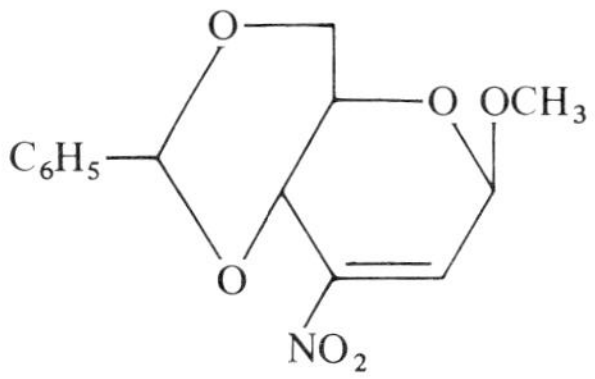	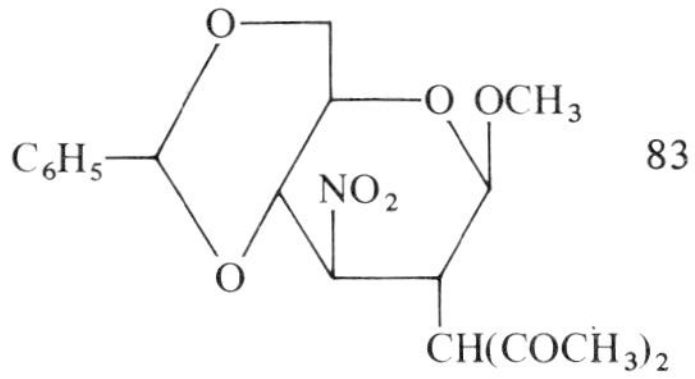 [product structure with C₆H₅, OCH₃, NO₂, CH(COCH₃)₂]	83	HDTBP	[74]
	[substrate structure with C₆H₅, OCH₃, NO₂, O, O]	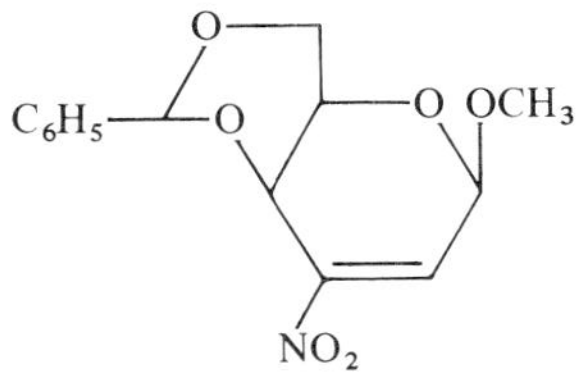	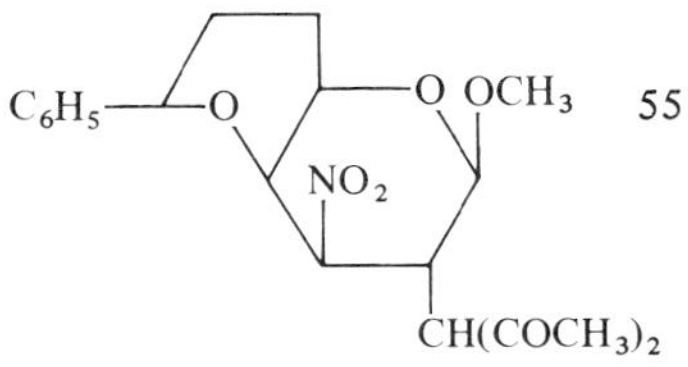 [product structure with C₆H₅, OCH₃, NO₂, CH(COCH₃)₂]	55	HDTBP	[74]

Table 10.3 (continued)

Number of carbons	Substrate	Electrophile	Product	% Yield	Catalyst	References
5	CH$_3$–C(=O)–CH$_2$–C(=O)–CH$_3$	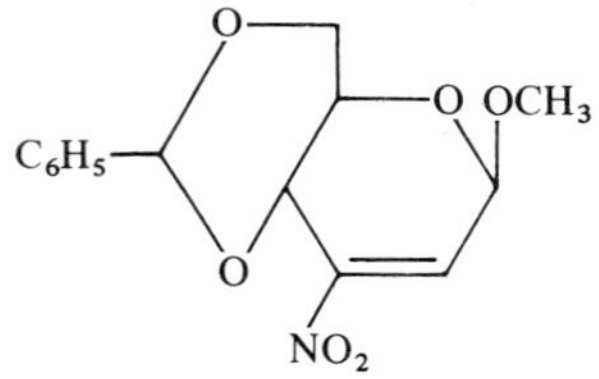 [pyranose with C$_6$H$_5$, OCH$_3$, NO$_2$]	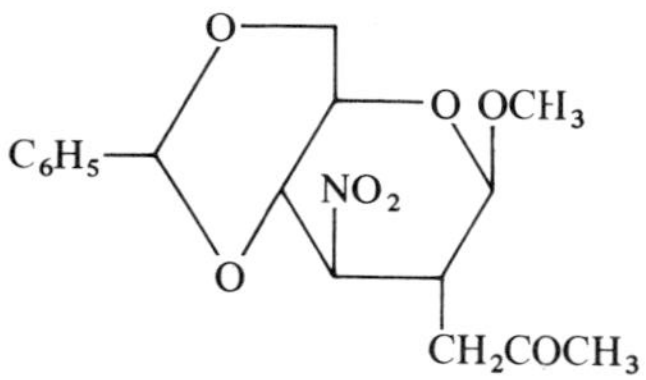 [pyranose product with CH$_2$COCH$_3$ substituent]	81	HDTBP	[74]
		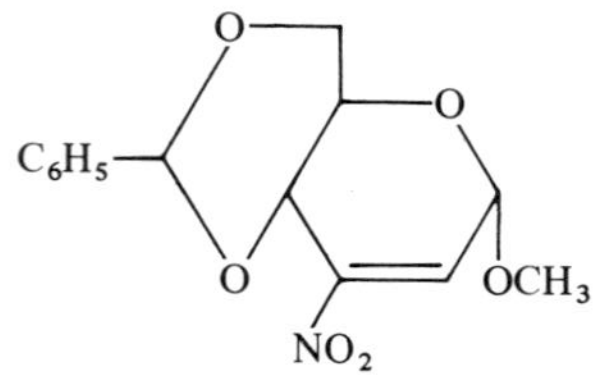 [pyranose with C$_6$H$_5$, OCH$_3$, NO$_2$]	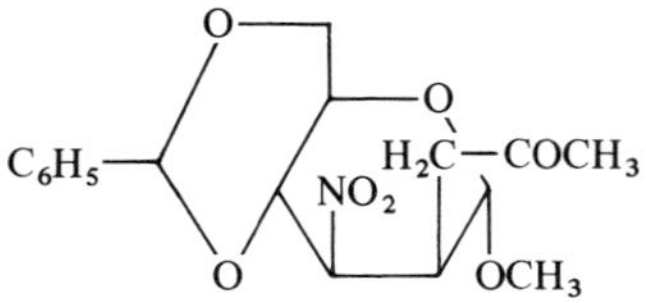 [pyranose product with CH(H$_2$C–COCH$_3$), OCH$_3$]	81	HDTBP	[75]
			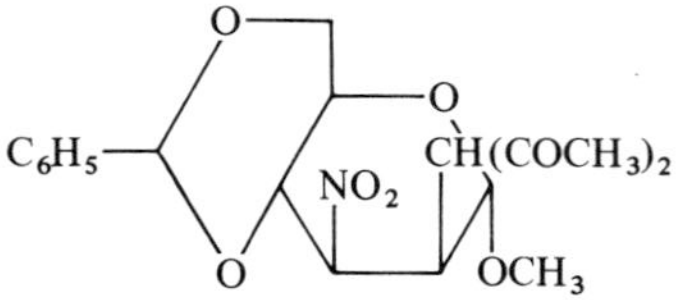 [pyranose product with CH(COCH$_3$)$_2$, OCH$_3$]	81	HDTBP	[75]
5	CH$_3$–C(=O)–CH$_2$CO$_2$CH$_3$	CH$_3$I	mono / di	80 / 10	TBA$^+$OH$^-$ (St)	[79]
		C$_2$H$_5$I	mono / di	84 / 9	TBA$^+$OH$^-$ (St)	[79]

Table 10.3 (continued)

Number of carbons	Substrate	Electrophile	Product	% Yield	Catalyst	References
5	CH$_3$-C(=O)-C$\underline{H}_2$-CO$_2$CH$_3$	i-C$_3$H$_7$I	mono O-alkylation	70 24	TBA$^+$OH$^-$ (St)	[79]
		CH$_2$=CH-CH$_2$Cl	mono	30	TCMAC	[10]
		CH$_2$=CH-CH$_2$Br	mono	85	TCMAC	[10]
		n-C$_4$H$_9$I	mono di	90 5	TBA$^+$OH$^-$ (St)	[79]
		(CH$_3$)$_2$C=CHCH$_2$Cl	mono	30	TCMAC	[10]
		C$_6$H$_5$CH$_2$Cl	mono	85	TCMAC	[10]
		Geranyl-Br	mono	85	TCMAC	[10]
5	NC-CH$_2$-C(=O)-O-C$_2$H$_5$	Br(CH$_2$)$_2$Br	⟨cyclopropane with CO$_2$H and CN⟩	86	BTEAC	[26]
6	CH$_3$-C(=O)-CH$_2$-CO$_2$C$_2$H$_5$	C$_2$H$_5$OTs	mono C	100	KOH/Benzene	[22]
			mono C O-alkylation	50 48	KOH/Benzene DC-18-C-6	[22]
			mono C	97	KOH/(C$_2$H$_5$)$_2$O	[22]
			mono C O-alkylation	54 41	KOH/(C$_2$H$_5$)$_2$O DC-18-C-6	[22]
			mono C O-alkylation	46 52	KOH/DME DC-18-C-6	[22]
			mono C	89	KOH/DME	[22]
		CH$_3$-C(Cl)=CH-CH$_2$Cl	mono	57	DBDMA	[78]

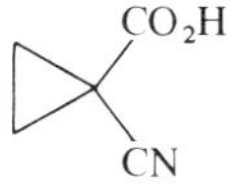

Table 10.3 (continued)

Number of carbons	Substrate	Electrophile	Product	% Yield	Catalyst	References
6	CH$_3$–C(=O)–CH$_2$–CO$_2$C$_2$H$_5$	Br(CH$_2$)$_2$Br	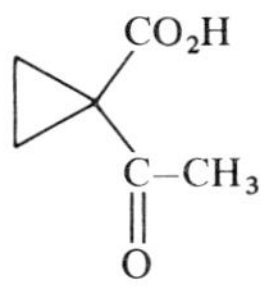 1-(CO$_2$H)-1-(C(=O)–CH$_3$)-cyclopropane	69	BTEAC	[26]
		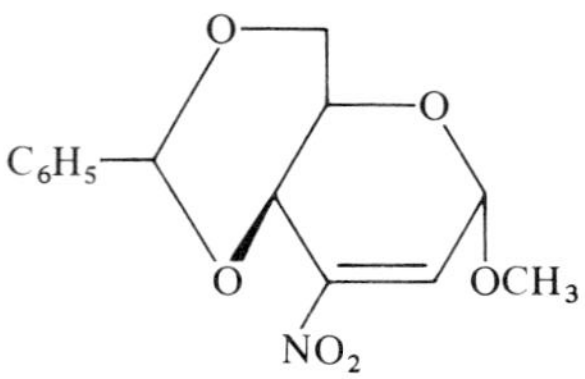 C$_6$H$_5$-(bicyclic sugar with OCH$_3$, NO$_2$)	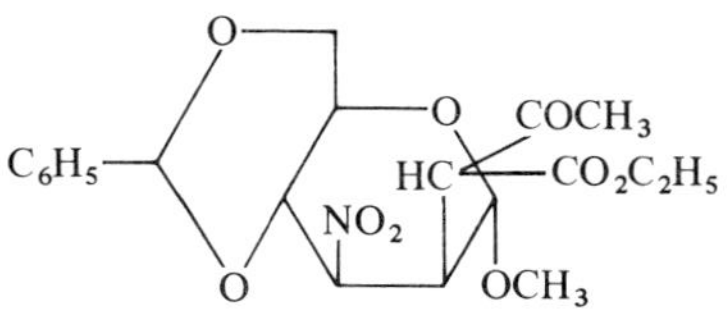 C$_6$H$_5$-(bicyclic sugar with COCH$_3$, CO$_2$C$_2$H$_5$, NO$_2$, OCH$_3$)	50	HDTBP	[75]
		4-CH$_3$C$_6$H$_4$SO$_2$N$_3$	CH$_3$–C(=O)–C(N$_2$)–CO$_2$C$_2$H$_5$	90	TBA$^+$OH$^-$ (St)	[80]
			H–C(N$_2$)–CO$_2$C$_2$H$_5$	53	TBA$^+$OH$^-$ (St)	[80]
6	(C$_2$H$_5$O)$_2$–P(=O)–CH$_2$CN	CH$_3$I	mono	80	TBA$^+$OH$^-$ (St)	[81]
		C$_2$H$_5$I	mono	70	TBA$^+$OH$^-$ (St)	[81]
		n-C$_3$H$_7$I	mono	70	TBA$^+$OH$^-$ (St)	[81]
		CH$_2$=CH–CH$_2$Br	mono	80	TBA$^+$OH$^-$ (St)	[81]
		C$_6$H$_5$CH$_2$Br	mono	30	TBA$^+$OH$^-$ (St)	[81]

Table 10.3 (continued)

Number of carbons	Substrate	Electrophile	Product	% Yield	Catalyst	References
6	[(CH$_3$)$_2$N]$_2$–P(=O)–CH$_2$CN	C$_2$H$_5$I	mono	94	TBA$^+$OH$^-$ (St)	[82]
		i-C$_3$H$_7$I	mono	87	TBA$^+$OH$^-$ (St)	[82]
		n-C$_4$H$_9$I	mono	94	TBA$^+$OH$^-$ (St)	[82]
		C$_6$H$_5$CH$_2$Cl	mono	71	BTEAC	[82]
7	CH$_2$(CO$_2$C$_2$H$_5$)$_2$	CH$_3$I	mono	86	TBA$^+$OH$^-$ (St)	[38]
		CS$_2$	[C$_2$H$_5$O$_2$C]$_2$C=C(SH)(S$^-$)	40	TBA$^+$OH$^-$ (St)	[73]
		C$_2$H$_5$I	mono	88	TBA$^+$OH$^-$ (St)	[38]
		i-C$_3$H$_7$I	mono	45	TBA$^+$OH$^-$ (St)	[38]
		n-C$_4$H$_9$I	mono	85	TBA$^+$OH$^-$ (St)	[38]
		Br(CH$_2$)$_2$Br	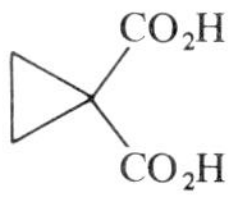 cyclopropane-1,1-dicarboxylic acid	75	BTEAC	[26]
	(benzylidene-protected sugar with NO$_2$, OCH$_3$)	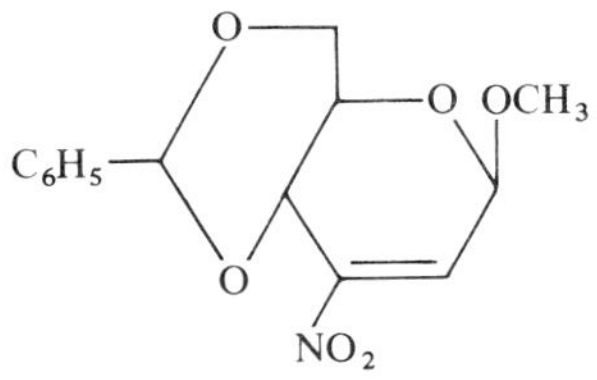	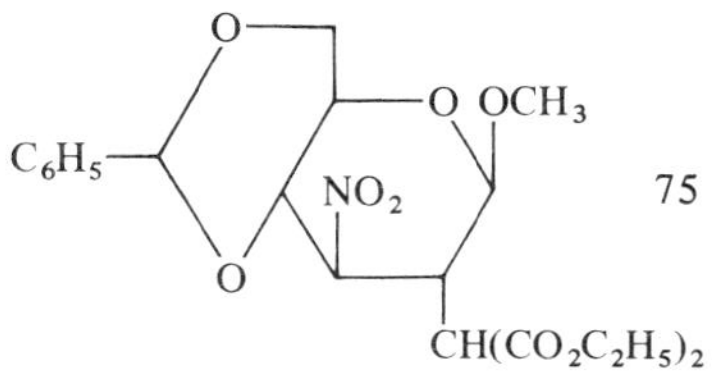 (alkylated sugar with CH(CO$_2$C$_2$H$_5$)$_2$)	75	HDTBP	[74]

Table 10.3 (continued)

Number of carbons	Substrate	Electrophile	Product	% Yield	Catalyst	References
7	$CH_2(CO_2C_2H_5)_2$	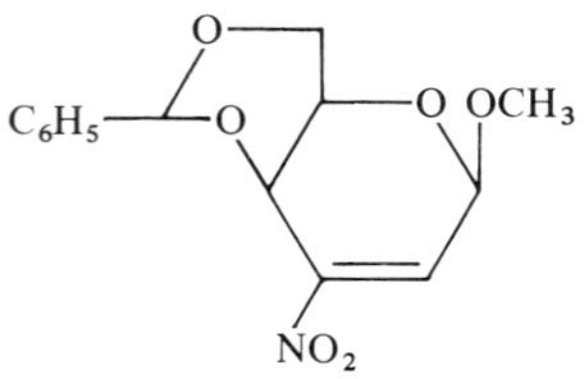 (bicyclic structure with C_6H_5, OCH$_3$, NO$_2$)	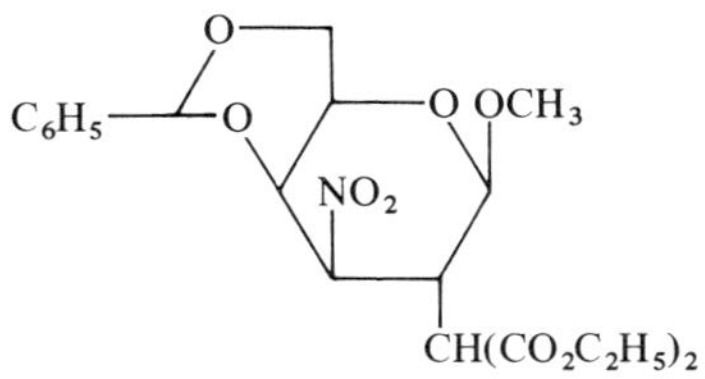 (bicyclic product with C_6H_5, OCH$_3$, NO$_2$, CH(CO$_2$C$_2$H$_5$)$_2$)	82	HDTBP	[74]
		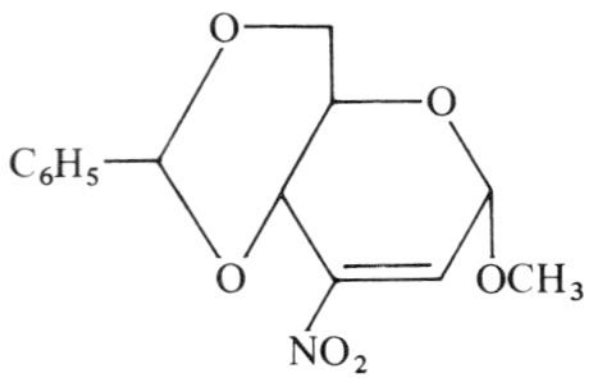 (bicyclic structure with C_6H_5, OCH$_3$, NO$_2$)	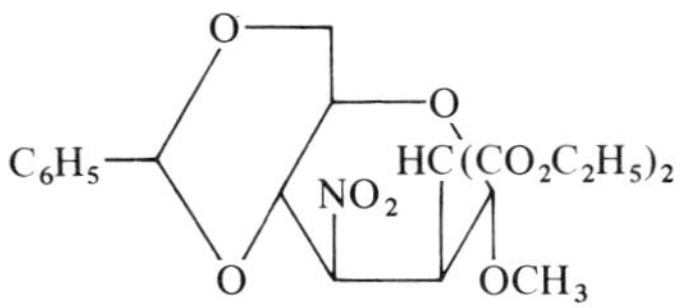 (bicyclic product with C_6H_5, OCH$_3$, NO$_2$, HC(CO$_2$C$_2$H$_5$)$_2$)	65	HDTBP	[75]
7	$[(CH_3)_2N]_2\overset{O}{\overset{\|}{P}}-\underline{C}H-CN$ CH_3	$ClCH_2CN$	mono	97	BTEAC	[82]
		$CH_2=CH-CH_2Cl$	mono	100	BTEAC	[82]
		$n-C_4H_9Br$	mono	100	BTEAC	[82]
		$C_6H_5CH_2Cl$	mono	78	BTEAC	[82]
		CH_2Cl_2	mono	89	BTEAC	[82]
8	$CH_3-\overset{O}{\overset{\|}{C}}-C\underline{H}_2-CO_2-t-C_4H_9$	$4-CH_3-C_6H_4SO_2N_3$	$CH_3-\overset{O}{\overset{\|}{C}}-\overset{N_2}{\overset{\|}{C}}-CO_2-t-C_4H_9$	92	TCMAC	[80]
			$\overset{N_2}{\overset{\|}{HC}}-CO_2-t-C_4H_9$	89	TCMAC	[80]

Table 10.3 (continued)

Number of carbons	Substrate	Electrophile	Product	% Yield	Catalyst	References
8	$C_6H_5-SCH_2-CN$	CH_3Br (excess)	di	75	BTEAC	[47]
		C_2H_5Br	mono	80	BTEAC	[47]
		$CH_2=CH-CH_2Cl$ (excess)	di	80	BTEAC	[47]
		$n\text{-}C_4H_9Br$	mono	82	BTEAC	[47]
		$C_6H_5CH_2Cl$ (excess)	di	82	BTEAC	[47]
		$Br(CH_2)_2Br$	cyclopropane	47	BTEAC	[47]
		$Br(CH_2)_3Cl$	$C_6H_5-S-CH-CN$ $\quad\quad\quad\quad\vert$ $\quad\quad\quad\quad(CH_2)_3Cl$	39	BTEAC	[47]
		$Br(CH_2)_4Br$	cyclopentane	69	BTEAC	[47]
		$Br(CH_2)_5Br$	cyclohexane	50	BTEAC	[47]
8	$[C_2H_5O]_2-\overset{\overset{\displaystyle O}{\|}}{P}-CH_2-CO_2C_2H_5$	CH_3I	mono	30	TBA^+OH^- (St)	[81]
8	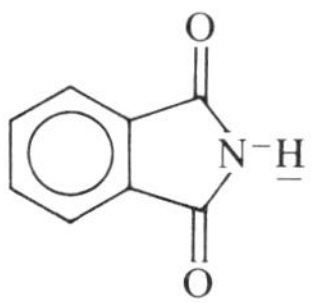 phthalimide (N-H)	$Br(CH_2)_2Br$	dimer	81	HDTBP	[83]
		$ClCH_2CO_2C_2H_5$	mono	91	HDTBP	[83]
		$Cl_3CHCO_2C_2H_5$ $\quad\quad\vert$ $\quad\quad Cl$	mono	90	HDTBP	[83]
		$(CH_3)_3C-CH_2Br$	mono	26	HDTBP	[83]
		$c\text{-}C_6H_{11}Cl$	cyclohexene	90	HDTBP	[83]
		$C_6H_5CH_2Cl$	mono	90	HDTBP	[83]
		$n\text{-}C_8H_{17}Cl$	mono	94	HDTBP	[83]
		$n\text{-}C_8H_{17}Br$	mono	90	HDTBP	[83]
		$n\text{-}C_8H_{17}O-SO_2CH_3$	mono	92	HDTBP	[83]

Table 10.3 (continued)

Number of carbons	Substrate	Electrophile	Product	% Yield	Catalyst	References
8	phthalimide (NH)	$n\text{-}C_6H_{13}\text{-}\underset{\underset{Cl}{|}}{CH}\text{-}CH_3$	mono	84	HDTBP	[83]
		$(R)\text{-}n\text{-}C_6H_{13}\text{-}\underset{\underset{Cl}{|}}{CH}\text{-}CH_3$	mono	84 (89% inversion)	HDTBP	[83]
		$n\text{-}C_6H_{13}\text{-}\underset{\underset{Br}{|}}{CH}\text{-}CH_3$	mono	86	HDTBP	[83]
		$n\text{-}C_6H_{13}\underset{\underset{O\text{-}SO_2CH_3}{|}}{CH}\text{-}CH_3$	mono	90	HDTBP	[83]
		$(R)\text{-}n\text{-}C_6H_{13}\underset{\underset{O\text{-}SO_2CH_3}{|}}{CH}\text{-}CH_3$	mono	85 (92% inversion)	HDTBP	[83]
		$n\text{-}C_{16}H_{33}Br$	mono	95	HDTBP	[83]
	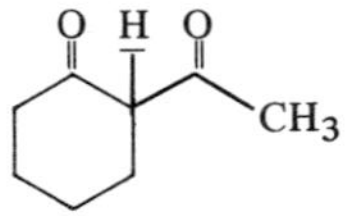 2-methyl-2-acetylcyclohexanone	CH_3I	mono	85	BMEB	[97]
		$CH_2\text{=}CH\text{-}CH_2Br$	mono	90	BMEB	[97]
		$CH_2\text{=}CH\text{-}CH_2Br$	mono	80	BMEB	[97]

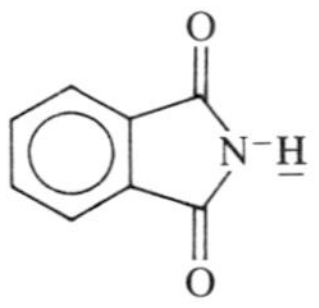

Table 10.3 (continued)

Number of carbons	Substrate	Electrophile	Product	% Yield	Catalyst	References
8	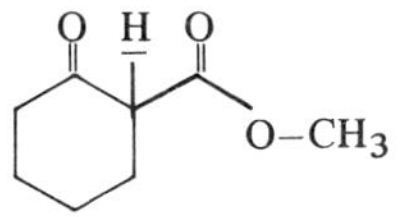 (cyclohexanone with CH(O)–O–CH₃ substituent)	$CH_2=CH-CH_2Br$	mono	75	BMEB	[97]
		$CH\equiv CH-CH_2Br$	mono	75	BMEB	[97]
9	$4\text{-}CH_3\text{-}C_6H_4\text{-}\underset{O}{\overset{O}{S}}\text{-}\underline{CH_2}\text{-}N{=}C$	CH_3I	mono	95	TBAI	[84]
		C_2H_5I	mono	90	TBAI	[84]
		C_2H_5Br	mono	80	TBAI	[84]
		$n\text{-}C_3H_7I$	mono	85	TBAI	[84]
		$i\text{-}C_3H_7I$	mono	40	TBAI	[84]
		$CH_2=CH-CH_2Cl$	mono	75	TBAI	[84]
		$n\text{-}C_4H_9I$	mono	75	TBAI	[84]
		$C_6H_5CH_2Br$	mono	80	TBAI	[84]
9	$C_6H_5\text{-}\underset{O}{\overset{O}{C}}\text{-}\underline{CH_2}\text{-}\underset{O}{\overset{O}{S}}\text{-}CH_3$	CH_3I	mono	81	TBA^+OH^- (St)	[85]
		C_2H_5I	mono	73	TBA^+OH^- (St)	[85]
		$CH_3\text{-}\underset{O}{\overset{O}{C}}\text{-}CH_2Br$	mono	81	TBA^+OH^- (St)	[86]
		$ClCH_2CO_2C_2H_5$	mono	74	TBA^+OH^- (St)	[85]
		$C_6H_5CH_2Cl$	mono	65	TBA^+OH^- (St)	[85]
		$C_6H_5\underset{O}{\overset{O}{C}}\text{-}CH_2Br$	mono	80	TBA^+OH^- (St)	[86]
		$4\text{-}Br\text{-}C_6H_4\text{-}\underset{O}{\overset{O}{C}}\text{-}CH_2Br$	mono	63	TBA^+OH^- (St)	[86]

Table 10.3 (continued)

Number of carbons	Substrate	Electrophile	Product	% Yield	Catalyst	References
9	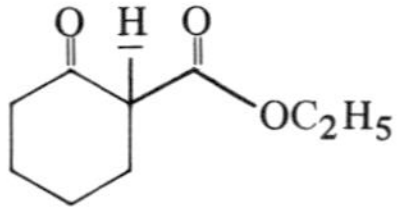 ![structure with OC$_2$H$_5$, H, O, cyclohexanone]	CH$_2$=CH–CH$_2$Br	mono	85	TBA$^+$OH$^-$ (St)	[97]
11	H$_2$C–[CO$_2$-t-C$_4$H$_9$]$_2$	C$_2$H$_5$Br	mono	47	BTEAC	[87]
			di	16		
		CH$_2$=CH–CH$_2$Cl (excess)	di	96	BTEAC	[87]
		n-C$_4$H$_9$Br (excess)	di	39	BTEAC	[87]
		C$_6$H$_5$CH$_2$Cl	mono	54	BTEAC	[87]
			di	16		
		C$_6$H$_5$CH$_2$Cl (excess)	di	86	BTEAC	[87]
		Br(CH$_2$)$_2$Br	cyclopropane	90	BTEAC	[87]
		Br(CH$_2$)$_4$Br	cyclopentane	75	BTEAC	[87]
		4-CH$_3$–C$_6$H$_4$SO$_2$N$_3$	N$_2$=C–[CO$_2$-t-C$_4$H$_9$]$_2$	87	TCMAC	[80]
12	C$_6$H$_5$CCH<u>I</u>C–OCH$_3$]$_2$	CH$_3$I	mono	100	TBA$^+$OH$^-$ (St)	[88]
		C$_2$H$_5$I	mono	54	TBA$^+$OH$^-$ (St)	[88]
			O-alkylation	46		
		i-C$_3$H$_7$I	mono	14	TBA$^+$OH$^-$ (St)	[88]
			O-alkylation	86		
		n-C$_4$H$_9$I	mono	47	TBA$^+$OH$^-$ (St)	[88]
			O-alkylation	53		
14	n-C$_6$H$_{13}$–C(O)–CH$_2$–SO$_2$C$_6$H$_5$	CH$_3$I	mono	78	TBA$^+$OH$^-$ (St)	[85]
		C$_2$H$_5$I	mono	81	TBA$^+$OH$^-$ (St)	[85]
		ClCH$_2$CO$_2$C$_2$H$_5$	mono	68	TBA$^+$OH$^-$ (St)	[85]
		C$_6$H$_5$CH$_2$Cl	mono	67	TBA$^+$OH$^-$ (St)	[85]

Table 10.3 (continued)

Number of carbons	Substrate	Electrophile	Product	% Yield	Catalyst	References
15	4-CH₃O–C₆H₄–C(=O)–CH₂–SO₂C₆H₅	CH₃I	mono	87	TBA⁺OH⁻ (St)	[85]
		C₂H₅I	mono	82	TBA⁺OH⁻ (St)	[85]
		ClCH₂CO₂C₂H₅	mono	78	TBA⁺OH⁻ (St)	[85]
		C₆H₅CH₂Cl	mono	74	TBA⁺OH⁻ (St)	[85]
15	C₆H₅–CH₂–C(=O)–CH₂–SO₂–C₆H₅	CH₃I	mono	86	TBA⁺OH⁻ (St)	[85]
		C₂H₅I	mono	76	TBA⁺OH⁻ (St)	[85]
		ClCH₂CO₂C₂H₅	mono	30	TBA⁺OH⁻ (St)	[85]
		BrCH₂CO₂C₂H₅	mono	90	TBA⁺OH⁻ (St)	[85]
		C₆H₅CH₂Cl	mono	59	TBA⁺OH⁻ (St)	[85]
		C₆H₅CH₂Br	mono	75	TBA⁺OH⁻ (St)	[85]
	C₆H₅–C(=O)–CH₂–C(=O)–C₆H₅	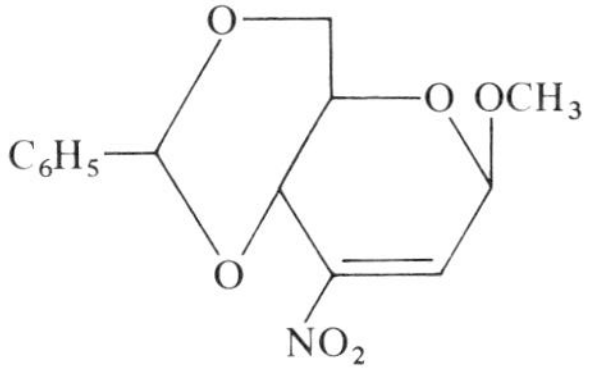 [pyranose with OCH₃, NO₂, C₆H₅]	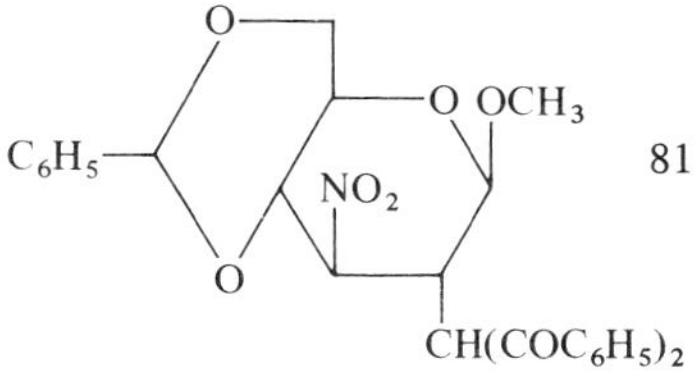 [pyranose product with OCH₃, NO₂, C₆H₅, CH(COC₆H₅)₂]	81	HDTBP	[74]
	C₆H₅–C(=O)–CH₂–C(=O)–C₆H₅	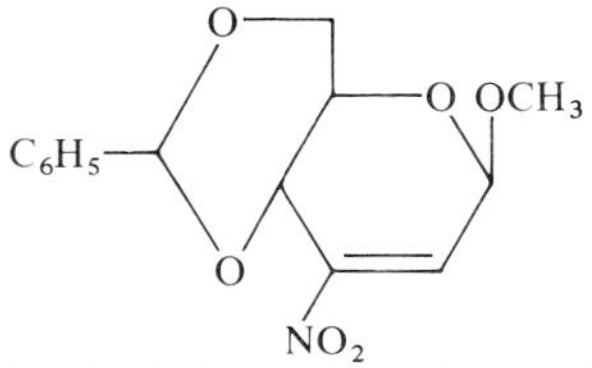 [pyranose with OCH₃, NO₂, C₆H₅]	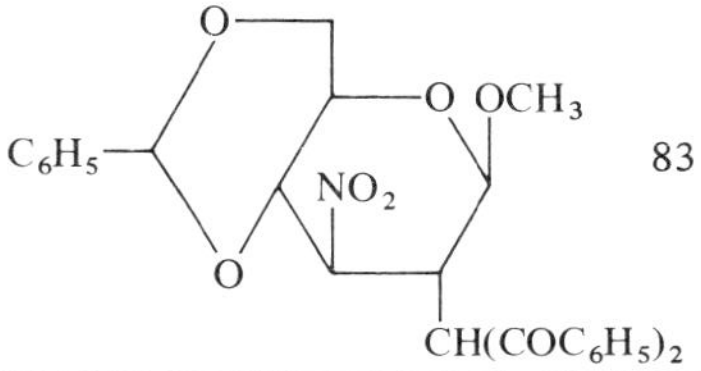 [pyranose product with OCH₃, NO₂, C₆H₅, CH(COC₆H₅)₂]	83	HDTBP	[74]

Table 10.3 (continued)

Number of carbons	Substrate	Electrophile	Product	% Yield	Catalyst	References
15	$C_6H_5-\overset{O}{\underset{\|}{C}}-CH_2-\overset{O}{\underset{\|}{C}}-C_6H_5$	(C₆H₅ glycal with OCH₃ and NO₂)	(glycoside product with HC(COC₆H₅)₂, OCH₃, NO₂, C₆H₅)	72	HDTBP	[75]

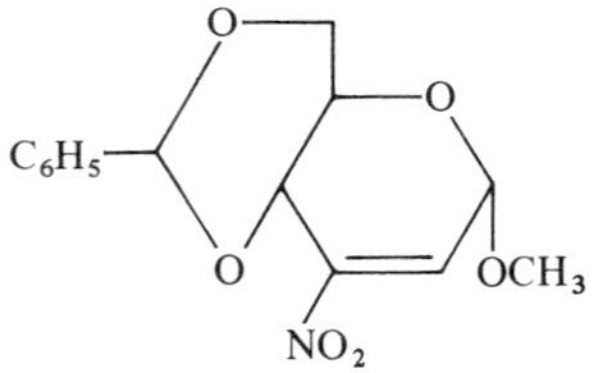

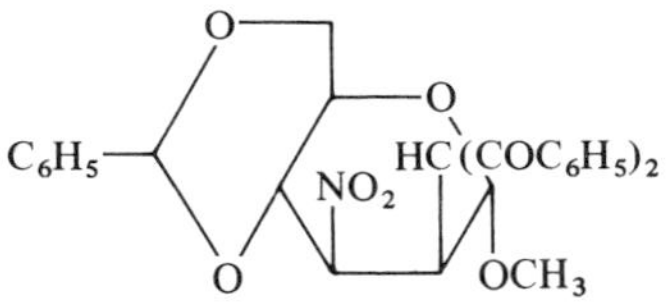

Table 10.4. Alkylation of hydrocarbons

Number of carbons	Substrate	Electrophile	Product	% Yield	Catalyst	References
1	CHCl₃	CH₃SCN	CH₃SCCl₃	61	BTEAC	[30]
		C₂H₅SCN	C₂H₅SCCl₃	64	BTEAC	[30]
		n-C₃H₇SCN	n-C₃H₇SCCl₃	67	BTEAC	[30]
		Cl(CH₂)₃SCN	Cl(CH₂)₃SCCl₃	77	BTEAC	[30]
		NC(CH₂)₃SCN	Cl(CH₂)₃SCCl₃	70	BTEAC	[30]

Table 10.4 (continued)

Number of carbons	Substrate	Electrophile	Product	% Yield	Catalyst	References
1	CHCl₃	 [structure with SCN, Cl, Cl]	 [structure with SCCl₃, Cl, Cl]	76	BTEAC	[30]
		n-C₅H₁₁SCN	n-C₅H₁₁SCCl₃	86	BTEAC	[30]
		C₆H₅SCN	C₆H₅SCCl₃	60	BTEAC	[30]
		C₆H₅CH₂SCN	C₆H₅CH₂SCCl₃	80	BTEAC	[30]
8	C₆H₅C≡CH	CH₃SCN	C₆H₅C≡C—S—CH₃	35	BTEAC	[89]
		C₂H₅SCN	C₆H₅C≡C—S—C₂H₅	42	BTEAC	[89]
		n-C₅H₁₁	C₆H₅C≡C—S—n-C₅H₁₁	45	BTEAC	[89]
		CCl₄	C₆H₅C≡C—Cl	45	BTEAC	[89]
9	Indene	CH₃Br	3-methyl	62	BTEAC	[90]
		n-C₃H₇Br	as above	68	BTEAC	[90]
		CH₂=CH—CH₂Cl	as above	73	BTEAC	[90]
		n-C₄H₉Br	as above	63	BTEAC	[90]
		(C₂H₅)₂N(CH₂)₃Br	as above	46	BTEAC	[90]
		Br(CH₃)₃Br	3-bromopropyl	45	BTEAC	[90]
		Br(CH₂)₄Br	1,1-spiropentane	no yield reported	BTEAC	[90]
		Br(CH₂)₅Br	3-bromopentyl	52	BTEAC	[90]
13	Fluorene	CH₂=CH—CH₂Cl	9,9-diallylfluorene	80	BTEAC/DMSO	[91]
		n-C₄H₉Br	mono and di	65	BTEAC/DMSO	[91]
		C₆H₅CH₂Cl	9,9-dibenzylfluorene	83	BTEAC/DMSO	[91]
		Br(CH₂)₄Br	9,9-spiropentane	64	BTEAC/DMSO	[91]
13	(C₆H₅)₂CH₂	C₆H₅CH₂Cl	mono		[2,2,2]-cryptate	[92]
19	(C₆H₅)₃CH	C₆H₅CH₂Cl	mono		[2,2,2]-cryptate	[92]

Table 10.5. Alkylation of amines

Number of carbons	Substrate	Electrophile	Product	% Yield	Catalyst	References
4	Pyrrole	$C_6H_5CH_2Cl$	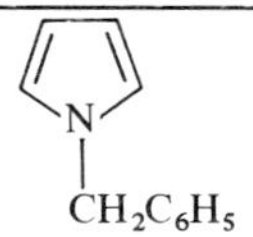 N-$CH_2C_6H_5$ pyrrole	60	BTEAC	[93]
6	$C_6H_5NH_2$	$4\text{-}CH_3\text{-}C_6H_4SO_2N_3$	$C_6H_5N_3$	94	BTEAC	[94]
6	$4\text{-}Br\text{-}C_6H_4NH_2$	$4\text{-}CH_3\text{-}C_6H_4SO_2N_3$	$4\text{-}Br\text{-}C_6H_4N_3$	68	BTEAC	[94]
6	$4\text{-}Cl\text{-}C_6H_4NH_2$	$4\text{-}CH_3\text{-}C_6H_4SO_2N_3$	$4\text{-}Cl\text{-}C_6H_4N_3$	46	BTEAC	[94]
6	$4\text{-}CH_3\text{-}C_6H_4NH_2$	$4\text{-}CH_3\text{-}C_6H_4SO_2N_3$	$4\text{-}CH_3\text{-}C_6H_4N_3$	18	BTEAC	[94]
8	Indole	$[CH_3O]_2SO_2$	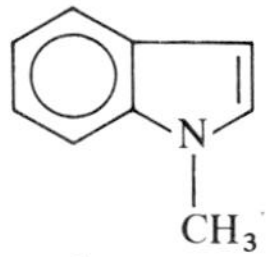 N-CH_3 indole	98	TBA^+OH^- (St)	[95]
		CH_3Cl	as above	93	TBA^+OH^- (St)	[95]
		C_2H_5Br	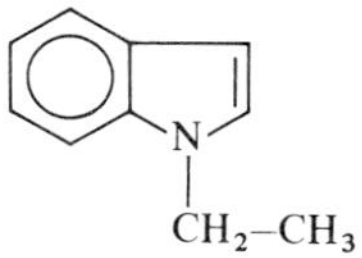 N-$CH_2\text{-}CH_3$ indole	87	BTEAC	[93]
		$[C_2H_5O]_2SO_2$	as above	95	TBA^+OH^- (St)	[95]
		C_2H_5I	as above	89	TBA^+OH^- (St)	[95]
		$n\text{-}C_4H_9Br$	as above	89	BTEAC	[93]
		$n\text{-}C_5H_{11}Br$	as above	78	TBA^+OH^- (St)	[95]
		$C_6H_5CH_2Cl$	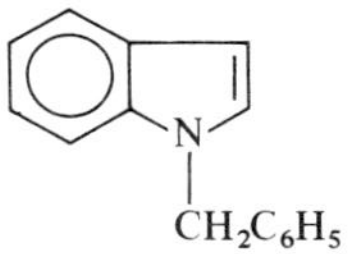 N-$CH_2C_6H_5$ indole	47	BTEAC	[93]

Table 10.5 (continued)

Number of carbons	Substrate	Electrophile	Product	% Yield	Catalyst	References
8	Indole	$C_6H_5CH_2Cl$	as above	96	BTMAC	[51]
		$C_6H_5CH_2Br$	as above	93	TBA^+OH^- (St)	[95]
10	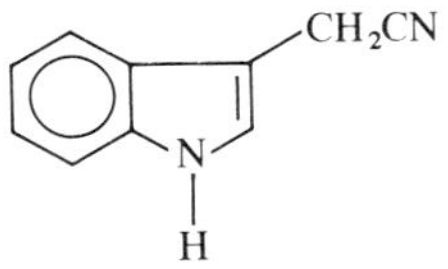 3-(cyanomethyl)-1H-indole	$C_6H_5CH_2Cl$	1-benzyl	no yield reported	BTMAC 60–70°	[51]
	carbazole	$C_6H_5CH_2Cl$	1,α,α-tribenzyl	no yield reported	BTMAC 100°	[51]
12	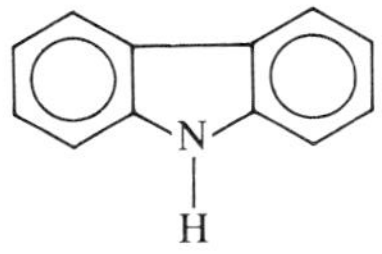	n-C_4H_9Br	mono	84	BTEAC	[93]
		$Br(CH_2)_4Br$	dimer	67	BTEAC	[93]
12	$(C_6H_5)_2NH$	$C_6H_5CH_2Cl$	$(C_6H_5)_2N$–$CH_2C_6H_5$	74	BTEAC	[93]
12	C_6H_5–NHNH–C_6H_5	CH_3I	mono	79	TBA^+OH^- (St)	[96]
		C_2H_5I	mono	42	TBA^+OH^- (St)	[96]
		C_2H_5Br	mono	7	TBA^+OH^- (St)	[96]
		n-C_3H_7Br	mono	13	TBA^+OH^- (St)	[96]
		i-C_3H_7Br	mono	13	TBA^+OH^- (St)	[96]
		CH_2=CH–CH_2Br	mono	79	TBA^+OH^- (St)	[96]
		$C_6H_5CH_2Br$	mono	79	TBA^+OH^- (St)	[96]
12	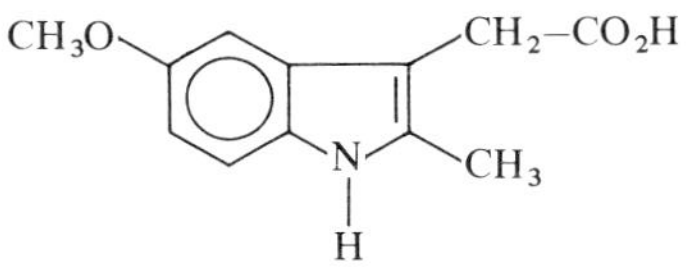 5-methoxy-2-methyl-3-(carboxymethyl)-1H-indole	$C_6H_5CH_2Cl$	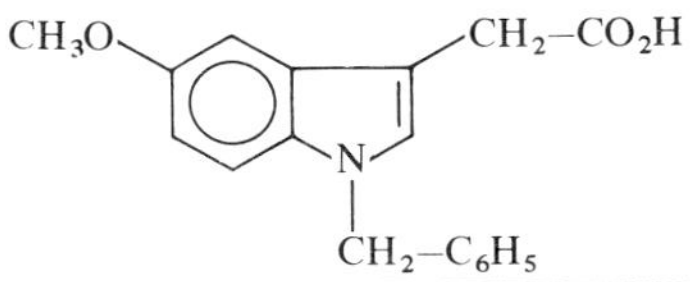 1-benzyl-3-(carboxymethyl)-5-methoxy-indole	no yield reported	BTMAC	[51]

References

1. Jarrousse, J.: Compt. Rend. *232*, 1424 (1951).
2. Makosza, M., Serafinowa, B.: Rocz. Chem. *39*, 1223 (1965).
3. Starks, C. M.: J. Am. Chem. Soc. *93*, 195 (1971).
4. Starks, C. M., Owens, R. M.: J. Am. Chem. Soc. *95*, 3613 (1973).
5. Herriott, A. W., Picker, D.: Tetrahedron Let. *1972*, 4517; J. Am. Chem. Soc. *97*, 2345 (1975).
6. Makosza, M., Kacprowicz, A., Fedorynski, M.: Tetrahedron Let. *1975*, 2119.
7. Makosza, M.: Pure Appl. Chem. *43*, 439 (1975).
8. Brandström, A.: Preparative Ion Pair Extraction. Lakemedel: Apotekarsocieteten, AB Hässle 1974.
9. Nakamura, E., Shimizu, M., Kuwajima, I.: Tetrahedron Let. *1976*, 1699.
10. Durst, H. D., Liebeskind, L.: J. Org. Chem. *39*, 3271 (1974).
11. Lange, J., Makosza, M.: Rocz. Chem. *41*, 1303 (1967).
12. Makosza, M., Goetzen, T.: Rocz. Chem. *46*, 1239 (1972).
13. Makosza, M., Serafinowa, B.: Rocz. Chem. *39*, 1595 (1965).
14. Makosza, M.: Rocz. Chem. *43*, 79 (1969).
15. Makosza, M., Serafinowa, B.: Rocz. Chem. *39*, 1805 (1965).
16. Makosza, M., Ludwikow, M., Urniaz, A.: Rocz. Chem. *49*, 297 (1975).
17. Brandström, A., Junggren, U.: Acta Chem. Scand. *23*, 2203 (1969).
18. Jonczyk, A., Fedorynski, M., Makosza, M.: Tetrahedron Let. *1972*, 2395.
19. Dietl, H. K., Brannock, K. C.: ibid *1973*, 1273.
20. Andreev, V. M., Bibicheva, A. I., Zhuravleva, M. I.: Zhur. Org. Khim. *10*, 1470 (1974).
21. Jonczyk, A., Serafin, B., Makosza, M.: Rocz. Chem. *45*, 1027 (1971).
22. Kurts, A. K., Dem'yanov, P. I., Beletskaya, I. P., Reutov, O. A.: Zhur. Org. Khim. *9*, 1313 (1973).
23. Makosza, M., Serafinowa, B.: Rocz. Chem. *39*, 1401 (1965).
24. Makosza, M., Serafinowa, B.: Rocz. Chem. *40*, 1647 (1966).
25. Makosza, M., Serafinowa, B.: Przem. Chem. *46*, 393 (1967).
26. Singh, R. K., Danishefsky, S.: J. Org. Chem. *40*, 2969 (1975).
27. Jonczyk, A., Serafin, B., Skulimowska, E.: Rocz. Chem. *45*, 1259 (1971).
28. Makosza, M., Serafinowa, B., Jawdosiuk, M.: Rocz. Chem. *41*, 1037 (1967).
29. Makosza, M., Serafinowa, B., Gajos, I.: Rocz. Chem. *43*, 671 (1969).
30. Makosza, M., Fedorynski, M.: Synthesis *1974*, 274.
31. Makosza, M.: Tetrahedron Let. *1966*, 5489.
32. Makosza, M., Jawdosiuk, M.: Bull. Acad. Polon. Sci. *16*, 589 (1968).
33. Makosza, M.: Tetrahedron Let. *1969*, 673.
34. Makosza, M., Ludwikow, M.: Bull. Acad. Polon. Sci. *19*, 231 (1971).
35. Gokel, G. W., DiBiase, S. A., Lipisko, B. A.: Tetrahedron Let. *1976*, 3495.
36. Jonczyk, A.: Bull. Acad. Polon. Sci. *22*, 849 (1974).
37. Jonczyk, A., Makosza, M.: Synthesis *1976*, 387.
38. Brandström, A., Junggren, U.: Tetrahedron Let. *1972*, 473.
39. Mikolajczyk, M., Grzejszczak, S., Zatorski, A., Montanari, F., Cinquini, M.: ibid. *1975*, 3757.
40. Reeves, W. P., Hilbrich, R. G.: Tetrahedron *32*, 2235 (1976).
41. Makosza, M., Jawdosiuk, M.: Bull. Acad. Polon. Sci. *16*, 597 (1968).
42. Netherland Pat.: 6,412,937, May 7, 1965.
43. Jonczyk, A., Serafinowa, B., Czyzewski, J.: Rocz. Chem. *47*, 529 (1973).
44. Rylski, L., Gajewski, F.: Acta. Polon. Pharm. *26*, 115 (1969).
45. Makosza, M.: Rocz. Chem. *43*, 333 (1969).
46. Lange, J.: Rocz. Chem. *42*, 1619 (1968).
47. Makosza, M., Bialecka, E., Ludwikow, M.: Tetrahedron Let. *1972*, 2391.
48. Makosza, M., Jagusztyn-Grochowska, J. M., Jawdosiuk, M.: Rocz. Chem. *45*, 851 (1971).
49. Makosza, M., Serafinowa, B.: Rocz. Chem. *40*, 1839 (1966).
50. Makosza, M., Serafinowa, B., Boleslawska, T.: Rocz. Chem. *42*, 817 (1968).

51. Suvorov, N. N., Smushkevich, Y. I., Velezheva, V. S., Rozhkov, V. S., Simakov, S. V.: Khim. Geterotsikl. Soedin *1976*, 191.
52. Makosza, M., Serafinowa, B.: Rocz. Chem. *39*, 1799 (1965).
53. Makosza, M., Goetzen, T.: Org. Prep. Proced. Int. *5*, 203 (1973).
54. Makosza, M.: Tetrahedron Let. *1969*, 677.
55. Gokel, G. W., Gerdes, H. M., Rebert, N. W.: ibid. *1976*, 653.
56. Merz, A., Märkl, G.: Angew. Chem., Int. Ed. *12*, 845 (1973).
57. Kuwajima, I., Nakamura, E.: J. Am. Chem. Soc. *97*, 3257 (1975).
58. Cederlund, B., Jesperson, A., Hörnfeldt, A. B.: Acta Chem. Scand. *25*, 3656 (1971).
59. Cederlund, B., Hörnfeldt, A. B.: Acta Chem. Scand. *25*, 3546 (1971).
60. Cardillo, G., Savoia, D., Umani-Ronchi, A.: Synthesis *1975*, 453.
61. Jończyk, A., Bańco, K., Makosza, M.: J. Org. Chem. *40*, 266 (1975).
62. Brehme, R.: Synthesis *1976*, 113.
63. Dryanska, V., Ivanov, C.: Tetrahedron Let. *1975*, 3519.
64. Jończyk, A., Serafin, B., Makosza, M.: Tetrahedron Let. *1971*, 1351.
65. Landini, D., Maia, A. M., Montanari, F., Pirisi, F. M.: Gazz. Chem. Ital. *105*, 863 (1975).
66. Cinquini, M., Montanari, F., Tundo, P.: J. Chem. Soc. Chem. Commun. *1975*, 393.
67. Jończyk, A., Serafin, B., Makosza, M.: Rocz. Chem. *45*, 2097 (1971).
68. Makosza, M., Fedorynski, M.: Rocz. Chem. *45*, 1861 (1971).
69. Jończyk, A., Pytlewski, T.: Rocz. Chem. *49*, 1425 (1975).
70. Boudeville, M. A., Abbayes, H. D.: Tetrahedron Let. *1975*, 2727.
71. Jończyk, A., Ludwikow, M., Makosza, M.: Rocz. Chem. *47*, 89 (1973).
72. Makosza, M., Jończyk, A., Serafinowa, B.; Mroczek, Z.: Rocz. Chem. *47*, 77 (1973).
73. Dalgaard, L., Kolind-Andersen, H., Lawesson, S. O.: Tetrahedron *29*, 2077 (1973).
74. Sakakibara, T., Yamada, M., Sudoh, R.: J. Org. Chem. *41*, 736 (1976).
75. Sakakibara, T., Sudoh, R.: J. Org. Chem. *40*, 2823 (1975).
76. Starks, C. M., Napier, D. R.: Brit. Pat. 1,227,144.
77. Brandström, A., Junggren, U.: Acta Chem. Scand. *23*, 3585 (1969).
78. Babayan, A. T., Indzhikyan, M. G.: Tetrahedron, *20*, 1371 (1964).
79. Brandström, A., Junggren, U.: Acta Chem. Scand. *23*, 2204 (1969).
80. Ledon, H.: Synthesis *1974*, 347.
81. D'Incan, E., Seyden-Penne, J.: Synthesis *1975*, 516.
82. Blanchard, J., Collignon, N., Savignac, P., Normant, H.: Synthesis *1975*, 655.
83. Landini, D., Rolla, F.: Synthesis *1976*, 389.
84. van Leusen, A. M., Bouma, R. J., Possel, O.: Tetrahedron Let. *1975*, 3487.
85. Samuelsson, B., Lamm, B.: Acta Chem. Scand. *25*, 1555 (1971).
86. Koutek, P., Pavlickova, L., Soucek, M.: Coll. Czech. Chem. Commun. *39*, 192 (1974).
87. Jończyk, A., Ludwikow, M., Makosza, M.: Rocz. Chem. *47*, 89 (1973).
88. Brandström, A., Junggren, U.: Acta Chem. Scand. *23*, 2536 (1969).
89. Makosza, M., Fedorynski, M.: Rocz. Chem. *49*, 1779 (1975).
90. Makosza, M.: Tetrahedron Let. *1966*, 4621.
91. Makosza, M.: Bull. Acad. Polon. Sci. *15*, 165 (1967).
92. Dietrich, B., Lehn, J. M.: Tetrahedron Let. *1973*, 1225.
93. Jończyk, A., Makosza, M.: Rocz. Chem. *49*, 1203 (1975).
94. Nakajima, M., Anselme, J. P.: Tetrahedron Let. *1976*, 4421.
95. Barco, A., Benetti, S., Pollini, G. P., Baraldi, P. G.: Synthesis *1976*, 124.
96. Nielsen, K. B., Bernatek, E.: Acta Chem. Scand. *26*, 4130 (1972).
97. Fiaud, J. C.: Tetrahedron Let. *1975*, 3495.
98. Makosza, M., Jonczyk, A.: Org. Syn. *55*, 91 (1976).
99. Makosza, M., Czyzewski, J., Jawdosink, M.: Org. Syn. *55*, 99 (1976).

11. Oxidation Reactions

11.1. Introduction

Oxidation is a reaction which can obviously assume many appearances. The conversion of an alcohol to a ketone or an aldehyde to an acid is unquestionably an oxidation, but the conversion of a primary amine to a nitrile or a bromide to an alcohol is just as much an oxidation process although it may be less obviously so. Included in this chapter are a wide variety of substrates and oxidizing reagents. A notable exception is the superoxide ion which can effect both displacement and oxidation. The reactions of superoxide ion in two-phase systems are treated separately in Chap. 8.

Most of the oxidation reactions discussed in this chapter are reactions of anions which are solubilized by the phase transfer method. The catalyst transports into solution an oxidizing anion which reacts stoichiometrically with a substrate. It is important to distinguish this kind of catalysis from catalytic oxidation in which the oxidizing agent is continually regenerated (Sect. 11.5). With the exception of carbanion oxidation (Sect. 11.7) and phosphorylation (Sect. 11.8), this chapter deals with ionic oxidizing reagents in nonpolar solutions.

11.2 Permanganate Ion

Gibson and Hosking were the first to report on phase transfer catalytic oxidation [1]. They found that it was possible to oxidize water insoluble substrates in aprotic media by ion-pairing the anion with a lipophilic cation. Permanganate anion was extracted from an aqueous reservoir of potassium permanganate by exchange of the chloride ion of methyltriphenylarsonium chloride. Methyltriphenylarsonium permanganate in chloroform was found to oxidize 1-octene, 1-propanol, 2-propanol, 4-heptanol, 1-nitropropane, 2-nitropropane, and formic acid but it did not oxidize t-butanol, toluene, ethyl acetate, diethyl ether, acetone, or dipropyl ketone. Gibson and Hosking outlined what appears to be the first statement of phase transfer principles but apparently did not extend their work.

Starks reported that although no reaction was observed at room temperature between aqueous neutral $KMnO_4$ and 1-octene, when a small amount of quaternary ammonium catalyst was added, an immediate reaction took place which generated "so much heat that the reaction mixture could not be contained in the flask" [2]. Under controlled conditions, 1-octene was converted quantitatively to heptanoic acid and 1-decene to nonanoic acid (91%) as shown in equation 11.1.

$$CH_3(CH_2)_7CH=CH_2 \xrightarrow[Q^+X^-]{KMnO_4} CH_3(CH_2)_7COOH \qquad (11.1)$$

Herriott and Picker [3] studied the extractability of permanganate ion from aqueous solution into benzene with various onium salts. Their findings for the tetrabutylammonium case are recorded in Table 11.1. In general, those salts known to be effective catalysts in other phase transfer reactions (i.e., benzyltriethylammonium chloride, tetrabutylammonium bromide, tetrabutylphosphonium chloride, cetyltrimethylammonium bromide, and tricaprylmethylammonium chloride) extracted permanganate ion to about the same extent. Tetramethylammonium chloride and sodium dodecyl sulfate were unsuccessful. As a representative catalyst (TBAB), the extraction data reported in Table 1 indicate that only a slight excess of catalyst over the amount of permanganate required need be added. Swern and Okimoto report similar results although these authors found evidence for some decomposition [4]. The two-phase aqueous permanganate-benzene system containing tricaprylmethylammonium chloride was used to oxidize several organic substrates [3]. The results are recorded in Table 11.2.

Table 11.1. Extraction of permanganate into benzene by tetrabutylammonium bromide

mmoles MnO_4^-	mmoles Q^+	mmoles MnO_4^- in benzene
1.00	0.298	0.30
1.00	0.79	0.71
1.00	1.54	0.95
1.00	3.12	0.97

Table 11.2. Catalytic permanganate oxidation

Starting material	Product	% Yield
1-octene	Heptanoic acid	81
E-stilbene	Benzoic acid	95
n-octanol	Octanoic acid	47
Benzyl alcohol	Benzoic acid	92
Phenylacetonitrile	Benzoic acid	86

Recently, the two-phase permanganate oxidation of piperonal to piperonylic acid (Eq. 11.2) has been carried out. A 65% yield was obtained when cetyltrimethylammonium bromide was present in the reaction mixture [5].

Oxidation Reactions

$$\text{(benzo[d][1,3]dioxole-5-carbaldehyde)} \longrightarrow \text{(benzo[d][1,3]dioxole-5-carboxylic acid)} \quad (11.2)$$

The two-phase permanganate oxidation of olefins generally affords products of oxidative cleavage. Weber and Shepherd found that when benzyltriethylammonium chloride was used as catalyst and the temperature was maintained near 0°C, internal olefins were oxidized by basic permanganate in dichloromethane to the corresponding cis-glycols in moderate yields (Eq. 11.3) [6]. Cyclohexene, cis-cyclooctene and trans-cyclododecene were dihydroxylated by this method in 15%, 50%, and 50% yields

$$R-CH=CH-R' \xrightarrow[Q^+X^-]{KMnO_4, \, 0\,°C} R-\underset{\underset{OH}{|}}{C}H-\underset{\underset{OH}{|}}{C}H-R' \quad (11.3)$$

respectively. If the product glycol is appreciably water soluble, overoxidation and cleavage become problems. Okimoto and Swern [4] have recently found that oleyl and elaidyl alcohols can be oxidized by a similar approach to respectively *erythro* and *threo*-9,10-dihydroxyoctadecanols.

Potassium permanganate can be readily solubilized in nonpolar media by complexing the potassium ion with a crown ether. Sam and Simmons [7] found that dicyclohexyl-18-crown-6 could solubilize solid potassium permanganate in benzene to the extent of about 0.06 molar. The resulting "purple benzene" solution was used to oxidize a number of organic substrates in good to excellent yield. Some of their results are recorded in Table 11.3. Similar results have recently been reported by other workers [8, 9].

18-Crown-6 catalyzes the solid-liquid permanganate oxidation of catechols in dichloromethane solution to *ortho*-quinones [10]. This appears to be a mild and effective method for generating these relatively unstable molecules (Eq. 11.4).

$$\text{3,5-di-}t\text{-butylcatechol} \xrightarrow[\substack{18-\text{Crown}-6 \\ CH_2Cl_2 \\ (97\%)}]{KMnO_4} \text{3,5-di-}t\text{-butyl-}o\text{-benzoquinone} \quad (11.4)$$

Table 11.3. Oxidation with DC-18-C-6 potassium permanganate complexes [7]

Substrate	Product	% Yield
Benzhydrol	Benzophenone	100
Toluene	Benzoic acid	100
Xylene	Toluic acid	78
Cyclohexene	Adipic acid	100
α-pinene	Pinonic acid	90
E-stilbene	Benzoic acid	100
n-heptanol	Heptanoic acid	70
Benzyl alcohol	Benzoic acid	100
Benzaldehyde	Benzoic acid	100

11.3 Chromate Ion

Potassium chromate in HMPT reacts with a variety of substrates in a crown catalyzed nucleophilic displacement. Both dicyclohexyl- and dibenzo-18-crown-6 catalyze the formation of chromate esters which decompose according to equation 11.5 to yield oxidized product [11].

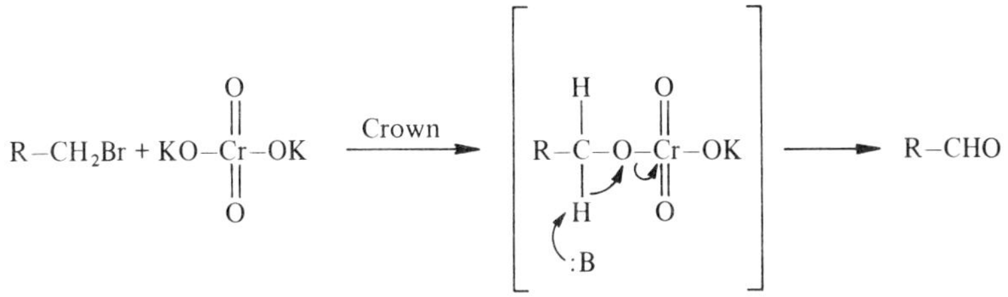

$$R-CH_2Br + KO-\underset{\underset{O}{\|}}{\overset{\overset{O}{\|}}{Cr}}-OK \xrightarrow{Crown} \left[R-\underset{\underset{H}{|}}{\overset{\overset{H}{|}}{C}}-O-\underset{\underset{O}{\|}}{\overset{\overset{O}{\|}}{Cr}}-OK \quad \overset{\curvearrowleft}{:B} \right] \longrightarrow R-CHO \quad (11.5)$$

Similar results have been obtained with a polymer bound ammonium chromate system [12]. The conversion of a number of bromides into carbonyl compounds is recorded in Table 11.4.

Table 11.4. Chromate ion oxidation

Substrate	% Yield	References
Ethyl 4-bromo-3-methylbutenoate	95	[12]
1-bromo-3-methyl-2-butane	78	[11]
Benzyl chloride	80	[11]
Benzyl chloride	95	[12]
Benzyl bromide	98	[12]
n-bromooctane	20	[11]
Geranyl bromide	82	[11]
Geranyl bromide	95	[12]
Diphenylmethyl bromide	95	[12]
9-bromofluorene	97	[12]
Farnesyl bromide	80	[11]

11.4 Hypochlorite Ion

Hypochlorite ion is an effective oxidizing agent for a variety of substrates when the reactions are conducted under phase transfer conditions [13]. Tetrabutylammonium ion catalyzes the reaction of hypochlorite with benzylic alcohols in dichloromethane or ethyl acetate solution to yield alcohols and ketones. Likewise, secondary alcohols are oxidized to ketones according to equation 11.6. The results of a series of such

$$Ar-CH_2-OH \xrightarrow{ClO^- Q^+} Ar-CHO \quad (11.6)$$

oxidations are recorded in Table 11.5. It is interesting that ethyl acetate seems to be a more effective solvent for this reaction than dichloromethane, a fact to which Lee and Freedman call attention [13]. Hypochlorite also oxidizes amines under phase transfer conditions, to either ketones (Eq. 11.7) or nitriles (Eq. 11.8). Apparently, an N-chloroimine is the intermediate which either hydrolyzes under the reaction conditions to a ketone or eliminates HCl to yield a nitrile, depending on whether the alpha carbon of the starting amine is primary or secondary.

$$R\text{–}\underset{\underset{R'}{|}}{C}H\text{–}NH_2 \xrightarrow{OCl^-} \left[R\text{–}\underset{\underset{R'}{|}}{C}=N\text{–}Cl\right] \xrightarrow{H_2O} R\text{–}\underset{\underset{R'}{|}}{C}=O \qquad (11.7)$$

$$R\text{–}CH_2\text{–}NH_2 \xrightarrow{OCl^-} \left[R\text{–}\underset{\underset{H}{|}}{C}=N\text{–}Cl\right] \xrightarrow{-HCl} R\text{–}C\equiv N \qquad (11.8)$$

Table 11.5. Hypochlorite oxidations [13]

Substrate	Product	Solvent	% Yield
$C_6H_5CH_2OH$	Aldehyde	CH_2Cl_2	76
$4\text{-}ClC_6H_4CH_2OH$	Aldehyde	CH_2Cl_2	82
$4\text{-}CH_3C_6H_4CH_2OH$	Aldehyde	CH_2Cl_2	78
	Aldehyde	EtOAc	100
$2\text{-}CH_3OC_6H_4CH_2OH$	Aldehyde	CH_2Cl_2	79
	Aldehyde	EtOAc	92
$4\text{-}CH_3OC_6H_4CH_2OH$	Aldehyde	CH_2Cl_2	79
	Aldehyde	EtOAc	92
$(C_6H_5)_2CHOH$	Ketone	CH_2Cl_2	82
9-fluorenol	Ketone	CH_2Cl_2	92
$c\text{-}C_7H_{13}OH$	Ketone	EtOAc	89
2-norbornanol	Ketone	EtOAc	36
$4\text{-}t\text{-}C_4H_9\text{-}c\text{-}C_6H_{10}OH$	Ketone	EtOAc	49
$c\text{-}C_6H_{11}NH_2$	Cyclohexanone	EtOAc	98
2-norbornylamine	2-norbornanone	EtOAc	84
$(C_6H_5)_2CHNH_2$	Benzophenone	EtOAc	94
$C_6H_5CH(CH_3)NH_2$	Acetophenone	EtOAc	98
$c\text{-}C_6H_{11}CH_2NH_2$	Cyclohexyl cyanide	EtOAc	76
$n\text{-}C_8H_{17}NH_2$	$n\text{-}C_7H_{15}CN$	EtOAc	60

11.5 Catalytic Oxidation

Starks has carried out an extensive study of olefin oxidation under phase transfer conditions. Unfortunately, with the exception of his work on permanganate ion which has been discussed, the results of these studies have appeared only in the patent literature [14, 15]. The oxidants used were 30% hydrogen peroxide or paraperiodic acid. Both of these are water soluble oxidants. The well known tendency of aqueous hydrogen peroxide to undergo metal catalyzed decomposition was circumvented by phase-transferring peroxide or paraperiodate into hydrocarbon solution. Starks found that tertiary amines, quaternary ammonium, and quaternary phosphonium salts could all be used in this application. The oxidations were co-catalyzed by osmium tetraoxide, ruthenium tetraoxide, and other metal salts. Osmium tetraoxide reacts with alkenes to yield cis-1,2-glycols and hydrogen peroxide reoxidizes the osmium. Cis-1,2-glycols are therefore the principal products of oxidation with 30% hydrogen peroxide and a catalytic amount of osmium tetraoxide, whereas, paraperiodic acid cleaves 1,2-glycols to aldehydes in addition to reoxidizing osmium. Ruthenium tetraoxide cleaves carbon-carbon double bonds to yield carboxylic acids. Ruthenium dioxide in this reaction can be reoxidized by paraperiodic acid or hydrogen peroxide to yield the reactive ruthenium tetraoxide. Examples of these systems are recorded in Tables 11.6 and 11.7.

Table 11.6. Catalytic oxidations under phase transfer conditions [14, 15]

Olefin	Oxidant	Metal	Catalyst	Product	% Yield
1-octene	Paraperiodic acid	OsO_4	TCMAC	Heptanal	86
				Heptanoic acid	12.8
1-octene	Paraperiodic acid	RuO_2	TCMAC	Heptanoic acid	99
1-octene	Paraperiodic acid	RuO_2	Tridodecylamine	Heptanoic acid	93

Table 11.7. Catalytic oxidation of cyclohexene with hydrogen peroxide

Metal compound	Yield of Products (mole %)			% H_2O_2 Decomposed	
	Cyclohexene Oxide	Cyclohexen-2-one & -2-ol	1,2-Cyclohexanediol	Presence of Q^+X^-	Absence of Q^+X^-
OsO_4	–	–	52.0	3.3	100
MoO_3	28.4	–	28.3	2.6	91
H_2WO_4	42.1	–	23.6	4.0	18
$SeOCl_2$	4.1	10.2	46.2	5.0	–
V_2O_5	–	88.6	–	5.6	95
Cr_2O_3	–	73.5	–	3.0	–
TiO_2	3.8	70.4	–	6.7	–
$CeSO_4$	2.3	65.8	–	33.5	100
NiO_2	3.5	59.0	–	5.6	–
$MnCl_2$	3.5	78.5	–	21.1	–
$CoCl_2$	0.9	69.5	–	20.1	–
PtO_2	4.9	74.4	–	44.9	100
$FeSO_4$	3.8	65.1	–	26.8	97
$Pb(OAc)_2$	4.3	76.2	–	5.0	–

Q^+ = tridecylmethylammonium.

Oxidation Reactions

The preparation of tetraethylammonium periodate and its chloroform solubility have been reported. Little other work has been done, however, with periodate anion as an oxidant in aprotic organic solvents [16].

11.6 Singlet Oxygen

The photochemical generation of singlet oxygen has suffered from the insolubility of those anionic dyes which are commonly used as sensitizers (see Fig. 11.1). It has recently been found that the photosensitizing dyes eosin and rose bengal can be solu-

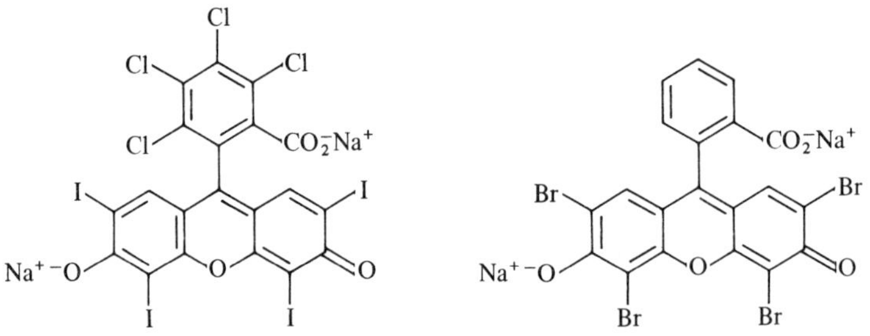

Rose bengal   Eosin   Figure 11.1

bilized in dichloromethane or carbon disulfide by complexation with 18-crown-6 or ion-pairing with the tricaprylmethylammonium ion [17]. By this method, singlet oxygen was generated which added to anthracene (Eq. 11.9) or to tetramethylethylene (Eq. 11.10) in good yield.

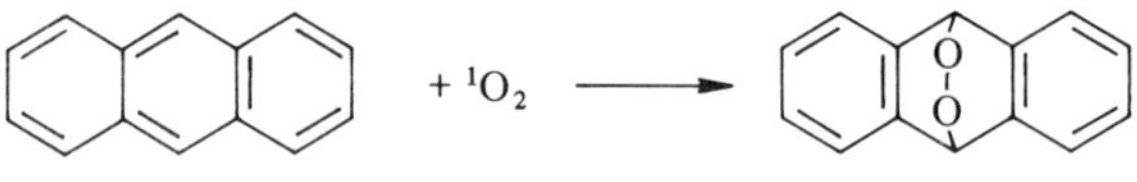

$$\text{anthracene} + {}^1O_2 \longrightarrow \text{anthracene endoperoxide} \tag{11.9}$$

$$\begin{array}{c}H_3C\\H_3C\end{array}\!\!C=C\!\!\begin{array}{c}CH_3\\CH_3\end{array} + {}^1O_2 \longrightarrow \begin{array}{c}H_2C\\H_3C\end{array}\!\!C\!-\!C\!\!\begin{array}{c}CH_3\\OOH\\CH_3\end{array} \tag{11.10}$$

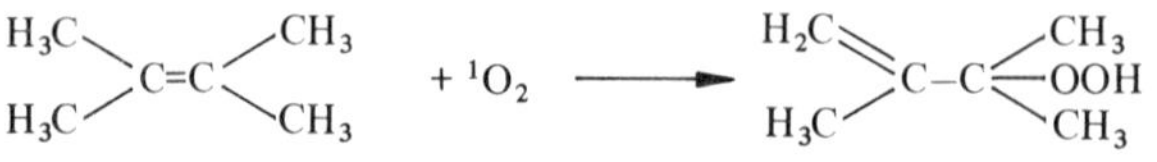

11.7 Oxidation of Anions

Carbanions have been oxidized under phase transfer conditions in the presence of both crown ethers and cryptates. The substrate which has been most studied is fluorene which undergoes phase transfer catalyzed air oxidation to yield fluorenone in high yield according to equation 11.11 [10, 18]. Crown complexed t-butoxide in

THF containing a small amount of water has also been found to effect oxidation followed by Haller-Bauer cleavage to give 2-carboxybiphenyl [19].

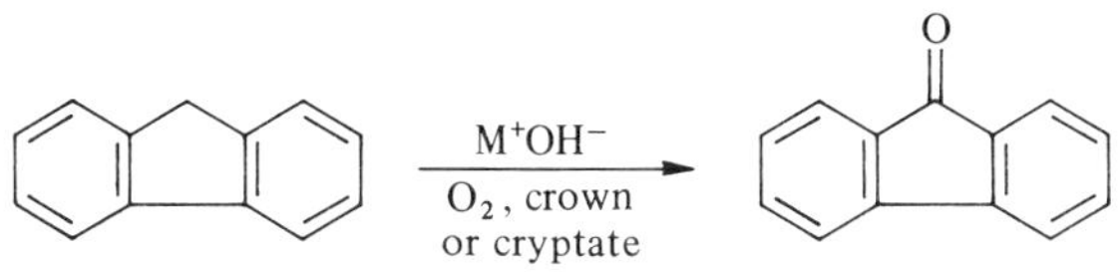

$$\text{fluorene} \xrightarrow[\substack{O_2,\text{ crown} \\ \text{or cryptate}}]{M^+OH^-} \text{fluorenone} \qquad (11.11)$$

11.8 Phosphorylation

Dialkyl phosphites have been halogenated to the corresponding dialkyl phosphorohalidates under basic phase transfer conditions by reaction with tetrahalomethanes [20–22]. The tetrahalomethane is reduced to the corresponding haloform in the reaction and the dialkyl phosphite is oxidized. The reaction is generally carried out under conditions such that the initial dialkyl phosphorohalidate [20] undergoes further reaction. Thus if the halogenation is carried out in the presence of an alcohol, the product is a trialkyl phosphate (Eq. 11.12 and Table 11.8) [21], whereas if the reaction is conducted in the presence of a primary or secondary amine, a phosphoroamidate results (Eq. 11.13 and Table 11.9) [22].

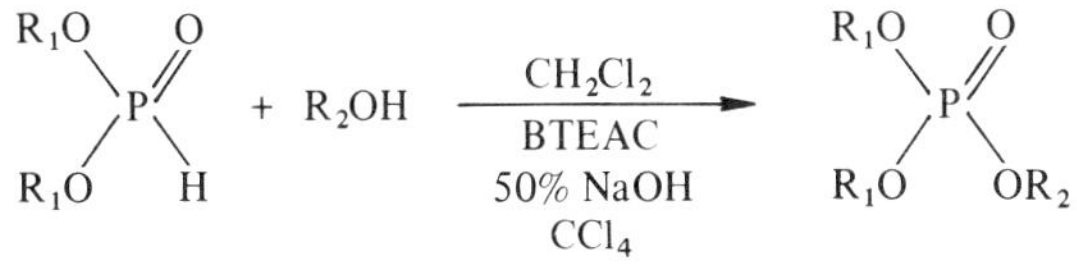

$$(R_1O)_2P(O)H + R_2OH \xrightarrow[\substack{\text{BTEAC} \\ 50\% \text{ NaOH} \\ CCl_4}]{CH_2Cl_2} (R_1O)_2P(O)(OR_2) \qquad (11.12)$$

Table 11.8. Phosphorylation of alcohols

R_1	R_2	% Yield
C_2H_5	$n\text{-}C_4H_9$	71
C_2H_5	$C_6H_5CH_2$	89
C_2H_5	$i\text{-}C_3H_7$	37
C_2H_5	$n\text{-}C_3H_7$	67
C_2H_5	$i\text{-}C_4H_9$	56
C_2H_5	C_6H_5	50
$n\text{-}C_4H_9$	C_2H_5	71
$n\text{-}C_4H_9$	$i\text{-}C_4H_9$	62
$n\text{-}C_4H_9$	$C_6H_5CH_2$	95
$n\text{-}C_4H_9$	$i\text{-}C_3H_7$	35
$n\text{-}C_4H_9$	C_6H_5	94

Oxidation Reactions

$$\underset{R_1O}{\overset{R_1O}{>}}\!\!\!\underset{H}{\overset{O}{P}} + CX_4 \xrightarrow[\underset{|}{R_2NH}]{\underset{20\% \text{ NaOH}}{\overset{CH_2Cl_2}{\text{BTEAC}}}} \underset{R_1O}{\overset{R_1O}{>}}\!\!\!\underset{N}{\overset{O}{P}}\!\!\underset{R_3}{\overset{R_2}{\diagdown}} \qquad (11.13)$$

Table 11.9. Phosphorylation of amines [22]

R_1	R_2	R_3	% Yield	X
C_2H_5	H	C_6H_5	77	Cl
C_2H_5	H	C_6H_5	84	Br
C_2H_5	H	$c\text{-}C_6H_{11}$	89	Cl
C_2H_5	H	$C_6H_5CH_2$	85	Cl
C_2H_5	C_2H_5	C_2H_5	86	Cl
C_2H_5	H	C_2H_5	81	Cl
$C_6H_5CH_2$	H	C_6H_5	78	Br
$C_6H_5CH_2$	H	$c\text{-}C_6H_{11}$	93	Cl
$C_6H_5CH_2$	H	$C_6H_5CH_2$	83	Cl
$C_6H_5CH_2$	C_2H_5	C_2H_5	91	Cl
$t\text{-}C_4H_9$	H	$c\text{-}C_6H_{11}$	90	Br
$t\text{-}C_4H_9$	H	$C_6H_5CH_2$	92	Br

References

1. Gibson, N. A., Hosking, J. W.: Aust. J. Chem. *18*, 123 (1965).
2. Starks, C. M.: J. Am. Chem. Soc. *93*, 195 (1971).
3. Herriott, A. W., Picker, D.: Tetrahedron Let. *1974*, 1511.
4. Okimoto, T., Swern, D.: private communication.
5. Menger, F. M., Rhee, J. U., Rhee, H. K.: J. Org. Chem. *40*, 3803 (1975).
6. Weber, W. P., Shepherd, J. P.: Tetrahedron Let. *1972*, 4907.
7. Sam, D. J., Simmons, H. E.: J. Am. Chem. Soc. *94*, 4024 (1972).
8. Landini, D., Montanari, F., Pirisi, F. M.: J. C. S. Chem. Commun. *1974*, 879.
9. Landini, D., Maia, A. M., Montanari, F., Pirisi, F. M.: Gazz. Chim. Ital. *105*, 863 (1975).
10. Durst, H. D.: private communication.
11. Cardillo, G., Orena, M., Sandri, S.: J. C. S. Chem. Commun. *1976*, 190.
12. Cardillo, G., Orena, M., Sandri, S.: Tetrahedron Let. *1976*, 3985.
13. Lee, G. A., Freedman, H. H.: Tetrahedron Let. *1976*, 1641.
14. Starks, C. M., Napier, D. R.: South African Pat. 71/1495, August 3, 1971.
15. Starks, C. M., Washecheck, P. H.: U. S. Pat. 3,547,962, December 15, 1970.
16. Qureshi, A. K., Sklarz, B.: J. Chem. Soc. (C) *1966*, 412.
17. Boden, R. M.: Synthesis *1975*, 783.
18. Dietrich, B., Lehn, J. M.: Tetrahedron Let. *1973*, 1225.
19. Gokel, G. W., DiBiase, S. A.: Eighth Central Regional ACS meeting, Akron, Ohio 1975.
20. Gajda, T., Zwierzak, A.: Synthesis *1976*, 243.
21. Zwierzak, A.: Synthesis *1976*, 305.
22. Zwierzak, A.: Synthesis *1976*, 507.

12. Reduction Techniques

12.1 Introduction

The two-phase catalytic method affords numerous advantages in many reactions because an inorganic salt dissolved in water can be phase-transferred into a nonpolar medium where it enjoys considerable solubility and frequently enhanced reactivity. The applicability of the phase transfer technique in reduction reactions has been limited because the nature of the reagents are such that there is little real advantage offered in executing most reduction reactions under two-phase conditions. The reducing reagents one commonly considers in organic applications are lithium aluminum hydride and borohydride, ion-paired with lithium, sodium or potassium cation. The phase transfer of lithium aluminum hydride from an aqueous system is obviously impractical and in any event, this reagent suffices in most organic applications and can be used in nonpolar media without complication [1].

Sodium and potassium borohydride can be dissolved in water and then phase-transferred into a nonpolar solution, but such solvents as benzene or toluene offer little advantage over alcohols which commonly dissolve organic substrates. One application in which the two-phase technique does afford special convenience is the two-phase reduction of acid chlorides, usually a heterogeneous reaction conducted in dioxane solution [2]. The hydride reductions and several other techniques are discussed in this chapter.

12.2 Borohydrides

Borohydride anion has been phase-transferred from an aqueous basic solution of sodium or potassium borohydride into benzene solution by a variety of catalytic systems including quaternary ammonium and phosphonium salts, crowns and cryptates. The rates of reaction of several quaternary onium borohydrides (under catalytic conditions) with 2-octanone (see Eq. 12.1) have been determined and give an appro-

$$CH_3-CO-(CH_2)_5-CH_3 \xrightarrow[C_6H_6]{Q^+BH_4^-} CH_3-CHOH-(CH_2)_5-CH_3 \qquad (12.1)$$

ximate measure of catalyst efficiency in these reactions [3]. 2-Octanone is reduced by aqueous borohydride to 2-octanol in benzene solution when tricaprylylmethylammonium chloride, (80% in 6.5 h) [3–5], N-dodecyl-N-methylephedrinium bromide (100%, 0.5 h) [5], or N-methyl-N-dodecyl-N,N-*bis*-hydroxyethylammonium bromide (100%, 6.5 h) is used as catalyst [5, 6].

Those catalysts containing a β-hydroxyl group appear to be more effective catalysts, allowing reduced reaction times and affording higher yields. The activating effect of the hydroxyl group is not understood (see also Sect. 2.3) but it may be the result of electrophilic catalysis involving a hydrogen bond from hydroxyl to carbonyl. When the hydroxylated catalyst N-dodecyl-N-methylephedrinium bromide is used as a phase transfer agent in the reaction of aqueous potassium borohydride with ketones in benzene solution, yields are good and reaction times short (see Eq. 12.2 and Table 12.1). Despite the fact that this quaternary salt is chiral, no asymmetric induction was witnessed in the reduction of either acetophenone or 2-octanone [5].

$$R^1-CO-R^2 \xrightarrow[C_6H_6/H_2O]{Q^+BH_4^-} R^1-CHOH-R^2 \qquad (12.2)$$

Table 12.1. Reduction of carbonyl compounds with borohydride under phase transfer conditions [5]

Carbonyl compound	% Yield
C_6H_5-CHO	95
$4\text{-}Cl-C_6H_4-CHO$	94
$i\text{-}C_3H_7-CO-C_3H_7\text{-}i$	86
$n\text{-}C_6H_{13}-CO-CH_3$	100
$C_6H_5-CO-CH_3$	97
$C_6H_5-CO-C_2H_5$	100
4-*t*-butylcyclohexanone	100
$C_6H_5-CO-C_6H_5$	100
$C_6H_5-CH_2-CO-CH_2-C_6H_5$	100

The crown ether catalyzed phase transfer reductions which have been reported are similar to the quaternary ion catalyzed reactions in the sense that if the proper catalyst is chosen the reactions yield well but appear to offer no special advantage over more traditional reduction methodology. The dibenzo-18-crown-6 catalyzed sodium borohydride reduction of several ketones in boiling toluene (5 h reaction time) has been reported [7]. By this method, acetophenone, cyclohexanone and 2-heptanone were reduced in 49%, 50% and 41% yields respectively.

The more lipophilic dicyclohexyl-18-crown-6 catalyzes the reaction of aqueous potassium borohydride with 2-octanone in refluxing benzene [6, 8]. After 2.5 h, 2-octanol was isolated in 92% yield. Likewise, this reaction can be effected by alkyl substituted [2.2.2]-cryptates in 97% yield [6].

12.3 Stoichiometric Reduction Systems

The desire for a mild reducing reagent like borohydride which would be soluble in nonpolar media has led to the development of several ammonium borohydrides. Both cetyltrimethylammonium borohydride $\{[C_{16}H_{33}(CH_3)_3N]BH_4\}$, a white granular solid, and tricaprylylmethylammonium borohydride $\{[(C_8H_{17})_3CH_3N]BH_4\}$, a grease, have been prepared and found to readily reduce aldehydes and acid chlorides in benzene, hexane or mineral oil [9]. These reagents did not effectively reduce nitriles or esters. It is interesting that the reactivity of these quaternary ammonium borohydrides differs substantially in nonpolar and alcoholic or aqueous solvents. For example, after two hours at 25 °C, cetyltrimethylammonium borohydride had reduced acetophenone to the corresponding alcohol in water (100%) or isopropanol (59%) but had failed to react in benzene even after 4 h. Similar results were obtained for tricaprylylmethylammonium borohydride [9], and even sodium borohydride in diglyme [10].

The reducing power of tetralkylammonium borohydrides in dichloromethane is similar to that observed in water or alcohol. A convenient preparation of tetrabutylammonium borohydride has recently been reported [11] and the utility of the reagent has been surveyed [12]. In dichloromethane solution, tetrabutylammonium borohydride readily reduces aldehydes, ketones and acid chlorides while reacting only very slowly with esters (see Table 12.2).

Brandström and coworkers have shown that tetrabutylammonium borohydride in dichloromethane solution will react with alkyl halides (methyl iodide, ethyl bromide or 1,2-dichloroethane) to yield solutions of diborane (see Eq. 12.3). Although diborane cannot be distilled from these solutions and its presence is therefore not

$$(n\text{-}C_4H_9)_4N^+BH_4^- + R\text{-}X \rightarrow (n\text{-}C_4H_9)_4N^+X^- + 1/2\, B_2H_6 + RH \qquad (12.3)$$

rigorously proved, the reactivity of these systems is such that the presence of diborane can be presumed. Aldehydes, esters, nitriles and even carboxylic acids are reduced by these "diborane"/dichloromethane solutions. Moreover, treatment of alkenes with these "diborane" solutions followed by oxidation with 30% hydrogen peroxide yields products expected from a hydroboration/oxidation sequence. Phenyl allyl ether, for example, affords 3-phenoxy-1-propanol (70%, see Eq. 12.4) under these conditions.

$$C_6H_5\text{-}O\text{-}CH_2CH\text{=}CH_2 \xrightarrow[CH_2Cl_2]{\text{"}B_2H_6\text{"}} \xrightarrow{[O]} C_6H_5\text{-}O\text{-}CH_2CH_2CH_2OH \qquad (12.4)$$

Several examples of reduction by tetrabutylammonium borohydride in the presence and absence of alkyl halide are recorded in Table 12.2.

Tetrabutylammonium cyanoborohydride has also been used under noncatalytic conditions [13]. In hexamethylphosphoric triamide (HMPA, HMPT), $(n\text{-}C_4H_9)_4N^+BH_3CN^-$ reduces primary alkyl iodides to alkanes in high yield. Primary alkyl bromides are about half as reactive as the iodides and primary chlorides and tosylates are virtually inert, as are the cyano, nitro and carbonyl groups.

Reduction Techniques

Table 12.2. Stoichiometric reductions of organic substrates in dichloromethane

Substrate	Product	Reductant	% Yield	Ref.
$(CH_3)_3C-CHO$	$(CH_3)_3C-CH_2OH$	Bu_4NBH_4	94	[12]
C_6H_5-CHO	$C_6H_5-CH_2OH$	Bu_4NBH_4	92	[12]
$4\text{-}Cl-C_6H_4-CHO$	$4\text{-}Cl-C_6H_4-CH_2OH$	Diborane	98	[11]
$C_6H_5-CH=CH-CHO$	$C_6H_5-CH=CH-CH_2OH$	Bu_4NBH_4	73	[12]
Pinacolone	$(CH_3)_3C-CHOH-CH_3$	Bu_4NBH_4	40	[12]
Cyclohexanone	Cyclohexanol	Bu_4NBH_4	75	[12]
$C_6H_5-CO-CH_3$	$C_6H_5-CHOH-CH_3$	Bu_4NBH_4	69	[12]
$C_6H_5-CO-Cl$	$C_6H_5-CH_2OH$	Bu_4NBH_4	100	[12]
C_6H_5-COOH	$C_6H_5-CH_2OH$	Diborane	98	[11]
$3\text{-}NO_2-C_6H_4-COOH$	$3\text{-}NO_2-C_6H_4-CH_2OH$	Diborane	90	[11]
$4\text{-}CH_3O-C_6H_4-CH_2COOH$	$4\text{-}CH_3O-C_6H_4-CH_2CH_2OH$	Diborane	90	[11]
$4\text{-}NO_2-C_6H_4-COOCH_3$	$4\text{-}NO_2-C_6H_4-CH_2OH$	Diborane	80	[11]
C_6H_5-CN	$C_6H_5-CH_2NH_2$	Diborane	95	[11]
phenacyl ethyl ester ($C_6H_5COCH_2CO_2C_2H_5$)	γ-phenyl-γ-butyrolactone	Bu_4NBH_4	70	[12]

The reducing power of this system is enhanced by the addition of acid. In the presence of 0.15 N sulfuric acid, aldehydes are rapidly reduced. If the concentration of acid is increased to 1.5 N, ketones are rapidly reduced. The addition of acid appears to have no influence on the halide reduction reaction. In Table 12.3 are recorded several applications of tetrabutylammonium cyanoborohydride as a reducing agent.

Table 12.3. Stoichiometric cyanoborohydride reductions in HMPA [13]

Substrate	Product	Acid conc.	% Yield
1-bromodecane	Decane	–	46
1-iododecane	Decane	–	81
1-tosyloxydecane	Decane	–	5
$CH_3(CH_2)_7CHO$	$CH_3(CH_2)_8OH$	0.15 N	92
$4\text{-}NC\text{-}C_6H_4CHO$	$4\text{-}NC\text{-}C_6H_4CH_2OH$	0.15 N	43
9-anthraldehyde	9-hydroxymethylanthracene	0.15 N	95
$CH_3-CO-(CH_2)_8CH_3$	$CH_3-CHOH-(CH_2)_8CH_3$	1.5 N	84
4-t-butylcyclohexanone	4-t-butylcyclohexanol	1.5 N	88
4-acetylbiphenyl	4-α-hydroxyethylbiphenyl	1.5 N	86
Cholestan-3-one	Cholestan-3-ol	1.5 N	94

12.4 Other Catalytic Reductions

Cyanoborohydride has been used under catalytic nonpolar conditions [14] as well as stoichiometrically in cation solvating media [13]. Sodium cyanoborohydride can be solubilized in dichloromethane by complexation with 18-crown-6. In dichloromethane O-methylsulfoxonium salts are reduced to the corresponding sulfides in good yield (see Sect. 13.8 and Eqs. 12.5 and 13.24). The methyl fluorosulfonate salts of dibutyl-, diphenyl-, and dibenzyl sulfoxides are reduced to the corresponding sulfides in 85%, 77% and 91% yields, respectively [14].

$$R_2S^{\pm}OCH_3 \; FSO_3^- \xrightarrow[CH_2Cl_2]{NaBH_3CN/18\text{-}C\text{-}6} R_2S \tag{12.5}$$

Formamidinesulfinic acid (obtained by the oxidation of thiourea with hydrogen peroxide) has been used to reduce disulfides to sulfides (see Eqs. 12.6 and 13.22) and N-tosylsulfilimines to sulfides (see Eqs. 12.7 and 13.23), under phase transfer catalytic conditions in the presence of hexadecyltributylphosphonium bromide [15]. Diphenyl, dibenzyl and dibutyldisulfides were reduced by this method to the corresponding sulfides in 72%, 62% and 90% yields respectively. Examples of the reduction of N-tosylsulfilimines are recorded in Table 12.4.

$$R-S-S-R + H_2N-C(=NH)-SO-OH \xrightarrow[Q^+X^-]{NaOH} R-S-R \tag{12.6}$$

$$\overset{\overset{\displaystyle N-Ts}{\|}}{R'-S-R} + H_2N-C(=NH)-SO-OH \xrightarrow[Q^+X^-]{NaOH} R'-S-R \tag{12.7}$$

Table 12.4. Reduction of N-tosylsulfilimines according to equation 12.7 [15]

R	R'	% Yield
CH_3	C_6H_5	93
CH_3	$2\text{-}Cl\text{-}C_6H_4$	63
CH_3	$4\text{-}Cl\text{-}C_6H_4$	82
CH_3	$4\text{-}CH_3\text{-}C_6H_4$	100
CH_3	$4\text{-}CH_3O\text{-}C_6H_4$	90
CH_3	$C_6H_5CH_2$	68
C_2H_5	C_6H_5	93
C_2H_5	$4\text{-}Cl\text{-}C_6H_4$	70
C_2H_5	$4\text{-}CH_3\text{-}C_6H_4$	73
$i\text{-}C_3H_7$	$4\text{-}Cl\text{-}C_6H_4$	60
C_6H_5	C_6H_5	26
C_6H_5	$C_6H_5CH_2$	83
$C_6H_5CH_2$	$C_6H_5CH_2$	67

12.5 Altered Reactivity

Pierre and Handel have studied the effect of [2.1.1]-cryptate on the lithium aluminum hydride reduction of cyclohexanone in diglyme [16]. The [2.1.1]-cryptate strongly complexes lithium ion and if sufficient cryptate is used to sequester all of the lithium ion, no reduction occurs. Apparently, lithium ion is needed as an electrophilic catalyst for the reduction to occur (see Eq. 12.8). Consistent with this interpretation is the observation that even in the presence of cryptate, reduction will occur if an excess of lithium iodide is also present. The relatively low reactivity of tetrabutylammonium borohydride in benzene solution may also reflect this property, at least in part [9]. Likewise, the β-hydroxyethyl quaternary ammonium ions may be better catalysts than non-oxygenated quaternary ions because the hydroxyl may hydrogen bond to carbonyl and provide electrophilic catalysis [5]. Similar, though less dramatic results, have been observed in the reduction of aromatic aldehydes and ketones by lithium aluminum hydride in the presence of [2.1.1]-cryptate [17].

$$\text{cyclohexanone} + \text{LiAlH}_4 \xrightarrow{[2.1.1]-\text{cryptate}} \text{no reaction} \qquad (12.8)$$

References

1. Horman, H.: "Newer Methods in Preparative Organic Chemistry" *2*, 213 (1963).
2. Chaikin, S. W., Brown, W. G.: J. Amer. Chem. Soc. *71*, 122 (1949).
3. Starks, C. M., Napier, D. R.: Brit. Pat. 1,227,144 filed Apr. 5, 1967.
4. Starks, C. M.: J. Amer. Chem. Soc. *93*, 195 (1971).
5. Colonna, S., Fornasier, R.: Synthesis *1975*, 531.
6. Cinquini, M., Montanari, F., Tundo, P.: J. C. S. Chem. Commun. *1975*, 393.
7. Matsuda, T., Koida, K.: Bull. Chem. Soc. Jap. *46*, 2259 (1973).
8. Landini, D., Maia, A. M., Montanari, F., Pirisi, F. M.: Gazz. Chim. Ital. *105*, 863 (1975).
9. Sullivan, E. A., Hinckley, A. A.: J. Org. Chem. *27*, 3732 (1962).
10. Brown, H. C., Mead, E. J., Subba Rao, B. C.: J. Amer. Chem. Soc. *77*, 6209 (1955).
11. Brandström, A., Junggren, U., Lamm, B.: Tetrahedron Let. *1972*, 3173.
12. Raber, D. J., Guida, W. C.: J. Org. Chem. *41*, 690 (1976).
13. Hutchins, R. O., Kandasamy, D.: J. Amer. Chem. Soc. *95*, 6131 (1973).
14. Durst, H. D., Zubrick, J. W., Kieczykowski, G. R.: Tetrahedron Let. *1974*, 1777.
15. Borgogno, G., Colonna, S., Fornasier, R.: Synthesis *1975*, 529.
16. Pierre, J. L., Handel, H.: Tetrahedron Let. *1974*, 2317.
17. Loupy, A., Seyden-Penne, J., Tchoubar, B.: Tetrahedron Let. *1976*, 1677.

13. Preparation and Reactions of Sulfur Containing Substrates

13.1 Introduction

A number of phase transfer reactions of sulfur containing substrates have been reported. These include substitution reactions in which sulfur is nucleophile, electrophile or an ancillary function. The synthesis of a variety of thioethers has been reported, as have the alkylation of numerous sulfur stabilized carbanions. In addition, reduction of both disulfides and N-tosylsulfilimines have been conducted under phase transfer catalytic conditions. The unifying theme of this chapter is, therefore, the synthesis and reactions of sulfur containing molecules. Inherent in this approach is some redundancy, but it is hoped that the ability to locate reactions of interest compensates.

13.2 Preparation of Symmetrical Thioethers

Symmetrical dialkyl sulfides have been prepared in high yield by the reaction of sodium or potassium sulfide with a primary or secondary halide under phase transfer conditions. Both quaternary ions [1] and crown ethers [2] have been successfully utilized in this reaction. It was found that at 70 °C, several different alkyl halides would condense with 0.6 to 1.0 equivalent of sodium sulfide to give generally excellent yields of symmetrical thioethers according to equation 13.1. Even the highly

$$R-X + Na_2S \xrightarrow{PTC} R-S-R + 2NaX \qquad (13.1)$$

hindered neopentyl bromide afforded product in good yield [1]. Likewise, symmetrical *bis*-nitrophenyl sulfides have been prepared under phase transfer conditions, although the yields are only moderate [3]. Several examples of the process described above are presented in Table 13.1.

Thiolacetic acid has also been used as a source of sulfur in the preparation of symmetrical sulfides [4]. Crotonaldehyde, for example, has been shown to condense with thiolacetic acid, hydrolyze, add to a second molecule of aldehyde, and then undergo aldol condensation to yield 4,6-dimethyl-5-thiacyclohexenecarboxaldehyde (81%), as shown in equation 13.2.

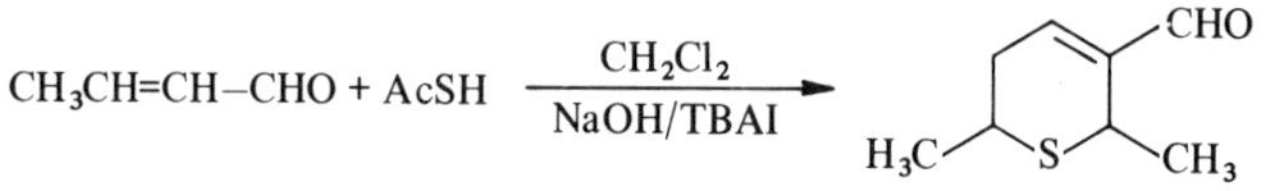

$$CH_3CH=CH-CHO + AcSH \xrightarrow[NaOH/TBAI]{CH_2Cl_2} \underset{H_3C\diagdown S\diagup CH_3}{\text{(ring with CHO)}} \qquad (13.2)$$

Table 13.1. Preparation of symmetrical thioethers

Alkyl halide	% Yield	Inorganic sulfide	Catalyst	References
n-$C_6H_{13}Cl$	90	Na_2S	HDTBP	[1]
n-$C_8H_{17}Cl$	91	Na_2S	HDTBP	[1]
$C_6H_5CH_2Cl$	94	Na_2S	HDTBP	[1]
n-$C_6H_{13}CHCl$ (CH$_3$)	99	Na_2S	HDTBP	[1]
n-$C_8H_{17}Br$	91	Na_2S	HDTBP	[1]
n-$C_8H_{17}Br$	64	K_2S	DC-18-C-6	[2]
n-$C_6H_{13}CHBr$ (CH$_3$)	91	Na_2S	HDTBP	[1]
$(CH_3)_3CCH_2Br$	81	Na_2S	HDTBP	[1]
4-$NO_2C_6H_4Cl$	40	Na_2S	DB-18-C-6	[3]

13.3 Preparation of Mixed Sulfides

Quaternary onium salts, crown ethers, and cryptates have all proved effective as catalysts in the synthesis of unsymmetrical thioethers under phase transfer conditions. Both primary and secondary alkyl bromides can be used to alkylate the mercaptan or thiophenol under basic conditions [5–9].

A detailed mechanistic study of the reaction of thiophenol with n-bromooctane in an alkaline two-phase system has been conducted [5, 6] (see Eq. 13.3) and the essential findings are discussed in Sect. 1.6. The reaction appears to be a true cata-

$$Br-(CH_2)_7CH_3 + C_6H_5SH \xrightarrow[Q^+X^-]{NaOH} C_6H_5-S-(CH_2)_7CH_3 \qquad (13.3)$$

lytic phase transfer process in which the observed reaction rates in the presence of either tetrabutylphosphonium chloride or dicyclohexyl-18-crown-6 are comparable. This similarity does not require in either case that benzene solubilized thiophenoxide anions be solvation free ("naked"). It means only that the degree of association and hydration are similar in both cases. The observation that tetrabutylammonium iodide is not a poison, but an effective catalyst in this reaction indicates that thiophenoxide is a "softer" anion than iodide [5]. Finally, the S_N2 character of the displacement is apparent from the reaction of optically active 2-chlorooctane with 4-thiocresoxide which yields 4-tolyl 2-octyl thioether (86%) which is 96% inverted (Eq. 13.4) [1].

$$4\text{-}CH_3C_6H_4\text{-}SH + CH_3\overset{*}{C}HCl(CH_2)_5CH_3 \xrightarrow[Q^+X^-]{NaOH}$$

$$4\text{-}CH_3C_6H_4\text{-}S\text{-}\overset{*}{C}H(CH_3)CH_2(CH_2)_4CH_3 \quad (13.4)$$

The phase transfer thioether formation reaction is successful for a broad range of thiols and alkylating agents as shown in Table 13.2. It is interesting that neopentyl phenyl sulfide can be prepared by this method in good yield [1]. In addition, dichloromethane is subject to reaction under these conditions [5]. The diphenylthioacetal of formaldehyde is thus produced (Eq. 13.5) in greater than 90% yield (see Table 13.2) [6].

$$2C_6H_5\text{-}SH + CH_2Cl_2 \xrightarrow[Q^+X^-]{NaOH} C_6H_5\text{-}S\text{-}CH_2\text{-}S\text{-}C_6H_5 \quad (13.5)$$

Table 13.2. Synthesis of mixed thioethers

Mercaptan	Alkyl halide	Catalyst	% Yield	Ref.
C_6H_5SH	CH_3I	TCMAC	93	[6]
C_6H_5SH	CH_3CH_2Br	TCMAC	93	[6]
C_6H_5SH	$i\text{-}C_3H_7Br$	TCMAC	88	[6]
C_6H_5SH	$i\text{-}C_4H_9Br$	TCMAC	85	[6]
C_6H_5SH	$(CH_3)_3CCH_2Br$	HDTBP	85	[1]
C_6H_5SH	$n\text{-}C_8H_{17}Br$	TCMAC	94	[5]
		DC-18-C-6	88	[5]
		HDTBP	100	[8]
		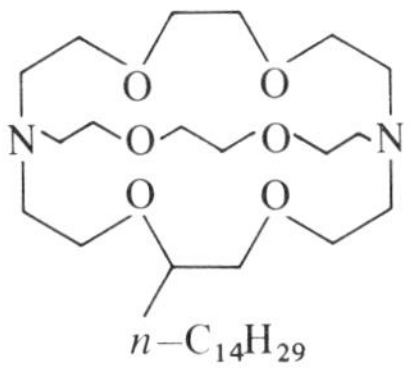 (cryptand with $n\text{-}C_{14}H_{29}$)	100	[8]
		HDTBP	91	[1]
		HDTBP	100	[7]

Preparation and Reactions of Sulfur Containing Substrates

Table 13.2 (continued)

Mercaptan	Alkyl halide	Catalyst	% Yield	Ref.
C_6H_5SH	$n\text{-}C_8H_{17}Br$	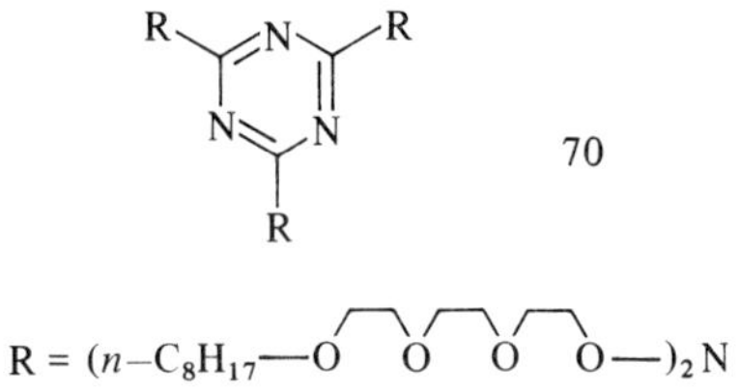 triazine with R substituents (see below)	70	[7]
		$R = (n\text{-}C_8H_{17}\text{—O}\frown\text{O}\frown\text{O}\frown\text{O—})_2N$		
C_6H_5SH	$n\text{-}C_8H_{17}Cl$	HDTBP	92	[1]
C_6H_5SH	$n\text{-}C_6H_{13}\overset{CH_3}{C}HCl$	HDTBP	90	[1]
C_6H_5SH	$n\text{-}C_6H_{13}\overset{CH_3}{C}HBr$	HDTBP	90	[1]
C_6H_5SH	$n\text{-}C_6H_{13}\overset{CH_3}{C}HBr$	N-dodecyl-N-methyl ephedrinium bromide	80	[9]
$p\text{-}CH_3\text{-}C_6H_4SH$	$n\text{-}C_6H_{13}\overset{CH_3}{C}HBr$	DC-18-C-6	85	[2]
$p\text{-}CH_3\text{-}C_6H_4SH$	$(R)\text{-}n\text{-}C_6H_{13}\overset{CH_3}{C}HCl$	HDTBP	86 (96% optical purity inversion)	[1]
C_2H_5SH	$n\text{-}C_8H_{17}Cl$	HDTBP	90	[1]
	$n\text{-}C_8H_{17}Br$	HDTBP	91	[1]
	$n\text{-}C_6H_{13}\overset{CH_3}{C}HCl$	HDTBP	88	[1]
	$n\text{-}C_6H_{13}\overset{CH_3}{C}HBr$	HDTBP	89	[1]
$n\text{-}C_4H_9SH$	CH_3CH_2Br	TCMAC	90	[6]
$i\text{-}C_4H_9SH$	$n\text{-}C_8H_{17}Br$	TCMAC	89	[6]
$i\text{-}C_4H_9SH$	CH_2Cl_2	TCMAC (di-substitution)	95	[6]
$n\text{-}C_4H_9SH$	CH_2Cl_2	TCMAC (di-substitution)	90	[6]
C_6H_5SH	CH_2Cl_2	TCMAC (di-substitution)	96	[6]

13.4 Preparation of Sulfides From Thiocyanates

An alternative to the synthesis of thioethers in which sulfur is a nucleophile, is a synthesis in which sulfur bears a leaving group. The reaction of a carbanion with a sulfenyl halide should, in principle, constitute a synthesis of thioethers. Such a reaction of sulfenyl halides has not yet been reported but the case in which sulfur is rendered electrophilic by attached cyano, has been [10]. Thus, benzylthiocyanate reacts with trichloromethide ion (generated from chloroform under phase transfer conditions) according to equation 13.6 to yield (80%) of benzyl trichloromethyl thioether [10].

$$C_6H_5-CH_2-S-CN + CHCl_3 \xrightarrow[Q^+X^-]{NaOH} C_6H_5-CH_2-S-CCl_3 + NaCN \quad (13.6)$$

The results of this and related reactions involving chloroform [10], phenylacetonitrile [10], and phenylacetylene [11] are recorded in Table 13.3.

Table 13.3. Preparation of sulfides from thiocyanates[a]

Thiocyanate	Carbon acid	% Yield	References
CH_3SCN	$CHCl_3$	61	[10]
	$C_6H_5C\equiv CH$	35	[11]
CH_3CH_2SCN	$CHCl_3$	64	[10]
	$C_6H_5C\equiv CH$	42	[11]
	$C_6H_5CH(CH_3)CN$	77	[10]
$n\text{-}C_3H_7SCN$	$CHCl_3$	67	[10]
$Cl(CH_2)_3SCN$	$CHCl_3$	77	[10]
$NC(CH_2)_3SCN$	$CHCl_3$	70	[10]
 Cl₂C=CH-CH₂-CH₂SCN (see structure)	$CHCl_3$	76	[10]
$n\text{-}C_5H_{11}SCN$	$CHCl_3$	86	[10]
	$C_6H_5C-C\equiv CH$	45	[11]
	$C_6H_5CH(CH_3)CN$	68	[10]
C_6H_5SCN	$CHCl_3$	60	[10]
$C_6H_5CH_2SCN$	$CHCl_3$	80	[10]
$(CH_2SCN)_2$	$C_6H_5CH_2CN$[b]	45	[10]

[a] According to equation 13.6.
[b] The product in this reaction is the ethylenedithioketal of benzoyl cyanide.

13.5 Preparation of Alkylthiocyanates

Alkylthiocyanates have been prepared in high yield by reaction of alkali metal thiocyanates with various primary and secondary alkyl halides under phase transfer conditions. Quaternary alkylammonium salts [12–14], crown ethers [15], cryptates [16], and tertiary amines [14] have all proved effective phase transfer catalysts for this reaction (Eq. 13.7 and Table 13.4). The mechanism of the thiocyanate displacement is probably similar to that of the cyanide displacement reaction (see Sect. 7.2).

$$R-X + M^+SCN^- \xrightarrow{PTC} R-SCN + M^+X^- \qquad (13.7)$$

Table 13.4. Preparation of alkyl thiocyanates

R–X	Catalyst	% Yield	References
C_2H_5Br	TCMAC or TBA	99	[14]
n-C_3H_7Br	TCMAC or TBA	99	[14]
n-C_4H_9Cl	TCMAC or TBA	100	[14]
s-C_4H_9Br	TCMAC	94	[14]
n-$C_5H_{11}Br$	TCMAC or TBA	97	[14]
s-$C_5H_{11}Br$	TCMAC or TBA	100	[14]
n-$C_6H_{13}Br$	TCMAC or TBA	99	[14]
n-$C_7H_{15}I$	TCMAC or TBA	100	[14]
n-$C_8H_{17}Br$	TCMAC or TBA	100	[14]
n-$C_8H_{17}Br$	DC-18-C-6	91	[2]
n-$C_{10}H_{21}Br$	TCMAC or TBA	99	[14]
n-$C_{10}H_{21}Br$	TCMAC	100	[12]
$C_6H_5CH_2Cl$	TCMAC or TBA	86	[14]
$C_6H_5CH_2Cl$	[2.2.2]-Cryptate	80	[16]
$C_6H_5CH_2OTs$	18-C-6	100	[15]

13.6 Sulfides Resulting From Michael Additions

Unsymmetrical sulfides possessing carbonyl groups beta to sulfur have been prepared by Michael addition of thiols to α,β-unsaturated aldehydes, ketones, and esters in tetrahydrofuran under phase transfer conditions [17] (see Eq. 13.8). This reaction is catalyzed by fluoride ion. In this application, tetrabutylammonium fluoride (TBAF)

$$R^1R^2C=CH-CO-R^3 \xrightarrow[THF, Q^+F^-]{R^4-SH} R^4S-C(R^1R^2)CH_2CO-R^3 \qquad (13.8)$$

was found to be a more effective catalyst than benzyltrimethylammonium fluoride (BTMAF). Crown ethers have been used to solubilize fluoride anion in other phase transfer reactions, but no report of crown ether catalysis of these reactions has yet appeared. The results of this approach to unsymmetrical sulfides are summarized in Table 13.5.

Table 13.5. Sulfides resulting from Michael addition [17]

R_4	R_1	R_2	R_3	Catalyst	% Yield
C_6H_5-	H	H	OCH_3	TBAF	99
	CH_3	H	OCH_3	BTMAF	83
	H	H	CH_3	TBAF	98
	CH_3	CH_3	CH_3	TBAF	94
	n-C_3H_7	H	H	TBAF	96
$C_6H_5CH_2-$	H	H	OCH_3	TBAF	87
	CH_3	H	OCH_3	BTMAF	100
	CH_3	CH_3	CH_3	BTMAF	93
	n-C_3H_7	H	H	TBAF	97
$\overset{O}{\underset{\|}{EtOCCH_2-}}$	H	H	OCH_3	TBAF	85
	CH_3	CH_3	CH_3	TBAF	80

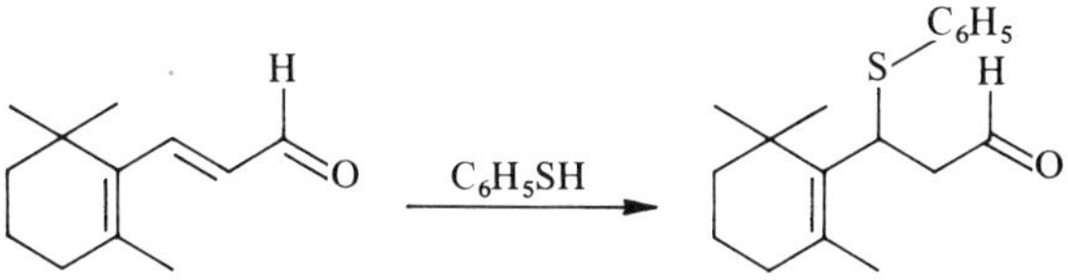

Reaction	Catalyst	% Yield
cyclohexenyl enal + C_6H_5SH → β-phenylthio aldehyde	BTMAF	87
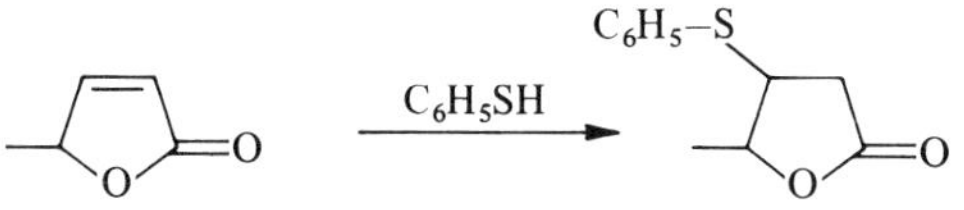 butenolide + C_6H_5SH → α-phenylthio-γ-butyrolactone	BTMAF	96

Acetylenic sulfides have also been used as Michael acceptors in phase transfer alkylation reactions (see also Chap. 10) [18]. A series of substituted benzyl cyanides have been condensed with n-butylthioacetylene as shown in equation 13.9 to give the products recorded in Table 13.6.

$$C_6H_5-\underset{R}{CH}-CN + HC\equiv C-S-n-C_4H_9 \xrightarrow{PTC} \quad (13.9)$$

$$C_6H_5-\underset{R}{C}(CN)-CH=CH-S-n-C_4H_9$$

Table 13.6. Alkylation of acetylenic sulfides [18]

R	% Yield	R	% Yield
CH_3	74	$C_6H_5CH(CH_3)$	64
$n\text{-}C_3H_7$	68	$(C_6H_5)_2CH$	96
$i\text{-}C_3H_7$	74	C_6H_5	55
$C_6H_5CH_2$	78		

13.7 Synthesis of α,β-Unsaturated Sulfur Compounds

The phase transfer Wittig-Horner-Emmons reaction has been used to prepare examples of α,β-unsaturated sulfides, sulfoxides, and sulfones [19]. Although these reactions are discussed in Sect. 14.3, we have summarized in Table 13.7 the results of transformations according to equation 13.10. Note that this reaction yields sulfur compounds on condensation with aldehydes only if the starting phosphonate is sulfur substituted.

$$(C_2H_5O)_2PO-CH_2X + R-CHO \xrightarrow[Q^+X^-]{NaOH} R-CH=CH-X \qquad (13.10)$$

α,β-Unsaturated sulfones have been synthesized directly by a Doebner-Knoevenagel condensation of aliphatic sulfones with aldehydes under alkaline phase transfer conditions [20]. (See Eq. 13.11). Several examples of this condensation are presented in Table 13.8.

$$R^1-SO_2-CH_2-R^2 + R^3-CHO \xrightarrow[Q^+X^-]{NaOH} R^1-SO_2-C(R^2)=CH-R^3 \qquad (13.11)$$

Table 13.7. Synthesis of α,β-unsaturated sulfur compounds [19]

X	Aldehyde	% Yield	E:Z Ratio
CH_3S	C_6H_5CHO	59	87:13
C_6H_5S	C_6H_5CHO	81	48:52
C_6H_5S	$4\text{-}(CH_3)_2NC_6H_4CHO$	40	80:20
CH_3SO	C_6H_5CHO	51	70:30
C_6H_5SO	C_6H_5CHO	54	83:17
C_6H_5SO	$4\text{-}ClC_6H_4CHO$	57	–
C_6H_5SO	$4\text{-}CH_3OC_6H_4CHO$	48	–
CH_3SO_2	C_6H_5CHO	85	100:0
CH_3SO_2	$4\text{-}ClC_6H_4CHO$	71	100:0
CH_3SO_2	$4\text{-}CH_3OC_6H_4CHO$	73	100:0

Table 13.8. Synthesis of sulfones by condensation reaction [20]

R_1	R_2	R_3	% Yield
C_6H_5	H	C_6H_5	86
C_6H_5	H	$4\text{-}Cl\text{-}C_6H_4$	98
C_6H_5	H	$C_6H_5CH=CH$	30
C_6H_5	H	β-naphthyl	85
C_6H_5	CH_3	C_6H_5	44
C_6H_5	C_6H_5	C_6H_5	49
C_6H_5	$(CH_3)_2C=CH$	C_6H_5	25
$(CH_3)_2N$	H	C_6H_5	58

3,5-Diphenyl-1,4-thiodioxane-1-oxide [21] and the corresponding sulfone (1,1-dioxide) [22] have been isolated in 12% and 48% yields respectively from phase transfer reactions containing sodium hydroxide, benzaldehyde, and either dimethyl sulfoxide or dimethyl sulfone. Apparently, an α,β-unsaturated sulfone is produced in the first condensation step and the β-hydroxysulfone which results from a second condensation step undergoes an intramolecular Michael addition as shown in equation 13.12.

$$CH_3SO_2CH_3 + C_6H_5-CHO \xrightarrow[Q^+X^-]{NaOH} CH_3SO_2CH=CH-C_6H_5 \xrightarrow[C_6H_5-CHO]{NaOH, Q^+X^-}$$

$$\underset{C_6H_5\overset{O^-}{C}H-CH_2-SO_2-CH=CHC_6H_5}{} \longrightarrow \underset{C_6H_5\diagdown O \diagup C_6H_5}{\overset{SO_2}{\diagup\diagdown}}$$

(13.12)

13.8 Other Phase Transfer Reactions of Sulfur Containing Substances

Alkyl p-tolyl sulfones have been prepared by nucleophilic addition of the p-tolyl sulfinate anion to primary, allylic, and benzylic halides in tetrahydrofuran [23]. Although these reactions were conducted in the presence of a stoichiometric amount of quaternary ammonium salt, it seems clear that the reaction could be catalytic and is therefore included here. The reaction is formulated in equation 13.13 and several examples of the condensation are shown in Table 13.9.

$$R-X + 4\text{-}CH_3-C_6H_4-SO_2^-\ ^+NBu_4 \xrightarrow{THF} 4\text{-}CH_3-C_6H_4-SO_2-R \quad (13.13)$$

Unsymmetrical α-haloalkyl p-tolyl sulfones have likewise been prepared under phase transfer conditions (see Eq. 13.14) [24]. Thus, α-halomethyl p-tolyl sulfones are deprotonated to form a carbanion which reacts with an alkyl halide to give product. α-Halo-α-sulfonylcarbanions apparently are not prone to undergo facile α-elimination to form carbenes despite the presence of chloride and sulfinate leaving groups. The carbanions are generally alkylated in 60–70% yield by alkyl bromides and chlorides [24]. Dihalomethyl *p*-tolyl sulfones are alkylated under similar conditions although in somewhat higher yields than are the monohalo sulfones (see Eq. 13.15).

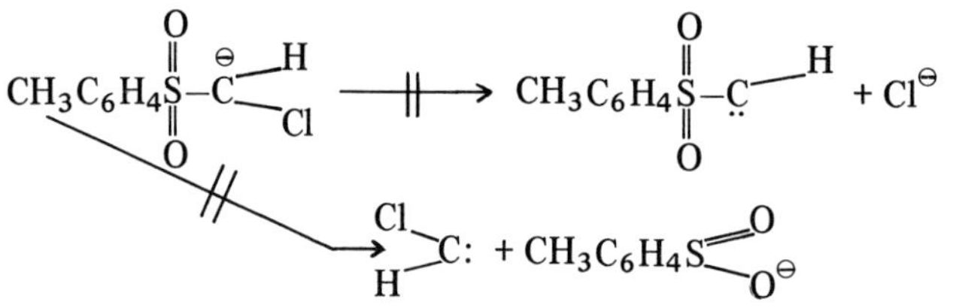

$$CH_3C_6H_4\overset{O}{\underset{O}{\overset{\|}{S}}}-\overset{\ominus}{C}\overset{H}{\underset{Cl}{{}}} \not\longrightarrow CH_3C_6H_4\overset{O}{\underset{O}{\overset{\|}{S}}}-\overset{H}{\underset{{}}{\ddot{C}}} + Cl^{\ominus} \qquad (13.14)$$

$$\longrightarrow \overset{Cl}{\underset{H}{{}}}\!\!\!\diagdown\!\!\!C\!: + CH_3C_6H_4\overset{O}{\underset{O^{\ominus}}{\overset{\diagup}{S}}}$$

$$CH_3C_6H_4\overset{O}{\underset{O}{\overset{\|}{S}}}-CH_2-X \xrightarrow[\text{BTEAC}]{\text{RX}}_{\text{50\% NaOH}} CH_3C_6H_4\overset{O}{\underset{O}{\overset{\|}{S}}}-\overset{}{\underset{X}{\overset{}{C}}}H-R$$

$$H_3C-\!\!\left\langle\!\!\bigcirc\!\!\right\rangle\!\!-\overset{O}{\underset{O}{\overset{\|}{S}}}-CHX_2 \xrightarrow{\text{RX}} H_3C-\!\!\left\langle\!\!\bigcirc\!\!\right\rangle\!\!-\overset{O}{\underset{O}{\overset{\|}{S}}}-CX_2-R \qquad (13.15)$$

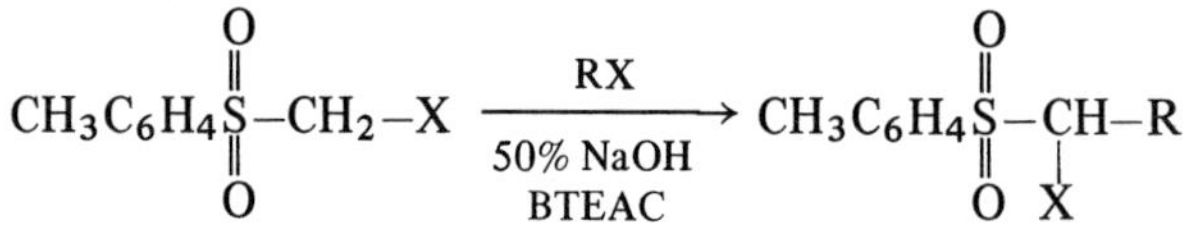

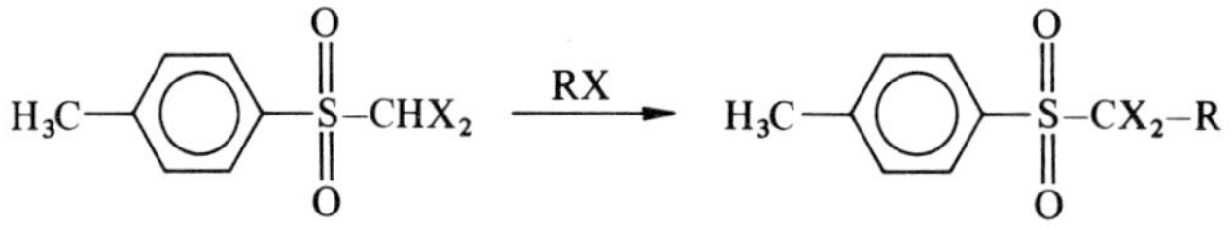

Table 13.9. Alkylation of sulfinate ion [23]

Alkyl halide		% Yield
CH$_3$I		93
ClCH$_2$Cl		85
CH$_3$OCH$_2$Cl		59
ClCH$_2$CCH$_2$Cl $\parallel$ O	(1 eq. sulfinate) (2 eq. sulfinate)	75 (mono) 75 (di)
H$_2$C=CHCH$_2$Br		80
i-C$_3$H$_7$Br		63
CH$_3$CH$_2$OCCH$_2$Br $\parallel$ O		80
4-ClC$_6$H$_4$CH$_2$Br		93
C$_6$H$_5$CCH$_2$Br $\parallel$ O		81
C$_6$H$_5$CH=CHCCH$_2$Cl $\parallel$ O		85
CH$_2$Br$_2$		89 (mono)
BrCH$_2$NO$_2$		100
BrCH$_2$CH$_2$Br	Vinyl *p*-tolylsulfone	45 24 (di)

α-Halomethyl p-tolyl sulfones also undergo base catalyzed phase transfer condensations with ketones and aldehydes to yield the corresponding α,β-epoxy sulfones. This reaction is similar to the Darzens condensation of α-halonitriles [25, 26]. The general equation is formulated in 13.16. Under phase transfer conditions chloromethyl p-tolyl sulfone has been found to condense and ring close to yield α,β-epoxysulfones of acetone (91%), benzophenone (90%), benzaldehyde (60%), isobutyraldehyde (65%), and cyclohexanone (90%).

$$H_3C-C_6H_4-\underset{\underset{O}{\|}}{\overset{\overset{O}{\|}}{S}}-CH_2-Cl \xrightarrow[50\% \text{ NaOH}, Q^+]{R_1-CO-R_2} \left[H_3C-C_6H_4-\underset{\underset{O}{\|}}{\overset{\overset{O}{\|}}{S}}-\underset{\underset{Cl}{|}}{\overset{\overset{H}{|}}{C}}-\underset{\underset{O^-}{|}}{\overset{\overset{R_1}{|}}{C}}-R_2 \right]$$

(13.16)

$$\longrightarrow H_3C-C_6H_4-\underset{\underset{O}{\|}}{\overset{\overset{O}{\|}}{S}}-HC\underset{O}{\overset{R_1}{\diagdown}}\overset{R_2}{\diagup}$$

A ring closure similar to that yielding α,β-epoxysulfones (Eq. 13.16) is expected in the phase transfer reaction of phenyl vinyl sulfone with α-chloropropionitrile. Under basic conditions, the nitrile is deprotonated and then adds in the Michael sense to the α,β-unsaturated sulfone. Protonation of the intermediate α-sulfonylcarbanion yields the simple Michael adduct, (Eq. 13.17), whereas intramolecular nucleophilic substitution would lead to sulfonylcyclopropanes (Eq. 13.18). The former process is favored over the latter to such an extent that less than a 10% yield of cyclopropanes are isolated in this reaction [27].

$$C_6H_5-SO_2-CH=CH_2 + CH_3CHClCN \xrightarrow[Q^+X^-]{NaOH} C_6H_5SO_2CH_2CH_2CCl(CH_3)CN \quad (13.17)$$

$$C_6H_5-SO_2-\bar{C}H-CH_2-CCl(CH_3)CN \longrightarrow C_6H_5-SO_2\triangle\overset{CN}{\underset{CH_3}{}} \quad (13.18)$$

Several alkylation reactions of sulfur containing species have also been reported in recent years. These reactions have all been conducted with quaternary ammonium cations present in stoichiometric quantities but it appears likely that such alkylations could be made catalytic. The tetrabutylammonium enethiol salt of thioacetoacetic ester is alkylated by allyl and propargyl bromides in 65% and 40% yields respectively, according to equation 13.19 [28]. Similarly, a number of β-hydroxydithiocinnamates

$$EtO_2C-CH=C(CH_3)-SH + CH_2=CH-CH_2Br \xrightarrow[\text{base}]{Q^+X^-} \quad (13.19)$$

$$EtO_2C-CH=C(CH_3)-S-CH_2CH=CH_2$$

have been alkylated in the presence of tetrabutylammonium hydroxide according to equation 13.20 [29]. By this method, β-hydroxydithiocinnamic acid was alkylated

$$
\text{Ar-C}\begin{array}{c}H\\ \diagup \\ C=C \\ \diagdown \\ O-H\end{array}\begin{array}{c}SH\\ \diagdown \\ \\ \diagup \\ S\end{array} \xrightarrow{Bu_4N^+\ ^-OH} \text{Ar-C}\begin{array}{c}H\\ \diagup \\ C=C \\ \diagdown \\ O-H\end{array}\begin{array}{c}S^-\\ \diagdown \\ \\ \diagup \\ S\end{array} \tag{13.20}
$$

$$
\xrightarrow{R-X} \text{Ar-C}\begin{array}{c}H\\ \diagup \\ C=C \\ \diagdown \\ O-H\end{array}\begin{array}{c}S-R\\ \diagdown \\ \\ \diagup \\ S\end{array}
$$

(methyl iodide) to give the corresponding thioester (46%). Nineteen other examples of this alkylation have been reported [29]. The related alkylation of the diethyl malonate-carbon disulfide condensation product has also been reported [30]. These alkylations are also stoichiometric in quaternary ion and in this case, the yields were low. *bis*-2-Carbethoxydithiolacetic acid is alkylated by dimethyl sulfate, ethyl bromide, and allyl bromide in 33%, 30%, and 29% yields respectively, according to equation 13.21 [30].

$$(C_2H_5OCO)_2CH-\overset{\overset{\displaystyle S}{\|}}{C}-S-H \xrightarrow[RX]{Bu_4N^+OH^-} (C_2H_5OCO)_2CH-\overset{\overset{\displaystyle S}{\|}}{C}-S-R \tag{13.21}$$

Phase transfer catalysis has been of value in several sequences involving reduction of sulfur containing substrates. Formamidine sulfinic acid proved to be an effective reducing agent for disulfides [31] (Eq. 13.22) and N-tosyl-sulfilimines [31] (Eq. 13.23). Yields in both cases were generally good (see Sect. 12.4 for details).

$$R-S-S-R + H_2N-C(=NH)-SO-OH \xrightarrow[Q^+X^-]{NaOH} R-S-R \tag{13.22}$$

$$R'-\overset{\overset{\displaystyle N-Ts}{\|}}{S}-R + H_2N-C(=NH)-SO-OH \xrightarrow[Q^+X^-]{NaOH} R'-S-R \tag{13.23}$$

The reduction of alkoxysulfonium salts (formed by alkylation of sulfoxides with methyl fluorosulfonate) with sodium cyanohydridoborate in dichloromethane is catalyzed by 18-crown-6 [32] (see Eq. 13.24). By this method, dibutyl, diphenyl, dibenzyl, and tetramethylene sulfoxides are reduced to the corresponding sulfides in 85%, 77%, 91%, and 87% yields, respectively.

$$R-\overset{\overset{\displaystyle O^-}{|}}{\underset{\displaystyle +}{S}}-R + CH_3OSO_2F \rightarrow R-\overset{\overset{\displaystyle OCH_3}{|}}{\underset{\displaystyle +}{S}}-R + FSO_3^- \xrightarrow[18\text{-}C\text{-}6]{NaBH_3CN} R_2S \tag{13.24}$$

References

1. Landini, D., Rolla, F.: Synthesis *1974*, 565.
2. Landini, D., Maia, A. M., Montanari, F., Pirisi, F. M.: Gazz. Chim. Ital. *105*, 863 (1975).
3. Kim, I. K., Noh, J. S.: Taehan Hwahak Hoechi *18*, 421 (1974) [Chem. Abstr. *82*, 124967h (1974)].
4. McIntosh, J. M., Khalil, H.: J. Org. Chem. *42*, 2123 (1977).
5. Herriott, A. W., Picker, D.: J. Am. Chem. Soc. *97*, 2345 (1975).
6. Herriott, A. W., Picker, D.: Synthesis *1975*, 447.
7. Fornasier, R., Montanari, F., Podda, G., Tundo, P.: Tetrahedron Let. *1976*, 1381.
8. Cinquini, M., Montanari, F., Tundo, P.: J. Chem. Soc. Chem. Commun. *1975*, 393.
9. Colonna, S., Fornasier, R.: Synthesis *1975*, 531.
10. Makosza, M., Fedorynski, M.: Synthesis *1974*, 274.
11. Makosza, M., Fedorynski, M.: Rocz. Chem. *47*, 1779 (1975).
12. Starks, C. M., Napier, D. R.: Brit. Pat. 1,227,144 (April 7, 1971).
13. Sugimoto, N., Fujita, T., Shigematsu, N., Ayada, A.: Chem. Pharm. Bull. *20*, 427 (1962).
14. Reeves, W. P., White, M. R., Hilbrich, R. G., Biegert, L. L.: Syn. Comm. *6*, 509 (1976).
15. Liotta, C. L., Grisdale, E. E., Hopkins, H. P.: Tetrahedron Let. *1975*, 4205.
16. Lehn, J. M.: Structure and Bonding, *16*, 1 (1973).
17. Kuwajima, I., Murofushi, T., Nakamura, E.: Synthesis *1976*, 602.
18. Makosza, M., Jawdosiuk, M.: Bull. Acad. Polon. Sci. *16*, 589 (1968).
19. Mikolajczyk, M., Grezejszczak, S., Midura, W., Zatorski, A.: Synthesis *1975*, 278.
20. Cardillo, G., Savoia, D., Umani-Ronchi, A.: Synthesis *1975*, 453.
21. Merz, A., Märkl, G.: Angew. Chem. Int. Ed. *12*, 845 (1973).
22. Gokel, G. W., Gerdes, H. M., Rebert, N. W.: Tetrahedron Let. *1976*, 653.
23. Vennstra, G. E., Zwaneburg, B.: Synthesis *1975*, 519.
24. Jończyk, A., Bańko, K., Makosza, M.: J. Org. Chem. *40*, 266 (1975).
25. Jończyk, A., Fedorynski, M., Makosza, M.: Tetrahedron Let. *1972*, 2395.
26. Makosza, M., Ludwikow, M.: Synthesis *1974*, 665.
27. Jończyk, A., Makosza, M.: Synthesis *1976*, 387.
28. Dalgaard, L., Lawesson, S. O.: Tetrahedron *28*, 2051 (1972).
29. Larsson, F. C. V., Lawesson, S. O.: Tetrahedron *28*, 5341 (1972).
30. Dalgaard, L., Kolind-Andersen, H., Lawesson, S. O.: Tetrahedron *29*, 2077 (1973).
31. Borgogno, G., Colonna, S., Fornasier, R.: Synthesis *1975*, 529.
32. Durst, H. D., Zubrick, J. W., Kieczykowski, G. R.: Tetrahedron Let. *1974*, 1777.

14. Ylids

14.1 Introduction

Over twenty years ago, Wittig found that alkyl substituted phosphonium salts could be deprotonated by strong bases to yield phosphonium ylids which in turn could react with aldehydes and ketones to yield olefins and the corresponding phosphine oxide [1]. The sequence is formulated in equations 1.1−1.3 for the reaction of methyltriphenylphosphonium halide with cyclohexanone in the presence of base to give methylenecyclohexane.

$$Ar_3P^+-CH_3 \ X^- \xrightarrow{Base} Ar_3P^+-CH_2^- + HX \qquad (14.1)$$

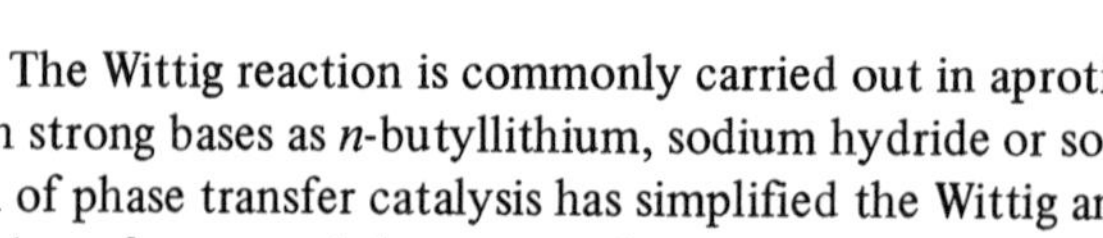

$$Ar_3P^+-CH_2^- + \underset{}{\bigcirc}=O \rightleftharpoons \underset{CH_2-P^+Ar_3}{\bigcirc}\!\!\!{}^{O^-} \qquad (14.2)$$

$$\longrightarrow \bigcirc=CH_2 + Ar_3P=O \qquad (14.3)$$

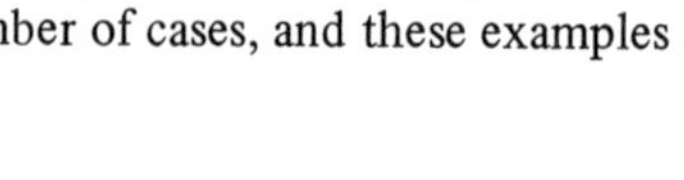

The Wittig reaction is commonly carried out in aprotic solvents in the presence of such strong bases as n-butyllithium, sodium hydride or sodamide [2]. The application of phase transfer catalysis has simplified the Wittig and related reactions in a number of cases, and these examples are presented in the chapter.

14.2 Phase Transfer Wittig Reactions

The major advantage of the phase transfer method in Wittig reactions is increasing the convenience of the reaction. Concentrated aqueous alkali is obviously easier to handle than is n-butyllithium or sodium hydride and such solvents as dichloromethane and benzene are more readily removed after the reaction is complete than is DMSO.

The phase transfer Wittig reaction can therefore be characterized by saying that the sequence 1.1–1.3 is altered only in the identities of the base and the solvent.

Numerous alkyltriphenylphosphonium salts have been shown to undergo the phase transfer Wittig reaction with a wide variety of aromatic and unsaturated aldehydes [3–7]. The results of these preparations are recorded in Table 14.1. Relatively few aliphatic aldehydes have been successfully condensed with phosphonium salts under these conditions, but the few successes point to the strength of the phase transfer method. Both formaldehyde [7] and glyoxal [5] have been used successfully under phase transfer conditions. In each case, the commercially available aqueous solution was used. The convenience of not having to obtain anhydrous solutions of formaldehyde or glyoxal is manifest. The Wittig reaction with formaldehyde led to styrenes in high yield (Eq. 14.4) whereas the condensation of pyridylphosphonium salts with aqueous glyoxal afforded more modest yields of butadienes (Eq. 14.5).

$$4-O_2N-C_6H_4-CH_2-{}^+P(C_6H_5)_3 \ X^- + CH_2O \xrightarrow{90\%} 4-NO_2-C_6H_4-CH=CH_2 \quad (14.4)$$

$$\text{Py}-CH=P(C_6H_5)_3 + OCH-CHO \xrightarrow{18\%} \text{Py}-(CH=CH)_2-\text{Py} \quad (14.5)$$

Table 14.1. Phase transfer Wittig reactions

Phosphonium salt	Aldehyde	% Yield	Ref.
$(C_6H_5)_3\overset{+}{P}-CH_3$	C_6H_5-CHO	80	[4]
	$4\text{-}Cl-C_6H_4-CHO$	95	[4]
	$4\text{-}CH_3-C_6H_4-CHO$	55	[4]
	$4\text{-}CH_3O-C_6H_4-CHO$	38	[4]
	$C_6H_5-CH=CH-CHO$	68	[4]
	furyl-CHO	63	[4]
	$n\text{-}C_7H_{15}-CHO$	73	[4]
$(C_6H_5)_3\overset{+}{P}-CH_2CH_3$	C_6H_5-CHO	20	[3]
	C_6H_5-CHO	46	[4]
	C_6H_5-CHO	93	[6]
	$C_6H_5-CH=CH-CHO$	51	[3]
$(C_6H_5)_3\overset{+}{P}-CH_2CH=CH_2$	C_6H_5-CHO	60	[4]
	furyl-CHO	15	[4]
	$C_6H_5-CH=CH-CHO$	30	[4]
	$n\text{-}C_7H_{15}-CHO$	40	[4]
$(C_6H_5)_3\overset{+}{P}-CH_2-C_6H_5$	CH_2O	87	[7]
	C_6H_5-CHO	80	[3]
	C_6H_5-CHO	81	[4]
	C_6H_5-CHO	96	[6]

Ylids

Table 14.1 (continued)

Phosphonium salt	Aldehyde	% Yield	Ref.
$(C_6H_5)_3\overset{+}{P}-CH_2-C_6H_5$	4-CH_3-C_4H_4-CHO	78	[3]
	4-NO_2-C_6H_4-CHO	72	[3]
	3-NO_2-C_6H_4-CHO	80	[3]
	C_6H_5-CH=CH-CHO	71	[4]
	C_6H_5-CH=CH-CHO	87	[3]
	furyl-CHO	60	[4]
	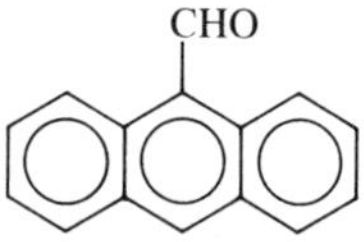 9-anthracenyl-CHO	65 (100% trans)	[3]
	C_2H_5-CHO	92	[6]
	n-C_7H_{15}-CHO	82	[4]
$(C_6H_5)_3\overset{+}{P}-CH_2-C_6H_4-NO_2$-4	CH_2O	92	[7]
$(C_6H_5)_3\overset{+}{P}-CH_2-C_6H_4-CN$-4	CH_2O	90	[7]
$(C_6H_5)_3\overset{+}{P}-CH_2-C_6H_4-Br$-4	CH_2O	98	[7]
$(C_6H_5)_3\overset{+}{P}-CH_2-C_6H_4-CH_3$-4	CH_2O	98	[7]
$(C_6H_5)_3\overset{+}{P}-CH_2-C_6H_4-OCH_3$-4	CH_2O	87	[7]
$(C_6H_5)_3\overset{+}{P}-n$-$C_5H_{11}$	C_6H_5-CHO	46	[4]
$(C_6H_5)_3\overset{+}{P}-CH(CH_3)_2$	C_6H_5-CHO	30	[3]
$(C_6H_5)_3\overset{+}{P}-CH_2-\overset{O}{\underset{\|}{C}}-C_6H_5$	C_6H_5-CHO	36	[4]
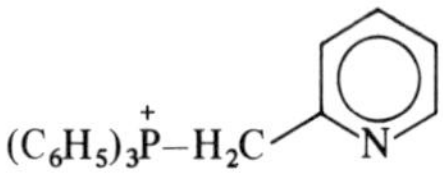 $(C_6H_5)_3\overset{+}{P}-H_2C$-(2-pyridyl)	OHC-CHO	18	[5]
$(C_6H_5)_3\overset{+}{P}-H_2C$-(2-pyridyl)	H-CO-CH=HC-(2-pyridyl)	71	[5]
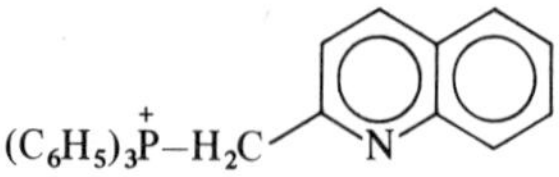 $(C_6H_5)_3\overset{+}{P}-H_2C$-(2-quinolyl)	OHC-CHO	13	[5]
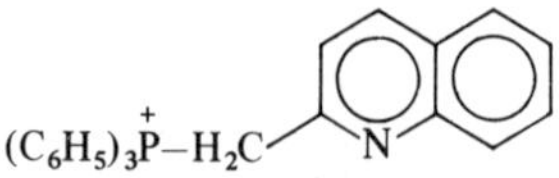 $(C_6H_5)_3\overset{+}{P}-H_2C$-(2-quinolyl)	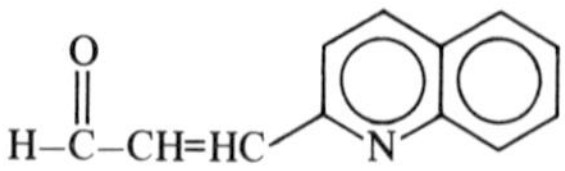 H-CO-CH=HC-(2-quinolyl)	not reported	[5]
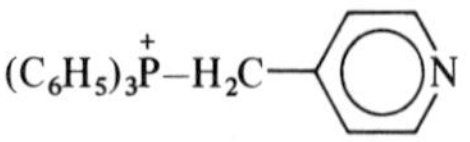 $(C_6H_5)_3\overset{+}{P}-H_2C$-(4-pyridyl)	OHC-CHO	23	[5]
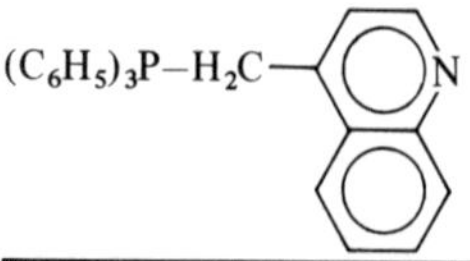 $(C_6H_5)_3P-H_2C$-(acridinyl)	OHC-CHO	15	[5]

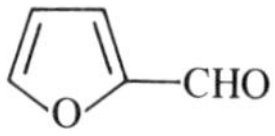

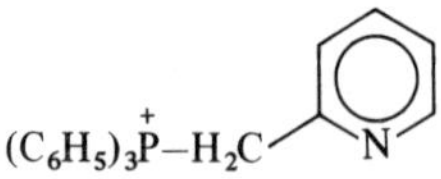

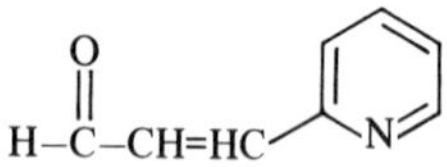

An interesting feature of the phase transfer Wittig reaction is that it succeeds in the absence of any added catalyst. Apparently, the phosphonium salts which are reactants in this system are also effective as phase transfer agents. It is less apparent why certain phosphonates should show the ability to catalyze these reactions and this is discussed in Sect. 1.3. In the phosphonium salt case, proton transfer must occur at the interface followed by dissolution of the ylid in the organic medium where it then undergoes the normal Wittig reaction.

It has recently been shown that crown ether catalyzed Wittig reactions give product distributions typical of the so-called "salt-free" reaction [6]. Using potassium *t*-butoxide or potassium carbonate as base, the yields of olefin were high and the isomer ratio was quite solvent dependent. The condensation of ethylidenetriphenylphosphine with benzaldehyde, for example, produced over 90% of β-methylstyrene in either tetrahydrofuran or dichloromethane, but the olefin was 85% *cis* in the former case and 78% *trans* in the latter [5].

14.3 The Wittig-Horner-Emmons Reaction

The Wittig reaction wherein a phosphonate is substituted for the more traditional phosphonium salt is called either the Wittig-Horner [8] or Wittig-Emmons [9] reaction. The synthesis of alkenes via base catalyzed phosphonate addition to an aldehyde or ketone has been accomplished under phase transfer conditions [10–13]. The general reaction is formulated in equation 14.6. An interesting feature of this system is that even with phosphonates, which are not obviously related to the quater-

$$(C_2H_5O)_2PO-CH_2R + R^1-CO-R^2 \xrightarrow{\text{base}} R^1R^2C=CH-R \quad (14.6)$$

nary ammonium or phosphonium salts, which have ordinarily been used as phase transfer catalysts, no additional catalyst is required. It seems likely that the phosphonate is coordinating a cation rather than itself ion-pairing with base. It is known that in phase transfer reactions involving benzaldehyde, the presence of catalytic amounts of ammonium salts prevents the Cannizzaro reaction [14]. In the presence of small amounts of crown (which coordinate cations rather than themselves ion-pairing an anion), the Cannizzaro is suppressed, but not sequestered. The observation that isonicotinic acid was produced in an attempted condensation with 2-formyl pyridine [11] suggests either a coordinating role for the phosphonate or an interfacial reaction. Examples of the phosphonate condensation under phase transfer conditions are presented in Table 14.2, including examples of a Wittig-Horner-Emmons carried out under ion pair extraction conditions [15].

Ylids

Table 14.2. Phase transfer Wittig-Horner-Emmons reactions

Phosphonate   Carbonyl compound

$$[C_2H_5O]_2-\overset{O}{\underset{}{P}}-CH_2R + R^1-\overset{O}{\underset{}{C}}-H \rightarrow R-CH=CH-R^1$$

R	R^1	% Yield	References
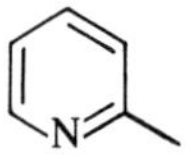 2-pyridyl	C_6H_5	71	[10]
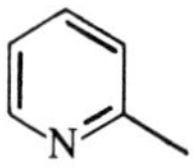 2-pyridyl	$C_6H_5-CH=CH$	68 75	[10] [12]
C_6H_5	$C_6H_5-CH=CH$	72	[12]
C_6H_4-Br-4	$C_6H_5-CH=CH$	81	[12]
$CH=CH-C_6H_5$	C_6H_5	70	[12]
$CH=CH-C_6H_5$	$4-O_2N-C_6H_4$	57	[12]
$CH=CH-C_6H_5$	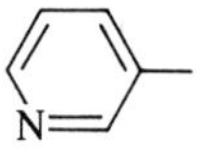 3-pyridyl	55	[12]
$CH=CH-C_6H_5$	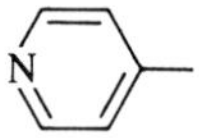 4-pyridyl	12	[12]
$CH=CH-C_6H_5$	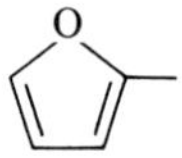 2-furyl	84	[12]
$CH=CH-C_6H_5$	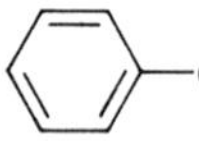 $C_6H_5-CH=CH$	80	[12]
CN	C_6H_5	77 80	[10] [13]
CN	CH_3	51	[10]
CN	Acetone	62	[10]
CO_2H	C_6H_5	95 100% E	[13]
$CO_2C_2H_5$	C_6H_5	56	[10]
$CO_2C_2H_5$	CH_3	54	[10]
$C_6H_5-\overset{O}{\underset{}{C}}-$	C_6H_5	55	[13]
	$4-Cl-C_6H_4$	65	[13]
	$4-Br-C_6H_4$	63	[13]

Table 14.2 (continued)

R	R^1	% Yield	References
[C$_2$H$_5$O]$_2$P(=O)–CH(CH$_3$)CN	C$_6$H$_5$	nr	[15]
CH$_3$S	C$_6$H$_5$	59	[11]
C$_6$H$_5$S	C$_6$H$_5$	81	[11]
		87	[13]
C$_6$H$_5$S	4-(CH$_3$)$_2$NC$_6$H$_4$	40	[11]
CH$_3$S(=O)–	C$_6$H$_5$	51	[11]
		51	[13]
C$_6$H$_5$S(=O)–	C$_6$H$_5$	54	[11]
		53	[13]
C$_6$H$_5$–S(=O)–	4-Cl–C$_6$H$_4$	57	[11]
C$_6$H$_5$–S(=O)–	4-CH$_3$O–C$_6$H$_4$	48	[11]
CH$_3$–S(=O)$_2$–	C$_6$H$_5$	85	[11]
CH$_3$–S(=O)$_2$–	C$_6$H$_5$	65	[13]
CH$_3$–S(=O)$_2$–	4-Cl–C$_6$H$_4$	71	[11]
CH$_3$–S(=O)$_2$–	4-CH$_3$O–C$_6$H$_4$	73	[11]
[C$_2$H$_5$O]$_2$P(=O)	C$_6$H$_5$	79	[13]
[C$_2$H$_5$O]$_2$P(=O)	4-Br–C$_6$H$_4$	74	[13]
[C$_2$H$_5$O]$_2$P(=O)	4-(CH$_3$)$_2$N–C$_6$H$_4$	76	[13]
[C$_2$H$_5$O]$_2$P(=O)	C$_6$H$_5$–CH=CH	70	[13]

14.4 Sulfur Stabilized Ylids

Trimethylsulfonium iodide undergoes ylid formation by reaction with 50% aqueous sodium hydroxide in the presence of catalytic tetrabutylammonium iodide [16]. The ylid thus formed reacts with aldehydes and ketones to form the corresponding epoxides (Eq. 14.7). The yields with aldehydes are considerably better than those with ketones. The fact that the reaction is slow (48 hours) may be due to the iodide of the catalyst. On the other hand, lauryldimethylsulfonium chloride undergoes reaction with ketones and aldehydes to yield epoxides under alkaline phase transfer conditions considerably more rapidly (6–10 hours). The enhanced rate of this methylene transfer reaction is probably due to the greater organic solubility of the lauryldimethylsulfonium cation [17]. Catalyst poisoning is observed with lauryldimethylsulfonium iodide. Similar reactions have been conducted under ion pair extraction conditions [18]. The data is presented in Table 14.3.

$$(CH_3)_3S^+ \xrightarrow[PTC]{OH^-} (CH_3)_2S=CH_2 + C_6H_5CHO \longrightarrow \underset{}{\overset{C_6H_5}{\vee\!\!\!-\!\!\!O}} \qquad (14.7)$$

Table 14.3. Synthesis of oxiranes

Sulfonium cation $(CH_3)_2\overset{+}{S}-R$ Anion $R_1-\underset{\|}{\overset{O}{C}}-R_2$

R	X$^-$	R$_1$	R$_2$	% Yield	References
CH$_3$	I	C$_6$H$_5$	H	90	[16]
n-C$_{12}$H$_{25}$	Cl	C$_6$H$_5$	H	86	[17]
CH$_3$	I	C$_6$H$_5$–CH=CH	H	85	[16]
n-C$_{12}$H$_{25}$	Cl	C$_6$H$_5$	CH$_3$	85	[17]
CH$_3$	I	C$_6$H$_5$	CH$_3$	36	[16]
n-C$_{12}$H$_{25}$	Cl	–(CH$_2$)$_5$–		88	[17]
n-C$_{12}$H$_{25}$	I	–(CH$_2$)$_5$–		27	[17]
CH$_3$	I	C$_6$H$_5$	C$_6$H$_5$	18	[16]

The reaction of trimethylsulfonium iodide with benzaldehyde under basic phase transfer conditions catalyzed by chiral quaternary ammonium salts such as (–)-N,N-dimethylephedrinium bromide has been reported to yield styrene oxide in high optical purity [19], which may be somewhat overestimated [20].

Likewise, trimethylsulfoxonium iodide undergoes ylid formation by reaction with 50% aqueous sodium hydroxide under phase transfer conditions. This ylid reacts with benzaldehyde to give styrene oxide, the expected product of methylene transfer in 20–30% yield (Eq. 14.8). The ylid, however, adds to α,β-unsaturated ketones to con-

$$(CH_3)_3 \overset{+}{S}=O\ I^- + C_6H_5CHO \xrightarrow{OH^-} C_6H_5-\overset{\overset{\displaystyle O}{\diagup\!\diagdown}}{CH-CH_2} \qquad (14.8)$$

vert the carbon-carbon double bond into a cyclopropane ring in much higher yields (86%, 78%, and 71% respectively for $X = H$, $X = CH_3$, and $X = OCH_3$ in Eq. 14.9) [16].

$$(CH_3)_3\overset{+}{S}=O \xrightarrow[PTC]{HO^-} (CH_3)_2SO=CH_2 + 4\text{-}X\text{-}C_6H_4CH=CH-CO-C_6H_5$$

$$\downarrow \qquad (14.9)$$

$$4\text{-}X\text{-}C_6H_4\underset{\underset{\displaystyle CH_2}{\diagdown\!\diagup}}{CH-CH}-CO-C_6H_5$$

References

1. Wittig, G., Schöllkopf, U.: Chem. Ber. *87*, 1318 (1954).
2. Wittig, G., Hesse, A.: Org. Syn. *50*, 66 (1970).
3. Märkl, G., Merz, A.: Synthesis *1973*, 295.
4. Tagaki, W., Inoue, I., Yano, Y., Okonogi, T.: Tetrahedron Let. *1974*, 2587.
5. Hunig, S., Stemmler, I.: ibid. *1974*, 3151.
6. Boden, R. M.: Synthesis *1975*, 784.
7. Broos, R., Anteunis, M.: Synthetic Commun. *6*, 53 (1976).
8. Horner, L., Hoffman, H., Wippel, H. G.: Chem. Ber. *91*, 61 (1958).
9. Wadsworth, W. S., Emmons, W. D.: J. Am. Chem. Soc. *83*, 1733 (1961).
10. Piechucki, C.: Synthesis *1974*, 869.
11. Mikolajczyk, M., Grzejszczak, S., Midura, W., Zatorski, A.: Synthesis *1975*, 278.
12. Piechucki, C.: Synthesis *1976*, 187.
13. Mikolajczyk, M., Grzejszczak, S., Midura, W., Zatorski, A.: Synthesis *1976*, 396.
14. Gokel, G. W., Gerdes, H. M., Rebert, N. W.: Tetrahedron Let. *1976*, 653.
15. D'Incan, E., Seyden-Penne, J.: Synthesis *1975*, 516.
16. Merz, A., Märkl, G.: Angew. Chem. Int. Ed. *12*, 845 (1973).
17. Yano, Y., Okonogi, T., Sunaga, M., Tagaki, W.: J. Chem. Soc., Chem. Commun. *1973*, 527.
18. Brändström, A., Lamm, B.: Acta Chem. Scand. *B28*, 590 (1974).
19. Hiyama, T., Mishima, T., Sawada, H., Nozaki, H.: J. Am Chem. Soc. *97*, 1626 (1975).
20. Hiyama, T., Mishima, T., Sawada, H., Nozaki, H:: J. Am. Chem. Soc. *98*, 641 (1976).

15. Altered Reactivity

15.1 Introduction

In a sense, all of phase transfer catalysis exemplifies altered reactivity. Much of the chemistry which has been successful in the phase transfer area involves reactions which have been envisioned, expected or attempted, but which have been unsuccessful or the results have been less than satisfactory. We attempt, in this chapter, to draw attention to those reactions which give a reactivity which is somewhat different from what might have been expected, or where there is an obvious advantage to using the phase transfer method instead of a more classical approach involving either expensive solvents or reagents.

Fundamental to any consideration of altered reactivity is a consideration of ion reactivity in nonpolar solutions. The majority of reactions which have been conducted under phase transfer conditions are of the S_N2 type and studies have been conducted regarding the reactivities of nucleophiles in this type of reaction under what we may loosely term "phase transfer conditions".

The classical nucleophilicity values for ions were reported by Swain and Scott [1]. The values reported for the halides are $I^- = 5.04$, $Br^- = 3.89$, $Cl^- = 3.04$ and $F^- = 2.0$. These values are, of course, logarithmic and the conclusion one draws from such values is that iodide is a much stronger nucleophile than bromide which is stronger than chloride and fluoride anion is the weakest of the four. Under "phase transfer conditions" the relative nucleophilicities seem to be reversed in nonpolar media [2, 3] and quite similar in acetonitrile solution [4]. Simmons and Sam [2] and Montanari and coworkers [3] have found that dissociated chloride in nonpolar solution is more nucleophilic (i.e. has a larger rate constant in a specific bimolecular reaction) than bromide and iodide reacts most sluggishly of the three. Liotta, Grisdale and Hopkins [4] have found that the reactivities of all four halide ions are quite similar in acetonitrile solution, the conductometric rates for a bimolecular process differing by less than a factor of two. The results are presented in Table 15.1. It should be borne in mind that the dielectric constants of the solvents involved, the temperatures and even the reactions differ in the cases presented, so results should be compared only in terms of gross trends.

Table 15.1. Rates of nucleophilic displacement reactions under phase transfer conditions

Nucleophile	Substrate	Solvent	Temp. (°C)	$10^3 k_2$	Ref.
$(n\text{-}C_4H_9)_4N^+Cl^-$	$n\text{-}C_4H_9\text{-}OBs$	Acetone	25.0	33.5	[2]
$(n\text{-}C_4H_9)_4N^+Br^-$	$n\text{-}C_4H_9\text{-}OBs$	Acetone	25.0	9.09	[2]
$(n\text{-}C_4H_9)_4N^+I^-$	$n\text{-}C_4H_9\text{-}OBs$	Acetone	25.0	1.68	[2]
DC-18-C-6/KBr	$n\text{-}C_4H_9\text{-}OBs$	Acetone	25.0	9.72	[2]
DC-18-C-6/KI	$n\text{-}C_4H_9\text{-}OBs$	Acetone	25.0	2.2	[2]
$BuP(C_{16}H_{33})_3^+Cl^-$	$n\text{-}C_8H_{17}\text{-}OMs$	C_6H_5Cl	60.0	19.7	[3]
$BuP(C_{16}H_{33})_3^+Br^-$	$n\text{-}C_8H_{17}\text{-}OMs$	C_6H_5Cl	60.0	8.1	[3]
$BuP(C_{16}H_{33})_3^+I^-$	$n\text{-}C_8H_{17}\text{-}OMs$	C_6H_5Cl	60.0	3.0	[3]
18-C-6/KF	$C_6H_5CH_2\text{-}OTs$	CH_3CN	30.0	135	[4]
18-C-6/KCl	$C_6H_5CH_2\text{-}OTs$	CH_3CN	30.0	130	[4]
18-C-6/KBr	$C_6H_5CH_2\text{-}OTs$	CH_3CN	30.0	120	[4]
18-C-6/KI	$C_6H_5CH_2\text{-}OTs$	CH_3CN	30.0	88	[4]

15.2 Cation Effects

Many organic reactions which involve ionic species involve the cations associated with the reactive anions, even though most of the attention is focused on the latter. Some reactions, like the displacement of bromide by cyanide in the synthesis of a cyanoalkane [5] (see Eq. 7.2), appear to involve the cation in only the most peripheral way. In the above example, replacing sodium cyanide by tetrabutylammonium cyanide [6, 7] or the 18-crown-6 complex of potassium cyanide does not alter the course of the reaction [8, 9], only its rate. There are, however, quite a few examples now available of alterations in the course of the reaction in the presence of crown ether, cryptate, or ammonium salts.

In many situations, crown ethers compete effectively with anions for the associated cations in solution. An example of this is found in the case of acetylacetonate ion [10]. In the absence of crown ether, the sodium ion appears to be positioned as one might anticipate for the Z,Z-conformation of the unperturbed (chelated) ion-pair. When 18-crown-6 is present, there is competition between the crown and oxygen anions for solvation of the cation and the Z,Z-conformation is destabilized relative to the E,Z-conformation to the extent the cation is dissociated (see Eq. 15.1).

$$\text{(15.1)}$$

A related phenomenon has been observed in the Cannizzaro reaction [11]. The potassium ion is apparently required to coordinate benzaldehyde and the benzaldehyde-hydroxide ion adduct which then undergoes hydride transfer via a cyclic six-membered

Altered Reactivity

transition state as shown in equation 15.2. When crown is present, there is competition for the potassium ion, and to the extent it is complexed by the polyether, the cation is dissociated from the Cannizzaro complex. The greater the amount of crown, the less likely is hydride transfer. Accordingly, the yield per unit time for the Cannizzaro reaction of benzaldehyde with potassium hydroxide was found to be inversely proportional to the amount of added 18-crown-6. Likewise, substitution of the bulky quaternary ammonium cation benzyltriethylammonium resulted in complete sequestration of the Cannizzaro reaction.

$$C_6H_5\text{-CHO} + KOH \rightleftharpoons C_6H_5\text{-}\underset{\underset{OH}{|}}{\overset{\overset{OK}{|}}{C}}\text{-H} \xrightarrow{C_6H_5\text{-CHO}} \cdots \quad (15.2)$$

$$\longrightarrow C_6H_5\text{-COOK} + C_6H_5CH_2OH$$

The presence of a cation is also important in the lithium aluminum hydride reduction of cyclohexanone [12, 13]. Lithium ion is apparently required as an electrophilic catalyst for the reduction of cyclohexanone and if an appropriately sized cryptate is added to the reducing medium, a retardation of the reduction is observed. Addition of [2.1.1]-cryptate to the reduction system in quantities sufficient to complex all of the available lithium ion leads to a total absence of reduction. An excess of lithium iodide, however, restores reactivity. This reaction is discussed in Sect. 12.5.

Although numerous complexes of macrocyclic ethers and cryptates exist, one of the more interesting and unusual examples is the complex between appropriately sized crown ethers and aryldiazonium compounds [14]. The insertion complex is formulated in equation 15.3 and the evidence for this structure has been reported [14]. Thus far, only 1:1 complexes have been isolated [15, 16].

Aryldiazonium tetrafluoroborates and hexafluorophosphates are poorly soluble in nonpolar media but the complexes are soluble due to lipophilization of the salt [14]. 4-t-Butylbenzenediazonium tetrafluoroborate is soluble in dichloromethane by virtue

$$(15.3)$$

of the alkyl residue and undergoes a homogeneous Schiemann reaction on standing for 10 days [17]. A mixture of the aryl fluoride and chloride is formed, presumably the result of aryl cation reaction with fluoride ion or solvent (see Eq. 15.4). In the presence of crown ether, the rate of the reaction is significantly retarded, indicating that the crown stabilizes the species to an appreciable extent [16]. In accordance with this observation is the stability of the solid 1:1 complex: the complex between

4-bromobenzenediazonium tetrafluoroborate and 18-crown-6 does not explode when placed on a steel plate and struck with a steel hammer [18].

$$4\text{-}t\text{-}C_4H_9\text{-}C_6H_4\text{-}N\equiv N^+BF_4^- \xrightarrow{CH_2Cl_2} 4\text{-}t\text{-}C_4H_9\text{-}C_6H_4F$$
$$+ 4\text{-}t\text{-}C_4H_9\text{-}C_6H_4Cl \quad (15.4)$$

Not only does crown complexation influence the reactivity of the diazonium ion directly, it has recently been shown to affect the reactivity of the ring. 4-Bromobenzenediazonium ion undergoes smooth *para*-halogen exchange in the presence of chloride ion in nonpolar solution (see Eq. 15.5), but the reaction is either retarded or sequestered in the presence of a sufficient concentration of 18-crown-6 [19]. This result has been interpreted in terms of heteroatom solvation of the diazoniom function with an attendant reduction in σ_p.

$$4\text{-}Br\text{-}C_6H_4\text{-}N_2^+X^- + Cl^- \rightarrow 4\text{-}Cl\text{-}C_6H_4\text{-}N_2^+X^- + Br^- \quad (15.5)$$

In work recently reported [20] it was found that reduction of aryldiazonium tetrafluoroborates in nonpolar solution by aqueous hypophosphorous acid and catalytic cuprous oxide, good to excellent yields of arenes were obtained under very mild conditions. The single exception in the group of salts studied was 4-methoxybenzenediazonium tetrafluoroborate which yielded only 67% of arene. Addition of 7.5 mole-% of 18-crown-6 increased the yield to 88%. The effect of crown in this system is not understood but believed to involve crown enhanced solubility of the salt (see Eq. 15.6) [20].

$$Ar\text{-}N_2^+BF_4^- + H_3PO_2 + Cu_2O \xrightarrow{CHCl_3} Ar\text{-}H \quad (15.6)$$

The crown catalyzed phase transfer of acetate ion into a nonpolar medium followed by reaction with aryldiazonium ion has also been shown to be an effective and mild method for the generation of aryl radicals [21]. Radicals thus generated can be reacted with deuteriochloroform or chloroform to yield deuterio- or protio-arenes. Reaction with bromotrichloromethane leads to the bromoarene and the reaction in benzene solution leads to unsymmetrical biphenyls in good yield [22]. These three processes are formulated in equation 15.7.

$$Ar\text{-}N_2^+ + AcO^-K^+ \xrightarrow{crown} Ar^{\cdot} \rightleftarrows \begin{matrix} Ar\text{-}C_6H_5 \\ Br\text{-}C_6H_5 \\ D\text{-}C_6H_5 \end{matrix} \quad (15.7)$$

In these reactions, crown apparently prefers to complex the potassium ion over the diazoniom function because the reactivity of the latter does not appear to be attentuated in its presence [20].

The reactions of carbenes depend appreciably on the means by which they are generated. Carbenes for example, which result from the photolysis of diazirines exhibit markedly different olefin selectivities from those generated by base catalyzed α-elimination [23]. The difference in reactivity is attributed to the difference between "free" and "complexed" carbenes. It is presumed that loss of nitrogen from a diazirine results in the "free" species, whereas those carbenes generated in the presence of salts tend to coordinate with them. Moss and coworkers [23–25] have shown that addition of crown ether which can complex the salt present in the base catalyzed generation technique results in "free" carbene species. This has been demonstrated by comparison of olefin selectivities of the free and complexed carbenes in the presence and absence of crown ether. The results are presented in Table 15.2 in which carbenes are generated according to equations 15.8–15.9. It is interesting to note that carbenes which are strongly substituent stabilized, exhibit selectivites characteristic of "free" species regardless of the means by which they are generated [25]. Thus, dichloro- and methylthio(chloro)carbenes exhibit "free" carbene behavior in the presence or absence of crown. The technique has also been applied to unsaturated carbenes [26].

$$C_6H_5-CHX_2 \xrightarrow{KO-t-C_4H_9} \text{carbenoid} + KX + C_4H_9OH \quad (15.8)$$

$$\downarrow \text{crown}$$

$$\underset{X}{\overset{C_6H_5}{>}}C=\underset{N}{\overset{N}{<}} \xrightarrow{h\nu} \text{free carbene} + N_2 \quad (15.9)$$

Table 15.2. Relative reactivities of alkenes towards phenylbromocarbene in competition with tetramethylethylene

Olefin	Rate ratios (k_{TME}/k_{olefin})		
	Base	Base/18-crown-6	Diazirine/hν
trimethylethylene	1.28	1.72	1.74
isobutylene	1.65	4.11	4.44
Z-butene	5.79	8.24	8.34
E-butene	11.3	17.1	17.5

15.3 Affected Anions

It was demonstrated some years ago that crown ethers complex cations and enhance the specific activity of the associated anions [27–34]. Considerable attention has been given to base catalyzed elimination reactions where the presence of crowns significantly affects the association of the base in solution and the differences in base species is reflected not only in the rates of elimination, but in the geometries and orientation of resulting olefins [35].

Such bases as potassium t-butoxide are associated in organic solutions. In general, the poorer the ability of the solvent to stabilize a cationic charge, the more associated will be the base. It is known, for example, that the conductivity of 0.1 M sodium t-butoxide in t-butanol is only 6% higher than the conductivity of the pure solvent [36]. In nonpolar media, sodium t-butoxide is tetrameric to approximately octameric, depending on the solvent. The associated base species is obviously much larger than the monomer, and the reactivity of the monomer differs from it. One means for assessing the degree of association and the related steric demand of the base is to observe product geometries in elimination reactions utilizing the base in question [35, 37]. In the case of potassium t-butoxide, for example, in a nonpolar medium where the base is essentially tetrameric, 2-iodobutane will undergo E-2 elimination to yield substantial amounts of 1-butene (see Eq. 15.10) [12]. In the presence of 18-crown-6 which is known to strongly complex potassium ion, the oligomeric base is (at least partially) dissociated (see Eq. 15.10). The dissociated base has a smaller steric demand and the preference for internal olefin in the butene system increases (see Eq. 15.11). In the equations formulated, it should be noted that the olefins indicated are not the exclu-

$$(t\text{-}C_4H_9\text{-}O^-K^+)_n + 18\text{-crown-6} \rightleftarrows (t\text{-}C_4H_9\text{-}O^-K^+)_{n-1} + t\text{-}C_4H_9\text{-}O^- \;\textcircled{K$^+$}$$
(15.10)

$$\textcircled{K$^+$} + t\text{-}C_4H_9\text{-}O^- + CH_3\text{-}CHI\text{-}CH_2\text{-}CH_3 \rightarrow KI + \text{crown} + ROH$$
$$+ CH_3CH\text{=}CHCH_3$$
(15.11)

$$(t\text{-}C_4H_9\text{-}O^-K^+)_n + CH_3\text{-}CHI\text{-}CH_2\text{-}CH_3 \rightarrow KI + ROH + CH_2\text{=}CH\text{-}CH_2\text{-}CH_3$$
(15.12)

sive products. In each case a mixture is obtained, but the product distribution varies according to the steric demand of the base. A detailed description of work in this area is available in a review by Bartsch which has recently appeared [35].

Ugelstadt suggested the term "self-solvating" some years ago to describe a base which could "wrap around" itself, thereby providing intramolecular solvation [38]. Several examples of this phenomenon are now known involving crown ethers and the corresponding open-chained systems. The solvent and solvation dependence of the base systems is illustrated by Table 15.3 which compares several self-solvating systems including the anions of polyethylene glycol (1) and monoaza-18-crown-6 (2) [39]. That 1 and 2 are self-solvating systems is apparent from the similarity in product distributions in a polar, cation-solvating medium and in a nonpolar solution.

$CH_3(OCH_2CH_2)_7OK$

1

2

Altered Reactivity

Table 15.3. Reactions of 2-Iodobutane with various bases at 50 °C

Base	DMSO Solvent		Toluene Solvent	
	% 1-Butene	E-2-Butene: Z-2-Butene	% 1-Butene	E-2-Butene: Z-Butene
t-BuOK	20	3.5	36	1.7
Et₃COK	21	3.8	47	1.8
$CH_3(CH_2)_{17}OK$	18	3.1	29	2.6
1	19	3.4	20	3.7
2	22	3.0	21	3.8

Another example of the well-known reactivity enhancement of certain anions in dipolar aprotic media in which the cations are solvated and the anions are relatively solvation free can be found in certain unimolecular reactions. Addition of crown ethers to solutions of the ion pairs appears to be an effective method for achieving the solvent (or ligand) separated ion pair whose reactivity is greater than that of the contact ion pair [27–34, 40]. Hunter and coworkers have shown that the decarboxylation of fluorenylacrylic acid salts occur at an enhanced rate in dipolar aprotic media, or in the presence of a crown ether (see Eq. 15.13) [41].

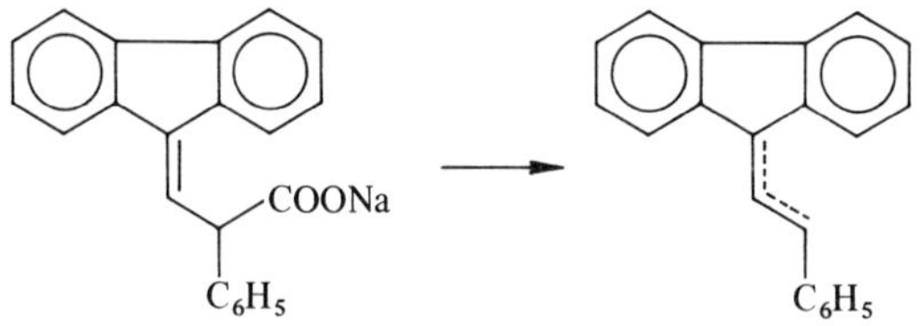

(15.13)

A combination of the enhanced decarboxylation and enhanced hydrolysis rates (see Sec. 9.6) has been utilized in an improved malonic ester synthesis as formulated in equation 15.14 [42].

$$R^1R^2C(CO_2C_2H_5)_2 \xrightarrow[RT]{KOH,\ 18\text{-}C\text{-}6} R^1R^2C(CO_2C_2H_5)COOK \xrightarrow{\Delta} R^1R^2\text{-}CH\text{-}CO_2C_2H_5 \quad (15.14)$$

Evans and coworkers have found a similar rate enhancement in the oxy-Cope rearrangement formulated in equation 15.15 [43].

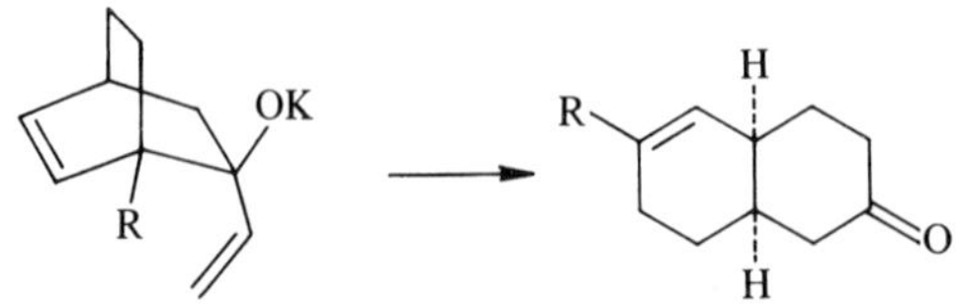

(15.15)

Reactivity enhancement has also been observed in the Smiles rearrangement involving non- or deactivated aromatic ring systems [44, 45]. In the presence of hexamethylphosphoric triamide (HMPA), sodium hydride effects the transformation of 2-aryloxyisobutyramides into N-aryl-2-hydroxyisobutyramides in good yield. Particularly noteworthy is the aryl ether to anilide conversion in 55% yield for the 4-methoxyphenyl compound. This reaction is formulated in equation 15.16.

$$CH_3O\text{-}C_6H_4\text{-}O\text{-}C(CH_3)_2\text{-}CO\text{-}NHR \xrightarrow[\text{HMPA}]{\text{NaH}} CH_3O\text{-}C_6H_4\text{-}NR\text{-}CO\text{-}C(CH_3)_2OH \quad (15.16)$$

15.4 Ambident Nucleophiles

Numerous studies of ambident anion alkylation have been conducted during the past twenty-five years. Among these, probably the most carefully studied species are the phenolates and the β-dicarbonyl compounds. The former were carefully studied by Kornblum and coworkers some years ago and found to afford different C/O alkylation ratios depending primarily on the solvent in which the reaction was conducted, the cation associated with the enolate and the alkylating agent [46–49]. In general, it appears that the more dissociated is the cation, the more accessible is oxygen for alkylation. This can be accomplished by using a dipolar solvent which enhances separation of cation and anion [50–53]; likewise, bulky, charge-diffuse cations which can separate more readily from anions lead to enhancement of the C/O ratio [54]. Finally, the exothermicity of the alkylation process influences the alkylation ratio: the greater the exothermicity of the alkylation, the greater the tendency for oxygen alkylation [50, 55, 56].

Admittedly, the above is a considerable simplification of what is a complex, multi-variable process, but some frame of reference seems required. The use of a quaternary ammonium salt instead of an alkali metal cation allows for the dissolution of ammonium enolate ion pairs in nonpolar media [54]. Whereas these ion pairs are relatively dissociated in such solvents as DMSO and give appreciable amounts of O-alkylation, in benzene or toluene, the C/O ratio is quite high. Brändström and Jünggren found that isopropylation of tetrabutylammonium acetylacetonate lead to a C/O ratio of 0.72 in DMSO and 13.8 in toluene solution (see Eq. 15.17) [54].

$$(CH_3\text{-}CO\text{-}CH\text{-}CO\text{-}CH_3)^- Q^+ + RX \rightarrow CH_3\text{-}CO\text{-}CHR\text{-}CO\text{-}CH_3$$
$$+ CH_3\text{-}CO\text{-}CH\text{=}COR\text{-}CH_3 \quad (15.17)$$

Durst and Liebeskind found that under phase transfer catalytic conditions, in the presence of a catalytic amount of tricaprylylmethylammonium chloride in benzene solution, acetoacetic ester alkylated exclusively on carbon within the limits of detection (see Eq. 15.18) [57]. The solid-liquid phase transfer process utilized in this alkyla-

Altered Reactivity

$$(CH_3-CO-CH-COOCH_3)^- Na^+ \xrightarrow[RX, C_6H_6]{TCMAC} CH_3-CO-CHR-COOCH_3$$

(15.18)

tion procedure was more effective when conducted in benzene or toluene than in hexane.

Another ambident anion species which shows cation and solvent dependent alkylation behavior is pyridone. Although N-alkylation is preferred thermodynamically, and benzylation of the sodium salt yields predominantly the N-alkylated product, O-alkylation of the anion is favored by dipolar aprotic solvents and soft cations [58]. In pentane or benzene, the silver salt of pyridone anion is O-benzylated. Likewise, under phase transfer conditions, where the anion is presumably paired with a quater-

$$\underset{N}{\bigcirc}\!\!-O^-M^+ \xrightarrow{R-X} \underset{N}{\bigcirc}\!\!-OR + MX \qquad (15.19)$$

nary ion, O-alkylation is preferred [59]. The process is illustrated in equation 15.19.

A number of other rather more detailed studies involving the reactions of enolate ions in the presence of crown ethers and cryptates have also been reported. The multivariable nature of these reactions is clearly demonstrated in these reports [60–62].

References

1. Swain, C. G., Scott, C. B.: J. Amer. Chem. Soc. 75, 141 (1953).
2. Sam, D. J., Simmons, H. E.: J. Amer. Chem. Soc. 96, 2252 (1974).
3. Landini, D., Maia, A. M., Montanari, F., Pirisi, F. M.: J. C. S. Chem. Commun. 1975, 950.
4. Liotta, C. L., Grisdale, E. E., Hopkins, H. P.: Tetrahedron Let. 1975, 4205.
5. Starks, C. M.: J. Amer. Chem. Soc. 93, 195 (1971).
6. Brändström, A.: "Preparative Ion Pair Extraction", Apotekarsocieteten, Hässle AB, Sweden 1974, p. 108.
7. Solodar, J.: Tetrahedron Let. 1971, 287.
8. Zubrick, J. W., Dunbar, B. I., Durst, H. D.: Tetrahedron Let. 1975, 71.
9. Cook, F. L., Bowers, C. W., Liotta, C. L.: J. Org. Chem. 39, 3416 (1974).
10. Noe, E. A., Raban, M.: J. Amer. Chem. Soc. 96, 6184 (1974).
11. Gokel, G. W., Gerdes, H. M., Rebert, N. W.: Tetrahedron Let. 1976, 653.
12. Pierre, J. L., Handel, H.: Tetrahedron Let. 1974, 2317.
13. Loupy, A., Seyden-Penne, J., Tchoubar, B.: Tetrahedron Let. 1976, 1677.
14. Gokel, G. W., Cram, D. J.: J. C. S. Chem. Commun. 1973, 481.
15. Haymore, B. L., Ibers, J. A., Meek, D. W.: Inorg. Chem. 14, 541 (1975).
16. Bartsch, R. A., Chen, H., Haddock, N. F., Juri, P. N.: J. Amer. Chem. Soc. 98, 6753 (1976).
17. Swain, C. G., Rogers, R. J.: J. Amer. Chem. Soc. 97, 799 (1975).
18. Korzeniowski, S. H., Gokel, G. W.: unpublished observation.
19. Gokel, G. W., Korzeniowski, S. H., Blum, L.: Tetrahedron Let. 1977, 1633.
20. Korzeniowski, S. H., Blum, L., Gokel, G. W.: J. Org. Chem. 42, 1469 (1977).
21. Korzeniowski, S. H., Gokel, G. W.: Tetrahedron Let. 1977, 1637.

22. Korzeniowski, S. H., Blum, L., Gokel, G. W.: Tetrahedron Let. *1977*, 1871.
23. Moss, R. A., Pilkiewicz, F. G.: J. Amer. Chem. Soc. *96*, 5632 (1974).
24. Moss, R. A., Mallon, C. B.: J. Amer. Chem. Soc. *97*, 344 (1975).
25. Moss, R. A., Joyce, M. A., Pilkiewicz, F. G.: Tetrahedron Let. *1975*, 2425.
26. Stang, P. J., Mangum, M. G.: J. Amer. Chem. Soc. *97*, 6478 (1975).
27. Hogen-Esch, T. E., Smid, J.: J. Amer. Chem. Soc. *91*, 4580 (1969).
28. Wong, K. H., Konizer, G., Smid, J.: J. Amer. Chem. Soc. *92*, 666 (1970).
29. Chan, L. L., Wong, K. H., Smid, J.: J. Amer. Chem. Soc. *92*, 1955 (1970).
30. Almy, J., Garwood, D. C., Cram, D. J.: J. Amer. Chem. Soc. *92*, 4341 (1970).
31. Staley, S. W., Erdman, J. P.: J. Amer. Chem. Soc. *92*, 3832 (1970).
32. Roitman, J. N., Cram, D. J.: J. Amer. Chem. Soc. *93*, 2231 (1971).
33. a) Svoboda, M., Hapala, J., Zavada, J.: Tetrahedron Let. *1972*, 265.
    b) Zavada, J., Svoboda, M., Pankova, M.: Tetrahedron Let. *1972*, 711.
34. Maskornick, M. J.: Tetrahedron Let. *1972*, 1797.
35. Bartsch, R. A.: Acct. Chem. Res. *8*, 239 (1975) and references therein.
36. Saunders, W. H., Bushman, D. G., Cockerill, A. F.: J. Amer. Chem. Soc. *90*, 1775 (1968).
37. Bartsch, R. A., Roberts, D. K.: Tetrahedron Let. *1977*, 321.
38. Ugelstad, J., Mork, P. C., Jenson, B.: Acta Chem. Scand. *17*, 1455 (1963).
39. Gokel, G. W., Garcia, B. J.: Tetrahedron Let. *1977*, 317.
40. Alunni, S., Baciocchi, E., Perucci, P.: J. Org. Chem. *41*, 2636 (1976).
41. Hunter, D. H., Lee, W., Sim, S. K.: J. C. S. Chem. Commun. *1974*, 1018.
42. Hunter, D. H., Perry, R. A.: Synthesis *1977*, 37.
43. Evans, D. A., Golob, A. M.: J. Amer. Chem. Soc. *97*, 4765 (1975).
44. Bayles, R., Johnson, M. C., Maisey, R. F., Turner, R. W.: Synthesis *1977*, 31.
45. Bayles, R., Johnson, M. C., Maisey, R. F., Turner, R. W.: Synthesis *1977*, 34.
46. Kornblum, N., Smiley, R. A., Blackwood, R. K., Iffland, D. C.: J. Amer Chem. Soc. *77*, 6269 (1955).
47. Kornblum, N., Berrigan, P. J., le Noble, W. J.: J. Amer. Chem. Soc. *85*, 1141 (1963).
48. Kornblum, N., Seltzer, R., Haberfield, P.: J. Amer. Chem. Soc. *85*, 1148 (1963).
49. Brieger, G., Pelletier, W. M.: Tetrahedron Let. *1965*, 3555.
50. le Noble, W. J., Puerta, J. E.: Tetrahedron Let. *1966*, 1087.
51. Kurts, A. L., Beletskaya, I. P., Macias, A., Reutov, O. A.:Tetrahedron Let. *1968*, 3679.
52. le Noble, W. J., Morris, H. F.: J. Org. Chem. *34*, 1969 (1969).
53. Kurts, A. L., Macias, A., Beletskaya, I. P., Reutov, O. A.: Tetrahedron Let. *1971*, 3037.
54. Brändström, A., Junggren, U.: Acta Chem. Scand. *25*, 1469 (1971).
55. Engemyr, L. B., Songstad, J.: Acta Chem. Scand. *26*, 4179 (1972).
56. House, H. O., Auerbach, R. A., Gall, M., Peet, N. P.: J. Org. Chem. *38*, 514 (1973).
57. Durst, H. D., Liebeskind, L.: J. Org. Chem. *39*, 3271 (1974).
58. Stein, A. R., Tan, S. H.: Can. J. Chem. *52*, 4050 (1974).
59. Freedman, H. H.: U. S. Pat. 3,969,360, July 13. 1976.
60. Kurts, A. L., Dem'yanov, P. I., Beletskaya, I. P., Reutov, O. A.: Zh. Org. Chem. *9*, 1313 (1973).
61. D'Incan, E., Viout, P.: Tetrahedron *31*, 159 (1975).
62. Cambillau, C., Sarthou, P., Bram, G.: Tetrahedron Let. *1976*, 281.

16. Addendum: Recent Developments in Phase Transfer Catalysis

Because phase transfer catalysis is such a rapidly moving field, this volume will inevitably lack some of the more recently published work. In an effort to minimize such omissions, we have included here articles which appeared after the body of the manuscript was complete. In some cases, this includes articles dating from the middle of 1976 due to the problems of mail service and library bindings. We have included with each article only a brief description of the work; a description which in many cases does not do the effort justice. It is our hope that a more detailed discussion can be included in a later edition.

Addendum to Chapter 1

Mechanism of phase-transfer catalysis. D. Landini, A. Maia and F. Montanari, J. C. S. Chem. Commun., 112 (1977)

Using an aqueous membrane system separating two organic phases, it is shown that transport from one organic phase to the other cannot occur with water insoluble quaternary 'onium salts such as hexadecyltributylphosphonium bromide, although this, among other 'onium salts, is an effective phase transfer catalyst. The implication seems to be that deep penetration of the quaternary 'onium compound into the aqueous phase is not a requisite for effective catalysis.

N-Alkylpentamethylphosphoramides: novel catalysts in two-phase reactions. M. Tomoi, T. Takubo, M. Ikeda and H. Kakiuchi, Chem. Let., 473 (1976)

Although hexamethylphosphoramide was not effective, monoalkyl or polymer bound (through nitrogen) pentamethylphosphoramides were found to be effective phase transfer catalysts for nucleophilic substitution and oxidation reactions.

Synthesis of alkyl-substituted crown ethers: efficient phase-transfer catalysts. M. Cinquini and P. Tundo, Synthesis, 516 (1976)

The synthesis of a variety of alkyl substituted (both at carbon and nitrogen) crown and azacrown ethers were reported and found to be efficient catalysts for both nucleophilic displacements and alkylation reactions. The more lipophilic the crowns, the more effective as catalysts they were found to be.

Phasentransfer-Katalyse durch offenkettige Polyäthyleneglykol-Derivate; I. Substitutionsreaktion von Benzylbromid mit Kaliumsalzen. H. Lehmkuhl, F. Rabet and K. Hauschild, Synthesis, 184 (1977)

Open-chained polyethylene glycols were found to be effective phase transfer catalysts in reactions of HS^-, NCS^-, N_3^-, AcO^-, NC^- and F^- ions with benzyl bromide as the substrate. The reactions were carried out in both benzene and acetonitrile and softer anions and longer-chained polyethers apparently were favored.

$$C_6H_5-CH_2-Br + KX \rightarrow C_6H_5-CH_2-X + KBr$$

On the theory of phase-transfer catalysis. J. E. Gordon and R. E. Kutina, J. Am. Chem. Soc. *99*, 3903 (1977)

The rate of nucleophilic displacement reactions run under PT-conditions is treated as the sum of the rate limiting homogeneous reaction $RX(org) + Y^-(org) \rightarrow RY(org) + X^-(org)$ and a rapid equilibrium exchange of anions between the two phases. This equilibrium depends on their selectivity coefficient. The various factors affecting the kinetics observed in PTC reactions are discussed and the expected behavior calculated.

Addendum to Chapter 2

Anwendungen der Phasentransfer-katalyse-4. Halogenaustausch und Reaktivitäten bei dibrom-, chlorbrom- und dichlorcarben. E. V. Dehmlow, M. Lissel and J. Heider, Tetrahedron *33*, 363 (1977)

Carbenes generated under phase transfer catalytic conditions from $CHBrCl_2$ and $CHBr_2Cl$ react with olefins to give all three possible dihalocyclopropanes. The implication is that the dihalocarbenes undergo halogen exchange. Experimental data is presented to demonstrate that dibromocarbene is more reactive to olefins than is dichlorocarbene, an observation which contradicts other literature reports.

Zur Bildung, Trennung und Zuordnung diastereomerer CCl_2-Bis-Adducte an Diolefine. W. Kuhn, H. Marschall and P. Weyerstahl, Chem. Ber. *110*, 1564 (1977)

Dichlorocarbene generated under phase transfer catalytic conditions is reacted with several dienes. The products are isolated and identified and stereochemistry is determined by a combination of partial resolution and magnetic resonance spectroscopy.

Synthesis and some properties of gem-dichlorocyclopropane carboxaldehyde acetals and cyclopropane carboxaldehyde acetals. A. Kh. Khusid, G. V. Kryshtal, V. A. Dombrovski, V. F. Kucherov, L. A. Yanovskaya, V. I. Kadentsev and O. S. Chizhov, Tetrahedron *33,* 77 (1977)

The phase transfer cyclopropanation technique was applied to and found to succeed with acrolein acetals and related systems where many other more traditional methods failed.

$$R-CH=CH-CH(OR)_2 \xrightarrow{:CCl_2(ptc)} R-CH\underset{}{\overset{}{-}}C(Cl)(Cl)\underset{}{\overset{}{-}}CH-CH(OR)_2$$

An improved route for ring expansion of five-membered heterocyclic compounds by use of phase transfer catalysts. F. De Angelis, A. Gambacorta and R. Nicoletti, Synthesis, 798 (1976)

Dichlorocarbene generated under phase transfer catalytic conditions adds to a variety of substituted indoles to give an adduct (see Sect. 2.7 and Eq. 2.25) which undergoes ring expansion resulting in the production of substituted chloroquinolines. Similar reactions with substituted pyrroles yield substituted chloropyridines.

Enantioselective addition of dichlorocarbene to olefins by use of chiral tertiary amines. Y. Kimura, Y. Ogaki, Y. Isagawa and Y. Otsuji, Chem. Let., 1149 (1976)

Dichlorocarbene generated in a two-phase system in the presence of a chiral amine catalyst, reacts with olefins to afford dichlorocyclopropanation products having small optical rotations. The reaction is discussed in terms of steric interactions in the transition state and is analogous to those discussed in section 2.2 and equations 2.9–2.11.

Addendum to Chapter 3

Phasentransfer-katalysierte Reaktionen; VI. Synthese von 1-Chloro-cyclohexancarbonsäure aus Cyclohexanon. P. Kuhl, M. Muhlstadt and J. Graefe, Synthesis, 825 (1976)

Cyclohexanone reacts with dichlorocarbene generated under phase transfer catalytic conditions to yield 1-chlorocyclohexane carboxylic acid [see Eq. 3.16 in Sect. 3.8].

Zur Reaktion des Makosza-Reagens mit Aldehyden und Ketonen. A. Merz and R. Tomahogh, Chem. Ber. *110,* 96 (1977)

Dichlorocarbene generated under phase transfer catalytic conditions is shown to react with aldehydes and ketones to give a dichloroepoxide. Ring opening of this epoxide leads to α-chloro- or α-hydroxy-acids [see Sect. 3.8 and Eq. 3.16].

Addendum to Chapter 4

Reactions of organic anions. Part LXVIII. An improved method of synthesizing dibromocyclopropane derivatives in a catalytic two-phase system. M. Makosza and M. Fedorynski, Rocz. Chem. *50,* 2223 (1976)

Experimental data is presented to support the observation that generation of dibromocarbene in dichloromethane solution using tri-*n*-butyl amine as catalyst is efficient and efficacious. A dozen olefins are dibromocyclopropanated in yields ranging from 60%–89% [see Eq. 4.1 and Sect. 2.2].

Addendum to Chapter 5

Monobenzylation of diols using phase-transfer catalysis. P. J. Garegg, T. Iversen and S. Oscarson, Carbohydrate Research *50,* C12 (1976)

Under phase transfer catalytic conditions, the yield of monobenzylation products (mixtures) from sugars is substantially improved over classical procedures.

Préparation d'éthers et d'acétals en série glucidique au moyen de la catalyse par transfert de phase. P. Di Cesare and B. Gross, Carbohydrate Research *48,* 271 (1976)

Benzyltriethylammonium chloride and tetrabutylammonium bromide have been utilized as catalysts in the synthesis of formaldehyde acetals from dichloromethane and ethers from several alkyl halides. Particular application is made to sugars [see Sect. 5.7].

Application of the macrocyclic polyether 18-crown-6 to the synthesis of N-alkyl- or -Aryl-2-(p-nitrophenoxy)ethylamines. A. C. Knipe, N. Sridhar and A. Loughran, J. C. S. Chem. Commun., 630 (1976)

The Smiles rearrangement of A is promoted by 18-crown-6 complexes potassium hydroxide in dichloromethane solution and compound B, itself usually unstable to Smiles rearrangement, can be isolated under these conditions in 28% yield. See Sect. 15.3.

$$O_2N-C_6H_4-SO_2-N(C_6H_5)CH_2CH_2OH \rightarrow O_2N-C_6H_4-O-CH_2CH_2NHC_6H_5$$
$$AB$$

Nucleophilic substitution in the aromatic series. LIII. Effect of solvents and macrocyclic ethers on the rate constant of p-nitrobromobenzene with potassium phenolate. V. V. Litvak and S. M. Shein, Zhur. Organich. Khim. *12,* 1723 (1976)

The rate of nucleophilic addition of potassium phenoxide to 4-bromonitrobenzene was found to be solvent dependent. In solvents of low polarity, the reaction was considerably slower then in polar solvents. The addition of crown ethers to the reaction mixture reversed this trend, having a greater effect in nonpolar media.

Addendum to Chapter 7

Emploi d'ether couronne 18-6 pour la photosubstitution par KCN en milieu anhydre de derives aromatiques substiues. R. Beugelmans, M.-T. LeGoff, J. Pusset and G. Roussi, Tetrahedron Let. 2305 (1976)

Photocyanation is enhanced by the presence of 18-crown-6 [see Sect. 7.8 and Eq. 7.16].

"Naked" cyanide-acetone cyanohydrin: a simple, efficient and stereoselective hydrocyanating reagent. C. L. Liotta, A. M. Dabdoub and L. H. Zalkow, Tetrahedron Let., 1117 (1977)

18-Crown-6 phase-transferred potassium cyanide adds, in either benzene or acetonitrile to several α,β-unsaturated ketones in the Michael sense. Acetone cyanohydrin serves as a proton source and cyanide carrier. This reaction is described in section 7.8 and equation 7.12.

Nucleophilic substitutions on hexachlorocyclotriphosphazene using 18-crown-6 ether complexes. E. J. Walsh, E. Derby and J. Smegal, Inorg. Chim. Acta *16*, L9 (1976)

Normally difficult nucleophilic displacements on hexachlorocyclotriphosphazene are facilitated by 18-crown-6 ether. Moreover, the displacement of Cl by CN apparently proceeds by formation first of the isonitrile which rearranges on standing to the nitrile.

Addendum to Chapter 8

Chemistry of superoxide ion. Reaction of superoxide ion with substrates having labile hydrogen. Y. Moro-Oka, P. J. Chung, H. Arakawa and T. Ikawa, Chem. Let., 1293 (1976)

The combination of potassium superoxide and 18-crown-6 was found to be useful in DMSO solution as an oxidizing agent for hydrocarbons bearing a readily abstracted allylic or benzylic hydrogen atom.

The reaction of superoxide with hydrazines, hydrazones and related compounds. C.-I. Chern and J. San Filippo Jr., J. Org. Chem. *42, 178 (1977)*

Potassium superoxide dissolved in benzene with the aid of 18-crown-6 was shown to oxidize hydrazines, hydrazones and related compounds into a variety of products. The products resulted from oxidation, dimerization and condensation. A mechanism is proposed to account for the products.

Addendum to Chapter 9

Phasentransfer-Katalyse durch offenkettige Polyathyleneglykol-Derivate; I. Substitutionsreaktion von Benzylbromid mit Kaliumsalzen. H. Lehmkuhl, F. Rabet and K. Hauschild, Synthesis, 184 (1977)

See abstract in Chap. 1 addendum.

Phase-transfer catalysis; preparation of alkyl azides. W. P. Reeves and M. L. Bahr, Synthesis, 823 (1976)

Alkyl halides (principally bromides) are shown to afford alkyl azides on treatment with azide ion and TCMAC. Six normal alkyl azides were prepared in good yield and one secondary alkyl azide was obtained in somewhat lower yield.

Protection of tryptophan in peptide synthesis. The use of crown ethers. M. Chorev and Y. S. Klausner, J. C. S. Chem. Commun., 596 (1976)

Crown ethers as catalysts of fluoride-anion-mediated reactions in peptide synthesis. Part 1. Protection of tryptophan by benzyloxycarbonyl and 2,4-dichlorobenzyloxycarbonyl groups. Y. S. Klausner and M. Chorev, J. C. S. Perkin I, 627 (1977)

Fluoride ion, crown ether phase-transferred into acetonitrile solution, apparently assists the deprotonation of the indole ring of tryptophan facilitating protection of nitrogen by benzyloxycarbonyl or 2,4-dichlorobenzyloxycarbonyl groups.

Phase transfer catalysis. Preparation of aliphatic and aromatic sulfonyl fluorides. T. A. Bianchi and L. A. Cate, J. Org. Chem. *42, 2031 (1977)*

Sulfonyl fluorides have been prepared in high (84–100%) yield by reaction of the corresponding sulfonyl chlorides with potassium fluoride/18-C-6 in acetonitrile or neat. Seven examples were given.

$$R-SO_2Cl \xrightarrow[KF]{18\text{-}C\text{-}6} R-SO_2F$$

Phase-transfer catalyzed substitution of group VI metal carbonyls. K.-Y. Hui and B. L. Shaw, J. Organomet. Chem. *124*, 262 (1977)

The displacement of one or two molecules of carbon monoxide by mono- or bidentate ligands such as triphenyl phosphine or 1,2-*bis*-(diphenylphosphino)ethane (diphos) respectively is facilitated by phase transferred hydroxide ion. Shorter reaction times and lower temperatures are possible with this technique.

$$Mo(CO)_6 + Ph_2PCH_2CH_2PPh_2 \xrightarrow[benzene]{NaOH/ptc} Mo(CO)_4(diphos)$$

Crown ether catalyzed reduction of aromatic nitro compounds to give amines. H. Alper, D. Des Roches and H. Des Abbayes, Angew. Chem. Int. Ed. *16*, 41 (1977)

Triirondodecacarbonyl in benzene in the presence of 1N KOH and 18-crown-6 was found to reduce two equivalents of an aromatic nitro compound to the corresponding aniline. Five cases are reported in yields ranging from 60%–84% (see Eq. 9.19).

Addendum to Chapter 10

Protection of tryptophan in peptide synthesis. The use of crown ethers. M. Chorev and Y. S. Klausner, J. C. S. Chem. Commun., 596 (1976)

Crown ethers as catalysts of fluoride-anion-mediated reaction in peptide synthesis. Part 1. Protection of tryptophan by benzyloxycarbonyl and 2,4-dichlorobenzyloxycarbonyl groups. Y. S. Klausner and M. Chorev, J. C. S. Perkin I, 627 (1977)

See abstract in Chap. 9 addendum.

Catalyse par transfert de phase en série hétérocyclique. N-alkylation des pyrazole et imidazole. H. J.-M. Dou and J. Metzger, Bull. Soc. Chim. Fr., 1861 (1976)

N-Alkylation of such heterocycles as imidazole and pyrazole is achieved under phase transfer catalytic conditions. Although the yields do not exceed about 80%, they appear to represent an improvement over traditional methodology.

Phase transfer catalysis in the N-alkylation of 2-chlorophenothiazine. J. Masse, Synthesis, 341 (1977)

N-Alkylation of 2-chlorophenothiazine has been accomplished using a two phase system (aq. NaOH/benzene) in the presence of TBAB or BTEAC. Eight examples are reported.

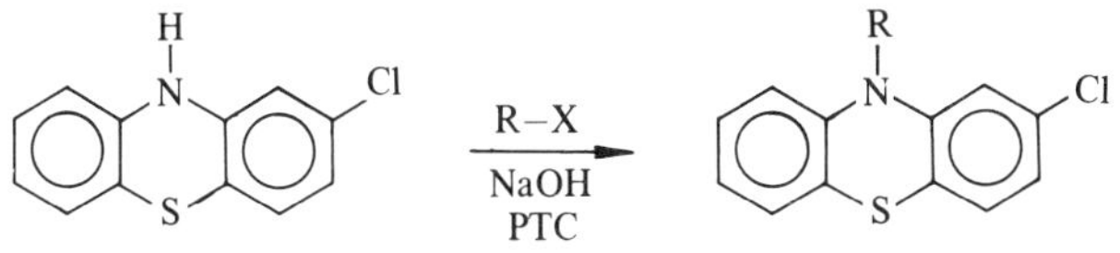

Reactions of organic anions. LXX. Catalytic N-alkylation of phenylhydrazones in aqueous medium. A. Jonczyk, J. Wlostowaska and M. Makosza, Synthesis, 795 (1976)

The phenylhydrazones of a number of aldehydes and ketones were found to alkylate in a two-phase system using 50% aqueous sodium hydroxide as base and tetrabutylammonium chloride as catalyst. Exclusive *N*-alkylation is observed in 43–98% yield for the thirteen examples reported.

$$R^1R^2C=N-NH-C_6H_5 \xrightarrow[\text{2. RX}]{\text{1. NaOH/TBAC}} R^1R^2C=N-NR-C_6H_5$$

N-Alkylation of tosylhydrazones through phase transfer catalysis. S. Cacchi, F. LaTorre and D. Misiti, Synthesis, 301 (1977)

Tosylhydrazones have been N-alkylated with methyl iodide or benzyl bromide by use of a two phase system (CH_2Cl_2/15% aq. NaOH) in the presence of BTMAC. Thirteen examples were reported.

$$\underset{R_2}{\overset{R_1}{>}}C=N-NH-Tos \xrightarrow[\substack{\text{NaOH}\\\text{PTC}}]{R_3X} \underset{R_2}{\overset{R_1}{>}}C=N-N\underset{Tos}{\overset{R_3}{<}}$$

The synthesis of amino acids by phase transfer reactions. M. J. O'Donnell Jr., J. Boniece and S. Earp, unpublished results, 1977

The Schiff's base derived from ethyl glycinate and benzophenone has been alkylated under both anhydrous and phase transfer conditions yielding, after hydrolysis, α-amino acids. Under phase transfer conditions, alanine, α-aminobutyric acid, valine, leucine and phenylalanine were prepared in 91%, 86%, 61%, and 55% yields, based on starting imine.

$$(C_6H_5)_2C=N-CH_2-CO_2C_2H_5 \xrightarrow[R-X]{\text{NaOH/ptc}} \xrightarrow{H_2O} H_2N-CHR-COOH$$

Synthetic applications of crown ethers; the malonic ester synthesis. D. H. Hunter and R. A. Perry, Synthesis, 37 (1977)

The crown ether enhanced decarboxylation of potassium salts is used to synthetic advantage in the malonic ester sequence. After traditional alkylation of malonic ester,

18-crown-6 is used to assist saponification and then loss of CO_2 when the mixture is heated. The one-pot reaction affords yields comparable to the traditional multi-step method and appears more convenient [see Eq. 15.14].

$$R^1R^2-CX-COOC_2H_5 \xrightarrow[2.\ rt \rightarrow 100°]{1.\ KOH/crown} R^1R^2-CX-H$$

$$X = -CO_2C_2H_5, -CO-R^3, -CN$$

Reactions of organic anions. LXXI. Reactions of enol esters with carbanions and dihalocarbenes in catalytic two-phase systems. M. Fedorynski, I. Gorzkowska and M. Makosza, Synthesis, 120 (1977)

In analogy to the previously reported addition of trihalomethyl anions to enol acetates (see Sect. 2.11 and Eq. 2.19), it is shown that a number of carbanions will likewise add to the double bond of such systems. Several substituted phenylacetonitriles add to vinyl acetate under phase transfer catalytic conditions to give the adducts shown in the equation below in 15–70% yields.

$$C_6H_5-CHR-CN + CH_2=CH-O-CO-CH_3 \rightarrow C_6H_5-\underset{R}{\overset{CN}{C}}-CH(CH_3)-O-CO-CH_3$$

Reactions of organic anions. Part LIX. Reactions of phenylacetonitrile and its derivatives with nitrobenzyl chlorides. M. Makosza and J. M. Jagusztyn-Grochowska, Rocz. Chem. 50, 1859 (1976)

The alkylation reactions of phenylacetonitrile, diphenylacetonitrile and 2-phenylbutyronitrile were studied in both homogeneous and two-phase systems using *o-*, *m-*, and *p-*nitrobenzyl chlorides as alkylating agents. Under typical phase transfer catalytic conditions, phenylacetonitrile was dialkylated by nitrobenzyl chloride although in some cases, products of electron transfer and self-condensation of the alkylating agent were observed.

Reactions of organic anions. Part LVIII. The problem of competetive nucleophilic substitution of ortho- and para-halogens in 2,4-dihalonitrobenzenes by some carbanions. M. Makosza, J. M. Jagusztyn-Grochowska and M. Jawdosiuk, Rocz. Chem. 50, 1841 (1976)

The question of where carbanions will attack 2,4-dichloronitrobenzene is addressed. Experimental evidence is presented that such carbon acids as alkylsubstituted phenylacetonitriles and ethylmalonate ester which contain a methine group prefer nucleophilic aromatic substitution in the *para*-position whereas active methylene compounds like desoxybenzoin, diethyl malonate, ethyl cyanoacetate and phenylacetonitrile prefer attack at the *ortho*-chlorine. A variety of conditions was used including strong base in DMSO but most of the cases are phase transfer alkylation conditions.

Addendum to Chapter 10

C$_6$H$_5$–CH$_2$–CN + [2-NO$_2$, 4-Cl chlorobenzene] $\xrightarrow[\text{HO}^-]{\text{PTC}}$ [2-NO$_2$-4-(CH(C$_6$H$_5$)CN)chlorobenzene] + [2-NO$_2$-4-Cl-1-(CH(C$_6$H$_5$)CN)benzene]

Reactions of organic anions. LXXIII. Alkylation of phenylacetonitrile at the interface with aqueous sodium hydroxide. M. Makosza and E. Bialecka, Tetrahedron Let., 183 (1977)

Using 50% aqueous sodium hydroxide as base, phenylacetonitrile can be alkylated by butyl iodide in the absence of any phase transfer catalyst. This observation is interpreted as evidence for interfacial generation of the carbanion which then presumably undergoes alkylation in the organic phase. Although chlorides are not effective alkylating agents under these conditions, addition of a small amount of the corresponding iodide gives an apparently co-catalytic reaction of the type Hennis and coworkers observed (see Eq. 6.2).

C$_6$H$_5$–CH$_2$–CN + R–I $\xrightarrow{\text{NaOH}}$ C$_6$H$_5$–CHR–CN

Reactions of organic anions. Part LXIX. Catalytic alkylation of some α-sulfur substituted ketones in aqueous medium. A. Jonczyk, M. Ludwikow and M. Makosza, Rocz. Chem. 51, 175 (1977)

This paper reports a study of the two-phase catalytic alkylation of α-S-phenylacetophenone and α-S-methylacetophenone.

Synthesis of 2-aminoindolenine derivatives via reductive cyclization of 2-(o-nitrophenyl)-2-phenylalkanenitriles. M. Jawdosiuk and M. Makosza, Rocz. Chim. 50, 857 (1976)

Utilizing previously developed alkylation technology which makes available substituted o-nitrophenylacetonitriles, the reductive cyclization of these substances is explored. Several different reducing agents including sodium alcoholate in alcohol, iron dust and HCl, sodium sulfide, sodium hydrosulfite and catalytic hydrogenation (Pd/C) lead to 2-aminoindolenine derivatives or the corresponding N-oxides.

[o-nitrophenyl-CH$_2$–CN] $\longrightarrow$ [indole-2-NH$_2$]

Synthesis of substituted 3,4-dihydroquinolines and their N-oxides via reductive cyclization of 2-phenyl-2-(o-nitrobenzyl)-alkanenitriles. M. Makosza, I. Kmiotek-Skarzynska and M. Jawdosiuk, Synthesis, 56 (1977)

Utilizing previously developed alkylation technology, the 2-alkyl-2-phenyl-3-(o-nitrophenyl)-propanenitriles shown in the equation below have become accessible. Reductive cyclization of these materials leads to substituted 3,4-dihydroquinolines or their N-oxides.

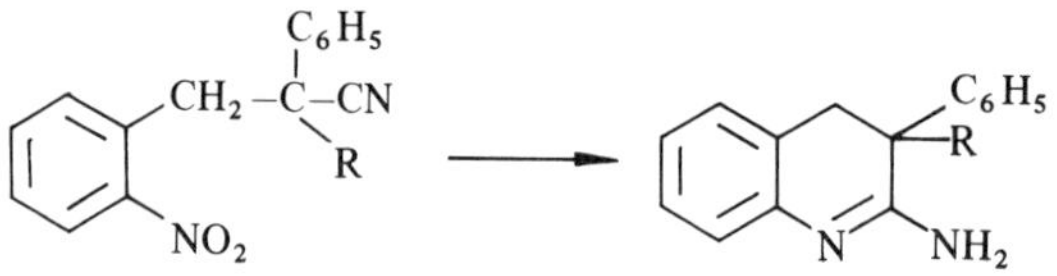

$$\underset{NO_2}{\underset{|}{C_6H_4}}-CH_2-\underset{R}{\underset{|}{\overset{C_6H_5}{\overset{|}{C}}}}-CN \longrightarrow \text{3,4-dihydroquinoline-2-amine with } C_6H_5, R \text{ substituents}$$

Naked fluoride-catalyzed Michael additions. I. Belsky, J. C. S. Chem. Commun., 237 (1977)

Potassium fluoride, solubilized in acetonitrile with the aid of 18-crown-6, was found to catalyze the Michael addition of several carbon acids to α,β-unsaturated nitriles or ketones. Nitromethane, for example, adds in 94% yield to chalcone under these conditions.

$$CH_3NO_2 + C_6H_5-CH=CH-CO-C_6H_5 \rightarrow C_6H_5-(O_2NCH_2)CH-CH_2-CO-C_6H_5$$

Fluoride catalyzed addition of silylacetylenes to carbonyl compounds. E. Nakamura and I. Kuwajima, Angew. Chem. Int. Ed., 15, 498 (1976)

Trimethylsilylphenylacetylene reacts with tetrabutylammonium fluoride to give the tetrabutylammonium salt of phenylacetylene. The ion pair readily reacts with carbonyl compounds to yield, after hydrolysis, acetylenic alcohols.

$$C_6H_5-C\equiv C-TMS \xrightarrow[2.\ RCOR']{1.\ TBAF} C_6H_5-C\equiv C-C(OH)RR'$$

Addendum to Chapter 11

The conversion in high yield of aromatic compounds to arene oxides using hypochlorite and phase transfer catalysts. S. Krishnan, D. G. Kuhn, H. F. Fonouni and G. A. Hamilton, Abstracts of the American Chemical Society Meeting, Chicago, August 1977

A variety of phenanthrenes have been oxidized to the K-region epoxides using a mixture of sodium hypochlorite, hydrogen peroxide, chloroform and tetrabutylammonium bisulfate at room temperature. Buffering to maintain pH in the range 8–9 was found efficacious and yields were generally in the range 70–90%.

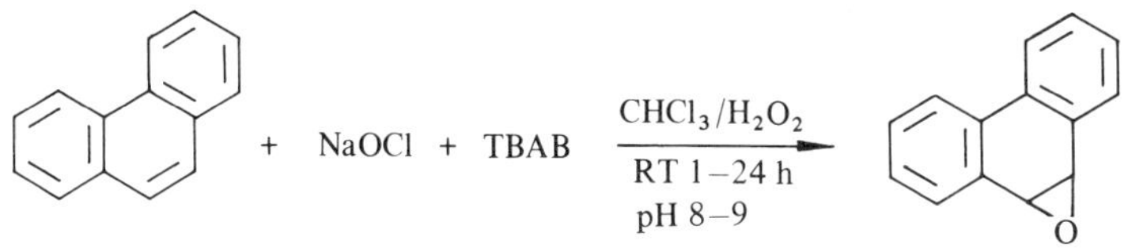

phenanthrene + NaOCl + TBAB $\xrightarrow[\text{pH 8-9}]{\text{CHCl}_3/\text{H}_2\text{O}_2, \text{RT 1-24 h}}$ phenanthrene-9,10-epoxide

Oxidation of organic compounds by aqueous hypohalites using phase transfer catalysis. G. A. Lee and H. H. Freedman, United States Patent no. 3,996,259

This is a patent which covers the work described in section 11.4.

Solid-liquid phase-transfer-catalyzed phosphorylation of hydrazine, preparation of dialkyl phosphorohydrazidates and N,N'-bis[dialkoxyphosphoryl]hydrazines.
A. Zwierzak and A. Sulewska, Synthesis, 835 (1976)

The previously reported phase transfer phosphorylation method (see Sect. 11.8) failed when applied to hydrazine. A solid-liquid phase transfer technique was successful, permitting the monophosphorylation of hydrazine with a variety of dialkyl phosphites. Phosphorylation of the second nitrogen could not be effected under phase transfer conditions and was achieved otherwise.

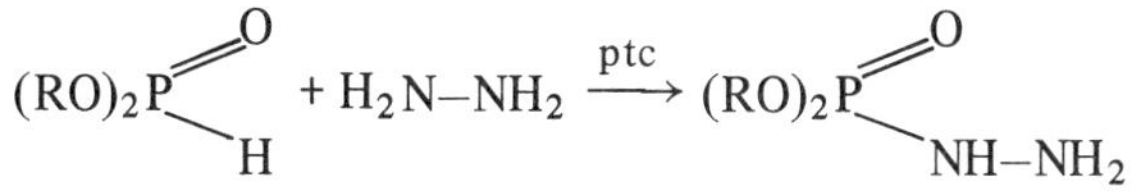

$(RO)_2P(=O)H + H_2N-NH_2 \xrightarrow{ptc} (RO)_2P(=O)NH-NH_2$

Two phase-dehydrogenation of 1,2-substituted hydrazines to 1,2-substituted diazenes (azo compounds). K. Dimroth and W. Tüncher, Synthesis, 339 (1977)

1,2-Substituted hydrazines were oxidized in CH_2Cl_2 by a lipophilic 2,4,6-triarylphenoxy radical to diazenes. 2,4,6-Triarylphenol thus produced is deprotonated by NaOH to the phenoxide anion which is oxidized in the aqueous layer by potassium ferricyanide to the phenoxy radical. Six examples are reported.

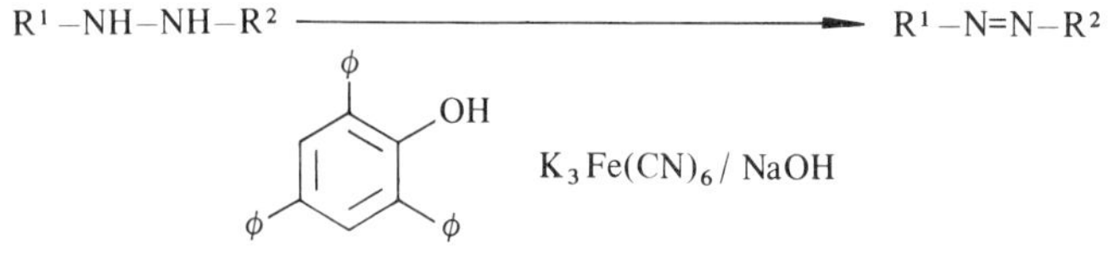

$R^1-NH-NH-R^2 \xrightarrow{\text{2,4,6-triarylphenol, } K_3Fe(CN)_6/NaOH} R^1-N=N-R^2$

Addendum to Chapter 12

Asymmetric reduction of ketones by phase transfer catalysis. J. P. Massé and E. R. Parayre, J. C. S. Chem. Commun., 438 (1976)

N-methyl-N-dodecyl- or N-methyl-N-hexadecylephedrinium bromides phase-transfer borohydride anion (from an aqueous reservoir of sodium borohydride) into

1,2-dichloroethane where modest amounts of asymmetric reduction of ketones to chiral alcohols is observed.

Addendum to Chapter 13

Reactions of organic anions. Part LXIX. Catalytic alkylation of some α-sulfur substituted ketones in aqueous medium. A. Jończyk, M. Ludwikow and M. Makosza, Rocz. Chem. *51*, 175 (1977)

See abstract in Chap. 10 addendum.

Addendum to Chapter 14

Catalyse par transfert de phase et extraction par paires d'ions. Stereoselectivite de la reaction de Horner-Emmons. E. d'Incan, Tetrahedron *33*, 951 (1977)

The Horner-Emmons reaction when conducted under phase transfer catalytic or ion pair extraction conditions shows a pronounced dependence on the phase transfer catalyst and the solvent. The similarity of results often observed in dipolar aprotic media such as HMPA is not observed in this particular case.

Addendum to Chapter 15

Metallation of weak hydrocarbon acids by potassium hydride-18-crown-6 polyether in tetrahydrofuran and the relative acidity of molecular hydrogen. E. Buncel and B. Menon, J. C. S. Chem. Commun., 648 (1976)

Carbanion mechanisms 6. Metallation of arylmethanes by potassium hydride/ 18-crown-6 ether in tetrahydrofuran and the acidity of hydrogen. E. Buncel and B. Menon, J. Am. Chem. Soc. *99*, 4457 (1977)

The basicity of potassium hydride is enhanced by complexation with 18-crown-6. The facile metallation of triphenylmethane, diphenylmethane, di-p-tolylmethane, and the partial metallation of di-2,4-xylylmethane with this system places the relative acidity of molecular hydrogen. The pKa value is estimated at 35.3.

Influence de complexant sur la reactivite des organocuprates. C. Ouannes, G. Dressaire and Y. Langlois, Tetrahedron Let., 815 (1977)

Addition of the complexing agent 12-crown-4 to lithium cuprates leads to altered reactivity of the reagent. 4-Methylcyclohexen-2-one normally undergoes

Addendum to Chapter 15

conjugate addition of lithium dimethylcuprate but in the presence of 12-crown-4, no reaction is observed. Addition of a large excess of LiI leads again to normal reactivity. Similar results are observed for the reaction of cuprate with hexanoyl chloride [see Sect. 12.5 and 15.2].

Nucleophilic substitutions on hexachlorocyclotriphosphazene using 18-crown-6 ether complexes. E. J. Walsh, E. Derby and J. Smegal, Inorg. Chim. Acta *16,* L9 (1976)

See abstract in Chap. 7 addendum.

Base-solvent systems for inducing clean bimolecular 1,2-eliminations. R. A. Bartsch, J. R. Allaway and J. G. Lee, Tetrahedron Let., 779 (1977)

Potassium *t*-butoxide in the presence of 18-crown-6 exhibits reactivity different from the highly associated base species usually present in solution (see Sect. 15.3). In the presence of this reagent, *l*-bornyl tosylate is found to give more unrearranged elimination product than under more ordinary conditions (i.e., more A than B or C).

Nucleophilic substitution in the aromatic series. LIII. Effect of solvents and macrocyclic ethers on the rate constant of p-nitrobromobenzene with potassium phenolate. V. V. Litvak and S. M. Shein, Zhur. Organisch. Khim. *12,* 1723 (1976)

See abstract in Chap. 5 addendum.

Étude de l'alcoylation de l'énolate de sodium dérivé de la 2-carbométhoxy cyclohexanone dans le DMSO en présence et en absence de coordinats macrocycliques. G. Nee and B. Tchoubar, C. R. Acad. Sc. Paris, Serie. C *283,* 223 (1976)

The effect of crown ether and cryptates on the alkylation (carbon/oxygen) ratio of 2-carbomethoxycyclohexanone is studied.

Alkylation d'anions ambidents en catalyse par transfert de phase. Cas des hydroxypyridines. H. J.-M. Dou, P. Hassanaly and J. Metzger, J. Heterocyclic Chem. *14,* 321 (1977)

Under phase transfer conditions (tetrabutylammonium bromide catalyst) butyl bromide was found to alkylate 2- and 4-hydroxypyridines at both O and N with alkylation at the latter predominating. A variety of alterations in reaction conditions (salts, solvent, temperature) did not profoundly affect the product distribution. This finding is in apparent contrast to the findings reported in section 15.4.

Addendum: Recent Developments in Phase Transfer Catalysis

Effect of crown ether complexation on asymmetric intramolecular nucleophilic attack. T. Wakabayashi and Y. Kato, Tetrahedron Let., 1235 (1977)

Asymmetric cyclization of (R)-6-α-methylbenzylaminocarbonylhex-2-enoate with $K^+t\text{-}BuO^-$ in chlorobenzene at $-40\,°C$ is affected by the presence or absence of 18-C-6.

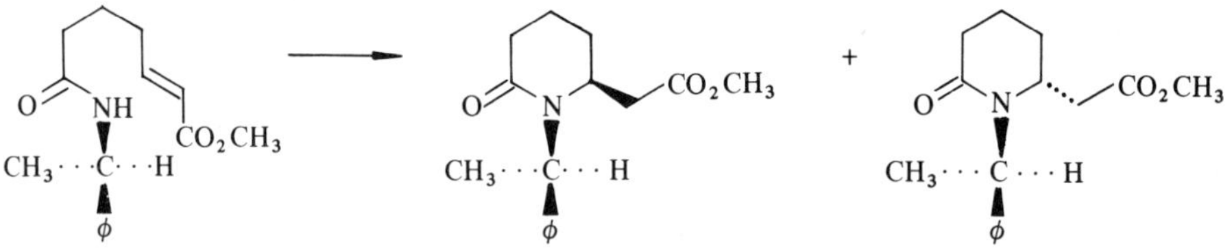

no 18–C–6	62%	38%
18–C–6	38%	62%

Author Index

Abbayes, H. D.   182, 184
Abskharoun, G. A.   58, 59, 60, 61
Adler, M.   28, 29
Agami, C.   3
Akabori, S.   93, 104
Allan, A. R.   60
Allaway, J. R.   265
Allcock, H. R.   134
Allen, R. W.   134
Almy, J.   246, 248
Alper, H.   132, 133, 258
Alunni, S.   248
Andreev, V. M.   138, 169, 170, 171, 172, 178
Andrews, G.   35
Anselme, J. P.   202
Anteunis, M.   34, 235, 236
Aoyama, Y.   45, 65
Arai, K.   104
Arakawa, H.   256
Arnold, C.   97
Arth, G. E.   1
Auerbach, R. A.   249
Ayada, A.   226

Babayan, A. T.   189, 191
Baciocchi, E.   248
Bahr, M. L.   257
Baird, M. S.   60
Baizer, M. M.   88, 89
Bańco, K.   82, 171, 172, 174, 229, 230
Baraldi, P. G.   202, 203
Barco, A.   202, 203
Bard, A. J.   109
Barlet, M. R.   62, 63, 64
Bartsch, R. A.   244, 246, 247, 265
Bashall, A. P.   79

Bass, R. G.   23
Battiste, M.   24
Bayles, R.   249
Beletskaya, I. P.   139, 160, 191, 249, 250
Belsky, I.   262
Benetti, S.   202, 203
Berge, A.   7, 75
Berguer, Y.   109
Bernatek, E.   203
Berrigan, P. J.   249
Beugelmans, R.   107, 256
Beyler, R. E.   1
Bialecka, E.   67, 150, 195, 261
Bianichi, T. A.   257
Bibicheva, A. I.   138, 169, 170, 171, 172, 178
Biegert, I. L.   8, 226
Blackwood, R. K.   249
Blanchard, J.   193, 194
Blum, L.   245
Blüme, G.   32, 33, 65, 66
Boche, G.   67
Bockum, K.   88
Boden, R. M.   212, 235, 236, 237
Boettger, H. G.   23, 24, 25, 26, 27
Boileau, S.   131
Boleslawska, T.   155, 161, 162
Böllert, V.   88
Boniece, J.   259
Borgogno, G.   219, 232
Boswell, R. F.   23
Boudeville, M. A.   182, 184
Bouma, R. J.   197
Bowers, C. W.   96, 97, 98, 99, 100, 106, 243
Bram, G.   250

Brandström, A.   2, 3, 7, 14, 88, 96, 103, 124, 125, 136, 146, 147, 149, 175, 188, 189, 193, 198, 217, 218, 240, 243, 249
Brannock, K. C.   138, 169, 172, 173
Brehme, R.   173
Brieger, G.   249
Bromels, M. J.   9, 130, 131
Broos, R.   235, 236
Brown, H. C.   217
Brown, W. G.   215
Brzozowski, Z. K.   132
Buddrus, J.   128, 129
Buncel, E.   264
Burger, U.   28, 61, 62
Bushman, D. G.   247

Cacchi, S.   259
Cainelli, G.   7, 93, 117, 118, 119, 120, 121, 122, 123, 125, 126, 127
Cambilau, C.   250
Cardilio, G.   171, 173, 178, 180, 181, 209, 228, 229
Carmichael, J. F.   2, 85
Carroll, G. L.   105, 106
Cassar, L.   134
Cate, L. A.   257
Cazes, B.   103
Cederlund, B.   169, 170
Chaikin, S. W.   215
Chan, L. L.   246, 248
Chau, L. V.   65, 66
Chen, H.   244
Chern, C.-I.   110, 111, 112, 113, 114, 257
Childs, M. E.   14, 102, 103
Chizhov, O. S.   254
Chorev, M.   257, 258

267

Author Index

Christensen, J. J. 9, 10
Chrumz, J. L. 88, 89
Chung, P. J. 256
Cihonski, J. C. 132
Cinquini, M. 7, 8, 11, 13, 87, 120, 122, 123, 146, 147, 172, 175, 216, 222, 223, 252
Clapp, M. A. 118, 122
Cockerill, A. F. 247
Colonna, S. 216, 219, 220
Collignon, N. 193, 194
Collings, J. F. 79
Collins, L. R. 3, 8, 13, 86
Colonna, S. 7, 99, 222, 224, 232
Connelly, S. A. 90
Cook, F. L. 11, 13, 96, 97, 98, 99, 100, 106, 243
Corey, E. J. 110, 111, 112
Cornelisse, J. 107
Cram, D. J. 11, 13, 244, 246, 248
Cretcher, L. H. 118, 122
Curtis, A. B. 109
Curtius, U. 88
Cuvigny, N. T. 8, 11, 86, 87, 90, 91
Czyzewski, J. 148, 153, 154, 158, 167, 176, 181, 182, 185

Daasvatn, K. 11
Dabdoub, A. M. 256
Dale, J. 11
Dalgaard, L. 187, 188, 192, 231, 232
Danishefsky, S. 140, 187, 189, 192, 193
De Angelis, F. 254
Deffieux, A. 131
Dehm, D. 90, 92
Dehmlow, E. V. 19, 22, 23, 25, 27, 28, 30, 33, 34, 35, 36, 37, 38, 39, 40, 41, 45, 46, 60, 61, 80, 253
Dehmlow, S. S. 34, 35, 60
Delay, F. 29
de los Heros, V. 28, 61, 62
Dem'yanov, P. I. 139, 160, 191, 250
Derby, E. 256, 265
Dervan, P. B. 126, 127

des Abbayes, H. 132, 133, 258
DeSmet, A. 34
des Roches, D. 132, 133, 258
DiBiase, S. A. 143, 145, 213
DiCesare, P. 255
Dietl, H. K. 138, 169, 172, 173
Dietrich, B. 9, 130, 201, 212
Dimroth, K. 263
Din, Z. U. 71
D'Incan, E. 192, 195, 237, 239, 250, 264
Dobinson, B. 128
Dombrovski, V. A. 253
Dou, H. J.-M. 134, 258, 265
Dressaire, G. 264
Dryanska, V. 174, 175
Dubois, R. A. 73, 74, 75
Dunbar, B. I. 14, 96, 97, 100, 107, 129, 243
Dunkelblum, E. 24, 25, 27, 61
Durr, G. 104, 105
Durst, H. D. 9, 10, 11, 14, 90, 91, 96, 97, 100, 107, 129, 137, 189, 208, 210, 219, 232, 243, 249

Earp, S. 259
Easterly, J. P. 3, 8, 13, 86
Eatough, D. J. 9, 10
Eberson, L. 107
Ebine, S. 33
Effenberger, F. 26
Eguchi, S. 30, 48, 52, 61, 62, 68, 69, 70
Ellingsen, T. 7, 75
Emmons, W. D. 237
Engemyr, L. B. 249
Erdman, J. P. 246, 248
Evans, D. A. 55, 105, 106, 248

Farrar, M. W. 87
Fedorynski, M. 8, 21, 22, 23, 39, 45, 46, 47, 56, 58, 59, 60, 64, 81, 136, 138, 142, 145, 146, 148, 150, 151, 178, 184, 186,
200, 201, 225, 231, 255, 260
Ferreira, D. 78
Fiaud, J. C. 76, 77, 80, 196, 197, 198
Finkelstein, H. 117
Foa, M. 134
Fonouni, H. F. 262
Fornasier, R. 8, 9, 11, 13, 99, 117, 122, 123, 216, 219, 220, 223, 224, 232
Freedman, H. H. 14, 15, 73, 74, 75, 94, 209, 210, 250, 263
Frensdorff, H. K. 10
Friedman, L. 98
Frimer, A. 114
Fritz, G. 88
Fröhlich, I. 51
Fujita, T. 226

Gajda, T. 213
Gajewski, F. 149
Gajos, I. 24, 25, 27, 31, 34, 38, 39, 141, 148, 158, 161
Galat, A. 7
Gall, M. 249
Gambacorta, A. 254
Gaoni, Y. 23
Garcia, B. J. 247
Gardano, A. 134
Garegg, P. J. 255
Garwood, D. C. 246, 248
Gerdes, H. M. 169, 229, 237, 243
Gesellchen, P. D. 92
Gibian, M. J. 110, 111
Gibson, N. A. 2, 206
Glazkov, Y. V. 34, 35
Goclawski, Z. 132
Goetzen, T. 137, 152, 153, 155, 157, 162, 163, 164
Gokel, G. W. 9, 10, 11, 13, 23, 24, 25, 26, 27, 50, 65, 143, 145, 169, 213, 229, 237, 243, 244, 245, 247
Goldschmidt, Z. 18, 27
Golob, A. M. 248
Gonzalez, T. 90
Gordon, J. E. 253
Gorgues, A. 128
Gorzkowska, I. 260

Graefe, J. 28, 29, 49, 51, 53, 54, 254
Green, G. E. 128
Greenside, H. S. 94
Greibrokk, T. 35, 36, 45, 58, 59, 60, 61
Grimm, K. G. 106
Grisdale, E. E. 226, 242, 243
Gromelski, S. J. 71
Gross, B. 255
Gross, M. 117, 119
Grushka, E. 90, 91
Grzejszczak, S. 8, 11, 13, 87, 146, 147, 172, 175, 228, 237, 238, 239
Guida, W. C. 217, 218
Gund, T. M. 45
Gustavii, K. 2, 14

Haberfield, P. 76, 249
Habner, H. 102
Haddock, N. F. 244
Hahn, R. C. 24
Hall, H. K. 117, 124, 125, 128
Hamilton, G. A. 262
Hamm, R. E. 109
Handel, H. 220, 244, 247
Hanessian, S. 121
Hansen, B. 13, 88, 89
Hapala, J. 246, 248
Hara, H. 105
Harris, H. P. 11, 13, 90, 117, 118, 119, 120, 121, 125, 126, 127
Hart, H. 30, 31
Harty, B. J. 24
Hass, E. C. 33, 34
Hassanaly, P. 265
Hauschild, K. 253, 257
Havinga, E. 107
Haymore, B. L. 244
Heider, J. 253
Heintzeler, M. 1
Helder, R. 82
Helgee, B. 107
Hemery, P. 137
Henrickson, J. B 90
Hennis, H. E. 3, 8, 13, 86, 98
Herriott, A. W. 5, 11, 13, 14, 132, 136, 207, 222, 223, 224, 225
Hesse, A. 234

Hessler, J. C. 128
Hetterely, R. M. 88
Hilbrich, R. G. 8, 147, 175, 226
Hinckley, A. A. 217, 220
Hiyama, T. 22, 23, 24, 26, 41, 42, 46, 47, 59, 60, 61, 82, 240
Hoffman, H. 237
Hoffman, J. M. 105, 106
Höfle, G. 39, 40, 41, 53, 54, 55
Hogen-Esch, T. E. 246, 248
Holmberg, K. 13, 88, 89
Holwick, J. L. 23, 24, 25, 26, 27
Hopkins, H. P. 226, 242, 243
Horman, H. 215
Horner, L. 237
Hörnfeldt, A. B. 169, 170
Hosking, J. W. 2, 206
House, H. O. 249
Hug, R. P. 76, 77, 80
Hui, K. Y. 258
Hummelen, J. C. 82
Hunig, S. 235, 236, 237
Hunter, D. H. 248, 259
Hurd, C. D. 13
Hutchins, R. O. 217, 218, 219

Ibers, J. A. 244
Iffland, D. C. 249
Ikan, R. 18, 27
Ikawa, T. 256
Ikeda, M. 31, 32, 62, 252
Indzhikyan, M. G. 189, 191
Inoue, I. 235, 236
Isagawa, K. 8, 21, 23, 52, 254
Ishiguro, I. 53, 54, 55
Ishimoto, S. 86, 92
Ivanov, C. 174, 175
Iversen, T. 255
Izatt, R. M. 9, 10

Jackman, L. M. 3
Jagusztyn-Grochowska, J. M. 152, 154, 157, 161, 164, 260
Jarrousse, M. J. 1, 8, 136, 139, 148
Jawdosiuk, M. 141, 142, 143, 147, 149, 150, 151,
152, 153, 154, 155, 156, 157, 160, 161, 163, 164, 165, 168, 227, 228, 260, 261
Jefford, C. W. 28, 29, 61, 62, 247
Jerzak, B. 46, 47
Jesperson, A. 169, 170
Johnson, E. L. 109
Johnson, M. C. 249
Johnson, R. A. 3, 11, 109
Johnson, R. P. 24
Jończyk, A. 81, 82, 138, 144, 145, 146, 147, 148, 154, 158, 159, 166, 167, 171, 172, 174, 175, 176, 177, 178, 179, 180, 181, 182, 183, 184, 185, 186, 198, 202, 203, 229, 230, 231, 259, 261, 264
Joshi, G. C. 23, 25, 26, 27
Joyce, M. A. 246
Julia, S. 68, 69, 70, 103
Junggren, U. 146, 147, 149, 175, 188, 189, 193, 198, 217, 218, 249
Juri, P. N. 244

Kacprowicz, A. 8, 21, 22, 28, 29, 49, 51, 59, 60, 136
Kadentsev, V. I. 254
Kaempf, B. 131
Kajigaeshi, S. 25
Kakiuchi, H. 252
Kandasamy, D. 217, 218, 219
Kapiack, L. 24
Kato, Y. 266
Kawashima, K. 53, 54, 55, 124
Kashiwaya, K. 129
Kendall, C. 90
Khalil, H. 221
Kheifets, V. C. 14
Khusid, A. Kh. 253
Kieczykowski, G. R. 219, 232
Kielkiewicz, J. 132
Kikta, E. J. 90, 91
Kim, I. K. 221, 222
Kimura, C. 8, 21, 23, 52, 129, 254
Kiriyama, T. 30, 48, 52, 61, 62

Author Index

Klabuhn, H. 33, 34
Klausner, Y. S. 257, 258
Klein, G. 18, 23, 24, 29, 39, 60, 61, 62
Klumpp, G. W. 28, 29
Kmiotek-Skarzynska, I. 261
Knipe, A. C. 255
Knöchel, A. 11, 13, 81
Knorr, E. 80
Kobayashi, M. 86, 92
Kobler, H. 98, 99, 100, 101
Koenig, K. E. 101, 103
Koenigs, W. 80
Koida, K. 216
Kolind-Andersen, H. 187, 188, 192, 232
Konizer, G. 246, 248
Kornblum, N. 76, 249
Korzeniowski, S. H. 245
Kostikov, R. R. 23, 25, 60, 61
Koutek, P. 197
Krasil'shchik, B. Ya. 14
Kraus, W. 18, 23, 24, 29, 60, 61, 62
Krishnan, S. 262
Kraus, W. 39
Kryshtal, G. V. 253
Kucherov, V. K. 254
Kuhl, P. 254
Kuhn, D. G. 262
Kuhn, W. 253
Kulinkovich, O. G. 34, 35
Kuroda, N. 25
Kurozumi, S. 86, 92
Kurts, A. K. 139, 160, 191
Kurts, A. L. 249, 250
Kurtz, W. 26
Kutina, R. E. 253
Kuwajima, I. 119, 120, 124, 137, 172, 173, 178, 179, 226, 227, 262
Kwantes, P. M. 28, 29
Kwon, S. 8, 21, 23, 31, 32, 52, 62

Laane, R. W. R. M. 82
Lacoste, J. 131
Lamm, B. 124, 125, 197, 198, 199, 217, 218, 240
Landini, D. 14, 23, 117, 118, 120, 121, 122, 123, 125, 126, 127, 175, 195, 196, 208, 216, 221, 222, 223, 224, 242, 243, 252

Lange, J. 137, 150, 151, 155, 160
Langlois, Y. 264
Larsson, F. C. V. 232
LaTorre, F. 259
Lavallee, P. 121
Lawesson, S. O. 187, 188, 192, 231, 232
LeBerre, A. 109
Le Coq, A. 128
Ledon, H. 177, 192, 194, 198
Lee, G. A. 14, 209, 210, 263
Lee, J. G. 265
Lee, W. 248
LeGoff, M-T. 256
Lehmkuhl, H. 253, 257
Lehn, J. M. 9, 11, 12, 130, 131, 201, 212, 226
le Noble, W. J. 249
Levinson, R. A. 132
Liang, C. 117, 123, 124, 125
Liebeskind, L. 137, 189, 249
Liebig, J. 102
Linstrumelle, G. 68, 69, 70
Liotta, C. L. 11, 13, 90, 96, 97, 98, 99, 100, 106, 117, 118, 119, 120, 121, 125, 126, 127, 226, 242, 243, 256
Lipisko, B. A. 143, 145
Lissel, M. 253
Litvak, V. V. 255, 265
Long, J. P. 3, 8, 13, 86, 98
Loughran, A. 255
Loupy, A. 220, 244
Ludwikow, M. 23, 37, 38, 81, 137, 143, 150, 152, 154, 155, 156, 157, 158, 159, 162, 163, 164, 165, 166, 167, 180, 185, 195, 198, 231, 261, 264

Machida, Y. 110, 111, 112
Macias, A. 249
Maerker, G. 2, 85
Magerlein, B. J. 81
Maia, A. M. 23, 175, 208, 216, 221, 222, 224, 226, 242, 243, 252
Maisey, R. F. 249

Makosza, M. 2, 3, 8, 13, 18, 20, 21, 22, 23, 24, 25, 27, 28, 29, 31, 34, 37, 38, 39, 45, 46, 47, 49, 51, 56, 58, 59, 60, 64, 67, 81, 82, 83, 136, 137, 138, 140, 141, 142, 143, 145, 146, 147, 148, 149, 150, 151, 152, 153, 154, 155, 156, 157, 158, 159, 160, 161, 162, 163, 164, 165, 166, 167, 168, 171, 172, 174, 175, 176, 177, 178, 179, 180, 182, 183, 184, 185, 186, 195, 198, 200, 201, 202, 203, 225, 227, 228, 229, 230, 231, 255, 259, 260, 261, 264
Mallon, C. B. 246
Manescalchi, F. 7, 93, 117, 118, 119, 120, 121, 122, 123, 125, 126, 127
Mangum, M. G. 246
Mansfield, R. C. 2, 7
Marschall, H. 253
Märkl, G. 82, 169, 229, 235, 236, 240, 241
Markus, A. 18, 27
Maryanoff, C. A. 90
Maskornick, M. J. 246, 248
Masse, J. 258
Massé, J. P. 263
Mathias, R. 65, 66,
Matsuda, T. 216
Matsumoto, G. 25
Matsushita, T. 93
Mayed, E. A. 109
McAdoo, D. J. 23, 24, 25, 26, 27
McDermott, M. 90
McIntosh, J. M. 221
McKillop, A. 76, 77, 80
Mead, E. J. 217
Meek, D. W. 244
Menger, F. M. 4, 97, 130, 207
Menon, B. 264
Merker, R. L. 7
Merz, A. 55, 76, 82, 169, 229, 235, 236, 240, 241, 254
Metzger, J. 258, 265
Michelot, D. 68, 69, 70
Midura, W. 8, 87, 228, 237, 238, 239

Mikolajczyk, M. 8, 11, 13, 87, 146, 147, 172, 175, 228, 237, 238, 239
Milano, M. 90
Miller, R. B. 24
Mills, R. H. 87
Mishima, T. 82, 240
Misiti, D. 259
Mislow, K. 90
Miura, S. 86, 92
Molchanov, A. P. 23, 25, 60, 61
Molinari, H. 7
Montanari, F. 7, 8, 9, 11, 13, 14, 23, 87, 99, 117, 118, 119, 120, 121, 122, 123, 125, 127, 146, 147, 172, 175, 208, 216, 221, 222, 223, 224, 226, 242, 243, 252
Morita, K. 53, 54, 55
Mork, P. C. 3, 247
Moro-Oka, Y. 256
Morris, H. F. 249
Morrisett, J. D. 130
Moss, R. A. 24, 67, 246
Mroczek, Z. 182, 183, 184, 185, 186
Mueller, A. C. 86
Mühlstädt, M. 28, 29, 51, 254
Müller, B. 24, 27, 33
Müller, C. 65, 66
Murai, K. 129
Murofushi, T. 226, 227

Nagashima, A. 25
Nakajima, M. 202
Nakamura, E. 119, 120, 124, 137, 172, 173, 178, 179, 226, 227, 262
Nakata, F. 68, 69, 70
Napier, D. R. 2, 3, 8, 18, 23, 83, 117, 118, 122, 130, 187, 210, 211, 216, 226
Nee, G. 265
Nef, J. U. 102
Nentwig, J. 88
Neumann, T. 25, 33
Newman, M. S. 71, 81, 117, 123, 124, 125
Nicolaou, K. C. 110, 111, 112

Nicoletti, R. 254
Nidy, E. G. 109
Nielsen, K. B. 203
Nishimura, Y. 31, 32, 62
Nitta, M. 30, 31
Noe, E. A. 243
Noh, J. S. 221, 222
Normant, H. 8, 11, 86, 87, 90, 91, 193, 194
Nozaki, H. 22, 23, 24, 26, 41, 42, 46, 47, 59, 60, 61, 82, 240

Oakwood, T. S. 102
Oberholzer, M. E. 78
O'Brien, J. P. 134
O'Donnell, Jr., M. J. 259
Oehler, J. 11, 13
Ogaki, Y. 254
Ogawa, T. 68, 69, 70
Ogura, F. 90
Ohno, M. 68, 69, 70
Ohtomi, M. 93, 104
Okawara, M. 124
Okimoto, T. 207, 208
Okonogi, T. 82, 134, 235, 236, 240
Orena, M. 209
Oscarson, S. 255
Otsu, T. 93
Otsuji, Y. 254
Ouannes, C. 264
Owens, R. M. 2, 5, 13, 19, 75, 96, 97, 136
Oxenrider, B. C. 88
Ozaki, Y. 42

Padwa, A. 90, 92
Palmertz, I. 124, 125
Pande, L. 23, 25, 26, 27
Pankova, M. 246, 248
Panunzio, M. 117, 118, 119, 120, 121, 122, 123, 125, 126, 127
Parayre, E. R. 263
Park, C. H. 11
Park, F. 88
Parker, A. J. 3
Patel, K. M. 130
Patrick, T. 68
Pavlickova, L. 197
Pearson, R. G. 7
Pedersen, C. J. 9, 10, 130, 131
Peet, N. P. 249

Pelletier, W. M. 249
Penszek, S. 131
Perry, R. A. 248, 259
Perucci, P. 248
Peter, F. 117, 119
Picker, D. 5, 11, 13, 14, 132, 136, 207, 222, 223, 224, 225
Piechucki, C. 237, 238
Pierre, J. L. 220, 244, 247
Pilkiewicz, F. G. 67, 246
Pirisi, F. M. 14, 23, 117, 118, 120, 121, 122, 123, 175, 208, 216, 221, 222, 224, 226, 242, 243
Pless, J. 121
Podda, G. 8, 9, 11, 13, 117, 122, 123, 222, 223, 224
Pollini, G. P. 202, 203
Pool, K.H. 109
Poos, G. I. 1
Port, W. S. 2, 85
Possel, O. 197
Pownall, H. J. 130
Puerta, J. E. 249
Pusset, J. 256
Pytlewski, T. 179

Quici, S. 117, 118, 121, 122, 123, 125, 126, 127
Qureshi, A. K. 212

Raban, M. 243
Raber, D. J. 217, 218
Rabet, F. 253, 257
Radcliffe, M. 48, 49
Rall, G. J. H. 78
Ranken, P. F. 24
Raynal, S. 131
Rebert, N. W. 169, 229, 237, 243
Reeves, W. P. 8, 98, 99, 100, 147, 175, 226, 257
Regen, S. L. 98
Reutov, O. A. 139, 160, 191, 249, 250
Rhee, H. K. 130, 207
Rhee, J. U. 130, 207
Ridgeway, R. W. 94
Roberts, D. K. 247
Roeske, R. W. 92
Rogers, R. J. 244
Roitman, J. N. 246, 248

Author Index

Rolla, F. 117, 118, 119, 120, 121, 122, 123, 125, 126, 127, 195, 196, 221, 222, 223, 224
Romano, L. J. 112
Rosenthal, I. 114
Rothenwöhner, W. 18, 23, 24, 29, 39, 60, 61, 62
Roussi, G. 107, 256
Roux, D. G. 78
Rozhkov, V. S. 155, 203
Rudolph, G. 11, 13, 81
Rylski, L. 149

Sadlo, H. 18, 23, 24, 29, 39, 60, 61, 62
Sakakibara, T. 187, 188, 190, 191, 192, 193, 199, 200
Sakito, Y. 52
Sam, D. J. 9, 78, 117, 118, 208, 242, 243
Samuelsson, B. 197, 198, 199
Sandri, S. 209
San Filippo, Jr., J. 110, 111, 112, 113, 114
Saraie, T. 53, 54, 55
Sarrett, L. H. 1
Sarthou, P. 250
Sasaki, T. 30, 48, 52, 61, 62, 68, 69, 70
Sato, M. 33
Saunders, W. H. 247
Savignac, P. 8, 11, 86, 87, 90, 91, 193, 194
Savoia, D. 171, 173, 178, 180, 181, 228, 229
Sawada, H. 22, 23, 24, 26, 41, 42, 59, 60, 61, 82, 240
Scherer, K. V. 61
Schleyer, P. V. R. 45
Schlosser, M. 65, 66
Schmidle, C. J. 2, 7
Schmidt, J. 80
Schneider, D. R. 67
Schnell, H. 88
Schöllkopf, U. 234
Schönefeld, J. 25, 27, 34, 37, 60, 61
Schué, F. 131
Schwarz, S. 94
Scott, C. B. 242
Scott, M. J. 7

Seltzer, R. 76, 249
Sepp, D. T. 51
Serafin, B. 2, 138, 175, 176, 177, 178, 179, 180, 181, 183, 184, 185
Serafinowa, B. 136, 137, 140, 141, 146, 147, 148, 149, 150, 151, 153, 154, 155, 156, 157, 158, 159, 160, 161, 162, 167, 176, 181, 182, 183, 184, 185, 186
Seyden-Penne, J. 192, 195, 220, 237, 239, 244
Seyferth, D. 20
Shaw, B. L. 258
Shechter, H. 98
Shein, S. M. 255, 265
Shepherd, J. P. 23, 24, 25, 26, 27, 208
Shibasaki, M. 110, 111, 112
Shigematsu, N. 226
Shimizu, M. 119, 120, 137, 172
Shimo, K. 7
Shiner, C. S. 110, 111, 112
Shippey, M. A. 126, 127
Sigwalt, P. 131
Sim, S. K. 248
Simakov, S. W. 155, 203
Simchen, G. 98, 99, 100, 101
Simmons, H. E. 9, 11, 78, 117, 118, 208, 242, 243
Singer, B. 24, 25, 27, 61
Singh, N. 23, 25, 26, 27
Singh, R. K. 140, 187, 189, 192, 193
Skattebøl, L. 58, 59, 60, 61, 62, 63, 64
Sklarz, B. 212
Skulimowska, E. 181
Slomkowski, S. 131
Smegal, J. 256, 265
Smid, J. 246, 248
Smiley, R. A. 97, 249
Smith, K. 90
Smudin, D. J. 24
Smushkevich, Y. I. 155, 203
Solodar, J. 96, 104, 243
Songstad, J. 249
Soucek, M. 197
Sparrow, J. T. 130
Spillane, W. J. 134

Sridhar, N. 255
Staley, S. W. 246, 248
Stang, P. J. 246
Starks, C. M. 2, 3, 4, 5, 8, 13, 14, 18, 19, 23, 75, 83, 96, 97, 99, 100, 117, 118, 121, 122, 123, 130, 134, 136, 187, 207, 210, 211, 216, 226, 243
Stein, A. R. 250
Stemmler, I. 235, 236
Stromquist, P. 48, 49
Subba Roa, B. C. 217
Sudoh, R. 187, 188, 190, 191, 192, 193, 199, 200
Sugimoto, N. 226
Sulewska, A. 263
Sullivan, E. A. 217, 220
Sunaga, M. 82, 240
Suvorov, N. N. 155, 203
Svoboda, M. 246, 248
Swain, C. G. 242, 244
Swern, D. 207, 208
Sweeney, A. 29
Sydnes, L. 62, 63, 64

Tabushi, I. 44, 45, 46, 47, 65
Tagaki, W. 82, 105, 134, 235, 236, 240
Takahashi, N. 44, 45, 46, 47
Takeishi, M. 124
Takubo, T. 252
Tamura, Y. 31, 32, 62
Tan, S. H. 250
Tanaka, T. 86, 92
Tavernier, D. 34
Tchoubar, B. 220, 244, 265
Thiem, J. 81
Thompson, L. R. 3, 8, 13, 86, 98
Tipson, R. S. 118, 122
Tishchenko, I. G. 34, 35
Tomahogh, R. 254
Tomoi, M. 252
Toru, T. 86, 92
Truesdale, L. K. 105, 106
Tsukanaka, M. 22, 23, 24, 26, 41, 42, 46, 47, 59, 60, 61
Tsunetsugu, J. 33
Tüncher, W. 263
Tundo, P. 7, 8, 9, 11, 13, 99, 117, 120, 122, 123,

175, 216, 222, 223, 224, 252
Turner, R. W. 249

Ugelstad, J. 3, 7, 75, 247
Ugi, I. K. 50, 65
Umani-Ronchi, A. 171, 173, 178, 180, 181, 228, 229
Umezawa, T. 134
Ungermann, T. 110, 111
Urniaz, A. 137, 158, 159, 162, 164, 166, 167

Valentine, J. S. 109, 110, 111, 112, 113, 114
van Leusen, A. M. 197
Velezheva, V. S. 155, 203
Vennstra, G. E. 229, 230
Verheijdt, P. L. 107
Viguier, M. 131
Vink, J. A. 107
Vinout, P. 250

Wadswoth, W. S. 237
Wagenknecht, J. H. 88, 89
Wakabayshi, T. 266
Wakamatsu, S. 7
Walsh, E. J. 256, 265
Washecheck, P. H. 210, 211
Wawrzyniewicz, M. 3, 8, 18, 20, 23
Weber, W. P. 14, 23, 24, 25, 26, 27, 48, 49, 50, 51, 65, 94, 101, 102, 103, 208
Weisberger, C. A. 102
Wetterling, M.-H. 1
Weda, E. 25
Weinkauff, O. J. 87
Weyerstahl, P. 24, 25, 27, 32, 33, 34, 65, 66, 253
White, M. R. 8, 98, 99, 100, 226
Widera, R. P. 50
Wiering, J. S. 82
Wippel, H. G. 237
Wittig, G. 1, 234
Wlostowaska, J. 259
Wöhler, F. 102
Wong, K. H. 246, 248
Wynberg, H. 82

Yakorlear, N. A. 14
Yamada, B. 93
Yamada, M. 187, 190, 192, 193, 199
Yano, Y. 82, 235, 236, 240
Yanovskaya, L. A. 254
Yasuda, Y. 93
Ykman, P. 117, 124, 125, 128
Yoshida, Z. 44, 45, 46, 47

Zalkow, L. H. 256
Zatorski, A. 8, 11, 13, 87, 146, 147, 172, 175, 228, 237, 238, 239
Zavada, J. 246, 248
Zhuravleva, M. I. 138, 169, 170, 171, 172, 178
Zubrick, J. W. 14, 96, 97, 100, 101, 129, 219, 232, 243
Zwaneburg, B. 229, 230
Zwierzak, A. 213, 214, 263

Subject Index

acetals 253, 255
acetate 7, 86, 89, 93, 245, 253
acetoacetic ester 249
acetone cyanohydrin 106, 256
acetonitrile 13, 87, 89, 90, 97, 143, 145
acetylacetonate 188, 243
acidity 138, 264
acyl azides 124
– nitriles (see cyanides)
acyloins 104
   (see also benzoin condensation)
adamantane 44
aldehydes 138, 144, 207, 222, 228, 235, 237, 240, 241, 244
–, reaction with chloroacetonitrile 81
aldol condensation 107, 138, 222
aldoximes 52
alkenes
–, addition of dichlorocarbene 23
alkyl azides 257
– chlorides from alcohols 46
– cyanides 99
   (see cyanides)
– nitrites (see nitroalkanes)
– thiocyanates 142, 226
alkylating agents 139
alkylation 136
–, C/O ratio 139, 141, 249, 265
–, early examples of 1
–, interfacial 261
– of acetoacetic ester 249
– of acetylacetonate 249
– of acetylenes 262
– of acetylenic sulfides 142, 151, 155, 156, 160, 163, 165, 168, 201, 227, 228
– of aldehydes 169
– of ambient ions 139, 141, 249, 265
– of carboxylates 2, 3, 85
– of chloroacetonitrile 81, 138, 145
– of 2-chlorophenothiazine 258
– of diactivated substrates 187

– of esters 136, 169
– of hydrocarbons 200
– of imidazole 258
– of ketones 136, 169
– of nitriles 137, 145
– of phenolates 249
– of phenylacetonitrile 137, 146, 260, 261
– of phenylhydrazones 259
– of pyrazole 258
– of pyridone 250, 265
– of Schiff's bases 259
– of sulfinate ion 230
– of sulfones 169
– of sulfoxides 232
– of α-sulfur ketones 261
– of tosylhydrazones 259
– with nitrobenzyl chloride 260
alkyne formation 128
allenes
–, carbene addition 35
–, dibromocarbene addition to 59
π-allyl complexes
allylic alcohols
–, dichlorocarbene addition to 41
allylic halides
–, dibromocarbene addition to 59
altered reactivity 220, 242
Amberlite IRA 93, 904
ambident anions 139, 141, 249, 265
amides
–, reaction with dichlorocarbene 52
amines
–, as catalysts 21
–, primary
–, –, isonitriles from 50
–, –, reaction with dichlorocarbene 50
–, secondary
–, –, reaction with dichlorocarbene 51
–, tertiary
–, –, reaction with dichlorocarbene 52

Subject Index

amidines
—, reaction with dichlorocarbene 52
amino acids 259
— polyethers 13
2-aminoindolenine 261
anilines 132
anionic polymerization 130
arene oxides 262
arylation 143, 245, 255, 260
aryldiazonium compounds 244
—, crown complexation 244
—, deuteration 245
—, halogen exchange 245
—, reduction 245
asymmetric induction
—, in carbene addition 42
—, in cyclizations 266
—, in dibromocarbene addition 61
azide 124, 253, 257
aziridines from imines 49
azobenzene
—, dimethylvinylidene carbene addition to 70

benzal chloride 141
benzodioxoles
—, formation 79
—, table 79
benzoin condensation 2, 104
benzyl cyanide (see phenylacetonitrile)
benzyl ether formation 255
benzyloxycarbonyl group 257
biphenyls 245
bisphenol A 88
bleach (see hypochlorite)
borohydride 215, 263
Brandström's alkylation method 136
Brandström's catalyst 7
bromide 117
t-butyl mesitoate 130
n-butylthioacetylene 142, 151, 155, 156, 160, 163, 165, 168, 227

Cannizzaro reaction 237, 243
carbenes 18, 44, 58, 253
—, amines as catalysts 8, 21
carbenoids 246
carbon tetrachloride 142
carbonylation 134
carbylamine reaction 50
catechols
—, oxidation 208
—, reaction with dihalomethanes 79
cation effects 243
catalysts 5
—, N-alkylpentamethylphosphoramides 252

—, alkyl substituted crowns 252
—, amines as 7
—, aminopolyethers as 13
—, Brandström's 7
—, charged 5
—, comparison 12
—, crown ethers 9
—, cryptands 11
—, efficiencies 5
—, Makosza's 7
—, phosphoryl sulfoxides 13
—, poisoning 7, 143
—, polyamines 7
—, polyethylene glycols 253
—, quaternary ions as 5
—, resins 2
—, Starks' 7
—, table of efficiencies 6
catalytic cycle (see Starks catalytic cycle)
— cyclopropanation 22
chalcones 114
chloride 117, 128
chloroacetonitrile 81, 138, 145
l-chlorocyclohexane carboxylic acid 254
chloroindoles 254
chloroiodocarbene 66
chloromethyl p-tolyl sulfone 82, 174
2-chlorophenothiazene 258
α-chloropropionitrile 146
chloroquinolines 254
β-chloro-α,β-unsaturated esters 125
chromate ion 209
cinnamonitrile 143
cobalt cluster compounds 133
cocatalysis 94, 261
conjugated olefins 33
cortisone 1
18-crown-6 10
—, preparations of 11
crown ethers 9
—, as catalysts 9, 10
cryptands 11
—, preparation of 12
cryptates (see cryptands)
cyanide
—, reactions 96
—, —, characteristics 97
—, —, mechanism 96
cyanide ion 96
cyanides 98, 101, 256
—, acyl 101, 102
—, alkyl 98
—, electrochemical generation 107
—, photochemical generation 107, 256
—, table 99

275

Subject Index

cyanoborohydride 219, 232
cyanoformates 102
cyanohydrin 103, 256
–, formation 2, 103
–, table 104
cyanosilylation 105, 106
cyclohexanone 254
cyclohexene 71, 83, 208
1,2-cyclohexanediol 83
cyclohexene oxide 83
cyclohexylidene carbene 71
cyclopentenones
 – from allylic alcohols 41
cyclopropanes 140

decarboxylation 248
dehalogenation 117, 125
dehydrogenation 263
deoxygenation 126
diacyl peroxides 113
dialkylation 139, 140
diamines
–, beta- 3
 – in ester formation 87
diazomethane 51
dibenzo-18-crown-6 10
dibromocarbene 255
–, addition to allenes 59
–, addition to allylic halides 59
–, addition to dienes 59
–, addition to double bonds, table 59
–, addition to enynes 59
–, addition to indoles 62
–, addition to olefins 58
–, addition to strained alkenes 61
–, addition to styrene 59
–, effect of alcohol 58
–, Michael addition 62
–, reaction with alcohols 65
–, reaction with amines 65
–, reaction with α,β-unsaturated carbonyl compounds 63
dibromocyclopropanes 58, 255
dicyclohexyl-18-crown-6 10
dicyclopropyl vinylidene carbene 70
dichloroacetylene 129
2,4-dichlorobenzyloxy carbonyl group 257
dichlorocarbene 253
–, addition followed by rearrangement 28, 48
–, addition to allenes 35
–, addition to allylic alcohols 41
–, addition to conjugated enynes 35
–, addition to conjugated olefins 27, 33, 34
–, addition to enamines 28

–, addition to enamines, table 29
–, addition to enol acetates 39
–, addition to furans 32
–, addition to hexamethyl-Dewar benzene 30
–, addition to imines 49
–, addition to indoles 31
–, addition to olefins 23
–, addition to phenols 42
–, addition to polycyclic aromatics 33
–, addition to polyolefins 26
–, addition to spirocyclopropylindene 31
–, addition to thiophenes 32
–, amine catalysis 8
–, asymmetric induction 42, 254
–, C-H insertions of 44, 45
–, catalytic scheme 20
–, dehydration with 52, 53
–, generation 18
–, interfacial reactions 20
–, reaction with alcohols 46, 47
–, reaction with aldehydes 254
–, reaction with aldoximes 52
–, reaction with allylic sulfides 55
–, reaction with amides 52
–, reaction with amidines 52
–, reaction with amines 50
–, reaction with benzaldehyde 55
–, reaction with cyclohexanone 254
–, reaction with dienes 253
–, reaction with glycols 48
–, reaction with hydrazine 51
–, reaction with indoles 254
–, reaction with ketones 254
–, reaction with non-olefinic substrates 44
–, reaction with olefins 18
–, reaction with phenylmercuric chloride 56
–, reaction with primary amines 50
–, reaction with pyrroles 254
–, reaction with secondary amines 51
–, reaction with thioamides 52
–, reaction with ureas
dichloromethane 141, 255
dichloroepoxides 254
dienes 253
–, dibromocarbene addition 59
–, dichlorocarbene addition 34
diesters 88, 89
diethers (see ethers)
difluorocarbene 65
dihalocarbenes 260
dihalomethanes 79
meso-dihydrobenzoin 49
3,4-dihydroquinolines 261
diiodocarbene 66

276

diiron dodecacarbonyl 132
4,4'-dimethoxybenzophenone 132
dimethyl sulfone 169, 229
– sulfoxide 229
4,6-dimethyl-5-thiacyclohexene-carboxaldehyde 222
4,4'-dimethylthiobenzophenone 132
dimethylvinylidene carbene 67
diols 49, 79, 83, 208, 211, 255
5,7-diphenyl-1,3-diazaadamantan-6-one 52
diphos 258
dipolar aprotic solvents 3
disulfides 219, 232
1,2-dithiocyantoethane 142
dithioketals 142
Doebner-Knoevenagel reaction 228

electrochemistry 107
elimination 125, 264
–, orientation 247
Emmons reaction (see Wittig-Horner-Emmons)
enol acetates 39
enamines 28, 29
enynes 35, 59
eosin 212
epichlorohydrin 85
epoxides 85, 240
–, deoxygenation 126
–, formation 81
–, from chloroacetonitrile 81
–, from sulfur ylids 240
α,β-epoxy ketones 83
α,β-epoxy nitriles 138
α,β-epoxy sulfones 81
ester hydrolysis 130, 141
esterification
–, crown catalyzed 89
–, cryptate catalyzed 93
–, resin catalyzed 93
esters 2, 3
–, crown catalyzed formation 89
–, fatty acid 2, 85
–, from carboxylate salts 85
–, non-catalytic formation of 87
–, phenacyl 90
–, –, table 91
–, phosphate 94
–, sulfonate 94
–, synthesis of 85
ethers 1, 73
–, di- 79
–, formation of methyl 76, 78
–, from phenols 78
–, mechanism of formation 73
–, methoxymethyl 78

–, methyl 76
–, phenyl 76, 78
–, preparation of alkyl aryl 77
–, preparation of unsymmetrical 74
–, rate enhancement in formation 75
ethyl isocyanide 65
ethylene dithioketals 142
ethylene oxide 128
extractive alkylation 2

fatty acid esters 2, 85
Finkelstein reaction 8, 118, 126
fluorene 201, 212
fluoride 117, 253, 257
fluorobromocarbene 66
fluorochlorocarbene 65
fluoroidocarbene 66
formaldehyde 235
formaldehyde acetals 255
formamides 51
formamidine sulfonic acid 219, 232
formates 46
"free" carbenes 246
2(5H)furanones 91
furans 32

glycidic nitriles 81, 138
1,2-glycols 48, 79, 83, 208, 211
glyoxal 235

halide ions 117
halomethanes
–, reaction with olefins, table 66
Hoffman carbylamine reaction 50, 65
Horner reaction (see Wittig-Horner-Emmons reaction)
hydrazine 51, 263
hydroboration 217
hydrocyanation 106, 256
hydrogen 264
– peroxide 83, 211
hydrolysis 130
β-hydroxydithiocinnamic esters 231
hydroxylation 208
5-hydroxymethyl-2-norbornene 30, 48
hypochlorite 209, 262, 263
hypohalite 263

imidazole 258
–, N-methyl 94
imines 49, 259
indoles 254
–, alkylation 202
–, carbene addition to 31
–, dibromocarbene addition to 62

277

Subject Index

cis-inositol hexaether 3
interfacial phenomena 5
iodide 74, 117
ion pair extraction 2
isonitriles 256
−, from dichlorocarbene 50
−, from dibromocarbene 65
−, from primary amines 50, 65
−, table of 50
isotopic exchange 20, 134

Jarrousse alkylation 1

ketones
−, reaction with chloroacetonitrile 81
Knoevenagel (see Doebner-Knoevenagel)
Koenigs-Knorr reaction 80
K-region epoxides 262

lithium aluminum hydride 220
lithium cuprates 264

macrocyclic polyethers (see crown ethers)
Makosza's alkylation method 136
− catalyst 7
"Makosza reagent" 254
malonic ester synthesis 248, 259
mandelic acid 55
mechanism
− of cyanide reactions 96
− of dichlorocarbene formation 19
− of ether formation 73
− of phase transfer catalysis 4
menthyl alcohol 47
metallation 132, 264
methyl cinnamate 40
methyl crotonate 40
methyl isocyanide 65
methyl mesitoate 130
methyl 3-methylcrotonate 40
methyl tetradecanoate 130
methylation (see ethers, methyl)
methylenation of carbonyl compounds 82
methylene diesters 89
methylenedithioacetals 223
methylthio(chloro)carbene 67, 246
methyltriphenylarsonium permanganate 2, 206
micellar effects 5
Michael Addition 31, 62
−, fluoride catalyzed 262
−, intramolecular 222, 226, 266
−, of lithium cuprates 264
− of sulfides 226
− of thiolacetic acid 222

− of trichloromethyl anion 37
−, possible 141
− to sugars 187, 199
molybdenum carbonyl 258
monoaza-18-crown-6 247

neopentyl phenyl sulfide 223
nitrile alkylation 145
nitriles, synthesis of 5, 96
nitrite esters 129
nitroalkanes 129
nitrobenzenes 132
p-nitrobromobenzene 255
norbornadiene 28
norbornene 29
nucleophiles (ambident) 249
nucleophilicity constants 242

octalenone 248
olefin formation 127
olefins 59
organocuprates 264
organometallics 132, 258
orthoformates 47
osmium tetraoxide 211
oxidation 206
−, catalytic 210
− of anions 212
− of arenes 262
− of carbanions 212
− of catechols 208
− of fluorene 212
− of hydrazines 257, 263
− of hydrocarbons 212, 216, 262
− of olefins 207
− of piperonal 207
−, solvent effect 209
− with hypochlorite 209, 262
− with hypohalite 263
− with metal oxides
− with permanganate 2, 206
− with potassium chromate 209
− with singlet oxygen 212
oxiranes 85, 240
oxy-Cope rearrangement 248
oxygen 212

paraperiodic acid 211
pentamethylenevinylidenecarbene 70
2,4-pentanedione 188, 243, 249
peptide synthesis 257
permanganate ion 2, 206
peroxides 109
−, diacyl 113

278

phase transfer catalysis
–, early examples 1
–, mechanism 5, 252, 253
–, principle of 3
–, role of water 14
–, solvents 13
phase transfer equilibria 4
phenacyl esters 90
phenols 42, 76, 78, 94, 208, 212, 263
phenylacetic acid 134
– esters 91
phenylacetonitrile 137, 147
phenylacetylene 128, 143, 163, 201
phenylbromocarbene 246
phenyl ethers (see ethers, phenyl)
phenylhydrazones 259
2-phenylpropionitrile 142
phenyl(trichloromethyl)mercury 56
phenylthiocarbene 66
phenylthio(chloro)carbene 66
phosphazine polymer 134
phosphazines 256
phosphonates 237
phosphonium salts 5, 235
phosphorylation 213
– of alcohols 213
– of amines 214
– of hydrazine 263
– of phenols 94
phosphorylsulfoxides 13
photochemical cyanation 107, 256
pinacol 49
piperonal 207
polycarbonates 88
polycyclic aromatics 33
polyethylene glycols 3, 253
polymerization 88, 130, 131
– of formaldehyde 130
polyolefins 26
polyvinyl azide 124
– chloride 124
potassium t-butoxide 247
– hydride 264
– permanganate 206
–, extraction of 207
prostaglandins 112
protecting group 257
pyrazole 258
pyridones 250, 265
pyrroles 254

quaternary ions (see catalysts)
quinolines
–, chloro 254
–, from indoles 31

quinones 105
–, ortho 208
rearrangement
– of carbene adducts 28, 62
–, oxy-Cope 248
–, Reissert's compound 203
–, sigmatropic 56, 70
–, Smiles 249, 255
reduction 215
–, altered reactivity in 220
–, asymmetric 263
– of aromatic nitro compounds 258
– of aryldiazonium compounds 245
– of disulfides 219
– of carbonyl compounds 215
– with borohydride
– with cyanoborohydride 218
– with diborane 217
– with formamidine sulfinic acid 219
– with LAH 220
– with tetrabutylammonium borohydride 217
reductive cyclization 261
Reimer-Tiemann reaction 42
Reissert's compound 144
ring expansion 254
rose bengal 212
ruthenium dioxide 211

Schiff's base 259
  (see also imines)
self-solvating bases 247
sigmatropic rerrangement 56, 70
silver carboxylates 85
silylacetylenes 262
silyl enol ethers 124, 237
singlet oxygen 212
Smiles rearrangement 249, 255
solvent effect 255
solvents 13
sodium thiosulfate 126
Starks' catalyst 7
– catalytic cycle 4
steroidal esters 28, 94
stirring speed 5, 75, 97
styrene
–, dibromocarbene addition to 59
–, reaction with dimethylvinylidene carbene 68
–, reaction with pentamethylenevinylidene carbene 70
sugars
–, alkylation of 81
sulfides
– by Michael addition

279

Subject Index

sulfides
— from thiocyanates 225
—, symmetrical 221
—, α,β-unsaturated 228
—, unsymmetrical 222
sulfones 139, 228
sulfonyl chlorides 94
— fluorides 257
sulfonylcyclopropanes 231
sulfoxides 228
superoxide 109, 256, 257
—, alcohol formation 110
—, aromatic substitution 115
—, cleavage reactions 113
—, ester cleavage 112
—, olefin formation 110
—, reaction with chalcones 114
—, stereochemistry 109
Swain-Scott parameters 242

tertiary amines
—, as catalysts 85
tetramethylethylene
—, reaction with methylthio(chloro)carbene 67
thioamides 52
thiobenzophenone 132
thiocyanate 142, 253
thiocyanates (see also alkyl thiocyanates)
thioethers 5, 142, 221
thiolacetic acid 221
thio-Micheler ketone 132
thiones 132
thiophenes 32
thiophenoxide 223

tosylhydrazones 259
tribromomethyl anion 63
trichloromethyl anion 37
trichloromethylbenzene 130
triirondodecacarbonyl 258
α-trimethylsilyldiazoacetate 107
trimethylsilyl potassium 126
trimethylsulfonium ion 82, 240
Triton B 1
tryptophan 257

uncharged catalysts 7
unsaturated carbenes 67
α,β-unsaturated sulfur compounds 228

vinyl acetate 260

water
— in cyanide reactions 97
— in ether synthesis 75
— in hydrolysis 130
—, role in ptc 14
Williamson reaction (see ethers)
Wittig-Horner-Emmons reaction 228, 237, 264
Wittig reaction 234
    (see also Wittig-Horner-Emmons reaction)
— of formaldehyde 235
— of glyoxal 235

ylids 234
— in carbene reactions 21
—, phosphorus 234
—, sulfur 239

Reactivity and Structure

Concepts in Organic Chemistry

Editors: K. Hafner, J.-M. Lehn, C. W. Rees,
P. v. Ragué Schleyer, B. M. Trost, R. Zahradník

Volume 1

J. TSUJI

Organic Synthesis by Means of Transition Metal Complexes

A Systematic Approach
4 tables. IX, 199 pages. 1975
ISBN 3-540-07227-6

Contents: Comparison of synthetic reactions by transition metal complexes with those by Grignard reagents. – Formation of σ-bond involving transition metals. – Reactivities of σ-bonds involving transition metals. – Insertion reactions. – Liberation of organic compounds from the σ-bonded complexes. – Cyclization reactions, and related reactions. – Concluding remarks.

Volume 2

K. FUKUI

Theory of Orientation and Stereoselection

72 figures, 2 tables. VII, 134 pages. 1975
ISBN 3-540-07426-0

Contents: Molecular Orbitals. – Chemical Reactivity Theory. – Interaction of Two Reacting Species. – Principles Governing the Reaction Pathway. – General Orientation Rule. – Reactivity Indices. – Various Examples. – Singlet-Triplet Selectivity. – Pseudoexcitation. – Three-species Interaction. – Orbital Catalysis. – Thermolytic Generation of Excited States. – Reaction Coordinate Formalism. – Correlation Diagram Approach. – The Nature of Chemical Reactions.

Volume 3

H. KWART, K. KING

δ-Orbital Involvement in the Organo-Chemistry of Silicon, Phosphorus and Sulfur

Approx. 220 pages. 1977
ISBN 3-540-07953-X

Contents: Introduction. – Theoretical Basis for d-Orbital Involvement. – Physical Properties Related to dp-π Bonding. – The Effects of dp-π Bonding on Chemical Properties and Reactivity. – Pentacovalency. – References. – Bibliography. – Author Index. – Subject Index.

Springer-Verlag
Berlin
Heidelberg
New York

Topics in Current Chemistry

Fortschritte der chemischen Forschung
Managing Editor: F. L. Boschke

Volume 65
Theoretical Inorganic Chemistry II
47 figures, 44 tables. IV, 153 pages. 1976
ISBN 3-540-07637-9

Contents: K. Bernauer: Diastereoisomerism and Diastereoselectivity in Metal Complexes (134 references)
M. S. Wrighton: Mechanistic Aspects of the Photochemical Reactions of Coordination Compounds (196 references)
A. Albini, H. Kisch: Complexation and Activation of Diazenes and Diazo Compounds by Transition Metals (119 references)

Volume 66
Triplet States III
10 figures. IV, 154 pages. 1976
ISBN 3-540-07655-7

Contents: P. J. Wagner: Chemistry of Excited Triplet Organic Carbonyl Compounds (198 references)
H. Dürr, B. Ruge: Triplet States from Azo Compounds (96 references)
H. Dürr, H. Kober: Triplet States from Azides (99 references)
G. Fischer: Spectroscopic Implications of Line Broadening in Large Molecules (69 references)

Volume 68
Theory
36 figures. IV, 156 pages. 1976
ISBN 3-540-07932-7

Contents: X. Chapuisat, Y. Jean: Theoretical Chemical Dynamics: A Tool in Organic Chemistry (285 references)
D. Papoušek, V. Špriko: A New Theoretical Look at the Inversion Problem in Molecules (81 references)
H. Schneider: Ion Solvation in Mixed Solvents (145 references)

Volume 70
Structural Theory of Organic Chemistry
By N. D. Epiotis, W. R. Cherry, S. Shaik, R. L. Yates, F. Bernardi
60 figures, 58 tables. VIII, 250 pages. 1977.
ISBN 3-540-08099-6

Contents: Theory. – Nonbonded Interactions. – Geminal Interactions. – Conjugative Interactions. – Bond Ionicity Effects. – Author Index. – Subject Index. (420 references)

Springer-Verlag
Berlin
Heidelberg
New York